Andreas Fink and Franz Rothlauf (Eds.)

Advances in Computational Intelligence in Transport, Logistics, and Supply Chain Management

Studies in Computational Intelligence, Volume 144

Further volumes of this series can be found on our homepage: springer.com

Vol. 123. Shuichi Iwata, Yukio Ohsawa, Shusaku Tsumoto, Ning Zhong, Yong Shi and Lorenzo Magnani (Eds.)
Communications and Discoveries from Multidisciplinary Data, 2008
ISBN 978-3-540-78732-7

Vol. 124. Ricardo Zavala Yoe
Modelling and Control of Dynamical Systems: Numerical Implementation in a Behavioral Framework, 2008
ISBN 978-3-540-78734-1

Vol. 125. Larry Bull, Bernadó-Mansilla Ester and John Holmes (Eds.)
Learning Classifier Systems in Data Mining, 2008
ISBN 978-3-540-78978-9

Vol. 126. Oleg Okun and Giorgio Valentini (Eds.)
Supervised and Unsupervised Ensemble Methods and their Applications, 2008
ISBN 978-3-540-78980-2

Vol. 127. Régie Gras, Einoshin Suzuki, Fabrice Guillet and Filippo Spagnolo (Eds.)
Statistical Implicative Analysis, 2008
ISBN 978-3-540-78982-6

Vol. 128. Fatos Xhafa and Ajith Abraham (Eds.)
Metaheuristics for Scheduling in Industrial and Manufacturing Applications, 2008
ISBN 978-3-540-78984-0

Vol. 129. Natalio Krasnogor, Giuseppe Nicosia, Mario Pavone and David Pelta (Eds.)
Nature Inspired Cooperative Strategies for Optimization (NICSO 2007), 2008
ISBN 978-3-540-78986-4

Vol. 130. Richi Nayak, Nikhil Ichalkaranje and Lakhmi C. Jain (Eds.)
Evolution of the Web in Artificial Intelligence Environments, 2008
ISBN 978-3-540-79139-3

Vol. 131. Roger Lee and Haeng-Kon Kim (Eds.)
Computer and Information Science, 2008
ISBN 978-3-540-79186-7

Vol. 132. Danil Prokhorov (Ed.)
Computational Intelligence in Automotive Applications, 2008
ISBN 978-3-540-79256-7

Vol. 133. Manuel Graña and Richard J. Duro (Eds.)
Computational Intelligence for Remote Sensing, 2008
ISBN 978-3-540-79352-6

Vol. 134. Ngoc Thanh Nguyen and Radoslaw Katarzyniak (Eds.)
New Challenges in Applied Intelligence Technologies, 2008
ISBN 978-3-540-79354-0

Vol. 135. Hsinchun Chen and Christopher C. Yang (Eds.)
Intelligence and Security Informatics, 2008
ISBN 978-3-540-69207-2

Vol. 136. Carlos Cotta, Marc Sevaux and Kenneth Sörensen (Eds.)
Adaptive and Multilevel Metaheuristics, 2008
ISBN 978-3-540-79437-0

Vol. 137. Lakhmi C. Jain, Mika Sato-Ilic, Maria Virvou, George A. Tsihrintzis, Valentina Emilia Balas and Canicious Abeynayake (Eds.)
Computational Intelligence Paradigms, 2008
ISBN 978-3-540-79473-8

Vol. 138. Bruno Apolloni, Witold Pedrycz, Simone Bassis and Dario Malchiodi
The Puzzle of Granular Computing, 2008
ISBN 978-3-540-79863-7

Vol. 139. Jan Drugowitsch
Design and Analysis of Learning Classifier Systems, 2008
ISBN 978-3-540-79865-1

Vol. 140. Nadia Magnenat-Thalmann, Lakhmi C. Jain and N. Ichalkaranje (Eds.)
New Advances in Virtual Humans, 2008
ISBN 978-3-540-79867-5

Vol. 141. Christa Sommerer, Lakhmi C. Jain and Laurent Mignonneau (Eds.)
The Art and Science of Interface and Interaction Design, 2008
ISBN 978-3-540-79869-9

Vol. 142. George A. Tsihrintzis, Maria Virvou, Robert J. Howlett and Lakhmi C. Jain (Eds.)
New Directions in Intelligent Interactive Multimedia, 2008
ISBN 978-3-540-68126-7

Vol. 143. Uday K. Chakraborty (Ed.)
Advances in Differential Evolution, 2008
ISBN 978-3-540-68827-3

Vol. 144. Andreas Fink and Franz Rothlauf (Eds.)
Advances in Computational Intelligence in Transport, Logistics, and Supply Chain Management, 2008
ISBN 978-3-540-69024-5

Andreas Fink
Franz Rothlauf
(Eds.)

Advances in Computational Intelligence in Transport, Logistics, and Supply Chain Management

Prof. Dr. Andreas Fink
Faculty of Economics and Social Sciences
Helmut-Schmidt-University Hamburg
Holstenhofweg 85
22043 Hamburg
Germany
Email: andreas.fink@hsu-hamburg.de

Prof. Dr. Franz Rothlauf
Lehrstuhl für Wirtschaftsinformatik
Johannes-Gutenberg-Universität Mainz
Jakob Welder-Weg 9
D-55099 Mainz
Germany
Email: rothlauf@uni-mainz.de

ISBN 978-3-540-69024-5 e-ISBN 978-3-540-69390-1

DOI 10.1007/978-3-540-69390-1

Studies in Computational Intelligence ISSN 1860949X

Library of Congress Control Number: 2008927876

Typeset & Cover Design: Scientific Publishing Services Pvt. Ltd., Chennai, India.

Printed in acid-free paper

9 8 7 6 5 4 3 2 1

springer.com

Preface

Logistics and supply chain management deal with managing the flow of goods or services within a company, from suppliers to customers, and along a supply chain where companies act as suppliers as well as customers. As transportation is at the heart of logistics, the design of traffic and transportation networks combined with the routing of vehicles and goods on the networks are important and demanding planning tasks. The influence of transport, logistics, and supply chain management on the modern economy and society has been growing steadily over the last few decades. The worldwide division of labor, the connection of distributed production centers, and the increased mobility of individuals lead to an increased demand for efficient solutions to logistics and supply chain management problems. On the company level, efficient and effective logistics and supply chain management are of critical importance for a company's success and its competitive advantage. Proper performance of the logistics functions can contribute both to lower costs and to enhanced customer service.

Computational Intelligence (CI) describes a set of methods and tools that often mimic biological or physical principles to solve problems that have been difficult to solve by classical mathematics. CI embodies neural networks, fuzzy logic, evolutionary computation, local search, and machine learning approaches. Researchers that work in this area often come from computer science, operations research, or mathematics, as well as from many different engineering disciplines. Popular and successful CI methods for optimization and planning problems are heuristic optimization approaches such as evolutionary algorithms, local search methods, and other types of guided search methods. Such methods do not enumerate all possible solutions but move "heuristically" through the search space searching for superior solutions. Heuristic optimization approaches must balance intensification, which focuses the search on high-quality solutions (exploitation), and diversification, which ensures that the search can escape from local optima and does not focus on small parts of the search space (exploration).

The book at hand presents a careful selection of relevant applications of CI methods for transport, logistics, and supply chain management problems. The chapters illustrate the current state-of-the-art in the application of CI methods in these fields and should help and inspire researchers and practitioners to apply and develop efficient methods. A few contributions in this book are extended versions of papers presented at *EvoTransLog2007: The First European Workshop on Evolutionary Computation in Transportation and Logistics* which was held in Valencia, Spain, in 2007. The majority of contributions are from additional, specially selected researchers, who have done relevant work in different areas of transport, logistics, and supply chain management. The goal is to broadly cover representative applications in these fields as well as different types of solution approaches. On the application side, the contributions focus on design of traffic and transportation networks, vehicle routing, and other important aspects of supply chain management such as inventory management, lot sizing, and lot scheduling. On the method side, the contributions deal with evolutionary algorithms, local search approaches, and scatter search combined with other CI techniques such as neural networks or fuzzy approaches.

The book is structured according to the application domains. Thus, it has three parts dealing with traffic and transportation networks, vehicle routing, and supply chain management.

The majority of the contributions of Part I focus on road traffic in urban settings. The first contribution on "Combined Genetic Computation of Microscopic Trip Demand in Urban Networks" by T. Tsekeris, L. Dimitriou, and A. Stathopoulos presents an approach to predicting the expected trip demand in urban networks. A proper estimation of dynamic origin-destination demands is fundamental for all network design and routing decisions. The authors formulate the multi-objective problem and present an evolutionary computation approach combined with a microscopic simulation model for estimating the expected traffic flow.

The next two contributions focus on design of road networks. The chapter "Genetically Optimized Infrastructure Design Strategies in Degradable Transport Networks" by the same authors builds on their first contribution. The authors take the origin-destination demands as given and formulate the stochastic equilibrium network design problem as a game-theoretic, combinatorial bi-level program. The design process of a transport network is considered as a game between the network designer and the network user. The system designer is the leader in a two-stage leader-follower Stackelberg game. He modifies the structure of the road network to optimize the system performance, while the users are the followers reacting to alternative design plans. The options of the system designer are either to build new roads or to add additional lanes to existing roads. To solve the problem, the authors again combine evolutionary algorithms with simulation.

The contribution "Genetic Algorithm for Constraint Optimal Toll Ring Design" by A. Sumalee focuses on design of toll rings. A toll ring is a ring of roads which encloses a specific area (e.g. city center) such that all vehicles travelling to the area must use a road that is part of the toll ring at least once. The author presents an evolutionary algorithm combined with a traffic simulator to find a toll ring such that the benefits gained less the costs of implementing the toll ring are maximized.

The final contribution in the context of road networks "Real Time Identification of Road Traffic Control Measures" by K. Almejalli, K. Dahal, and M. A. Hoasain deals with control of road traffic and presents a system for determining appropriate road traffic control options. The options are chosen according to different traffic states. The performance of the control actions is determined by travel time, fuel consumption, and average length of traffic jams. To solve the problem, the authors combine several CI approaches, namely fuzzy logic, neural networks, and evolutionary algorithms.

The contribution "Simultaneous Airline Scheduling" by T. Grosche and F. Rothlauf forms the bridge between Part I and Part II of the book. It addresses both design of transportation networks and routing of vehicles (airplanes) on those networks. The authors present a novel and integrated approach for airline scheduling which allows airlines simultaneously to determine an optimal structure of the flight network, routing of the airplanes on the flight network, and scheduling of flights. The optimization goal is to maximize the revenue of an airline. Since the problem is too complex to be efficiently solved by classical optimization methods, different types of heuristic optimization methods such as evolutionary algorithms and threshold accepting are studied. Important for a high performance of the heuristic optimization methods are the proper choice of the representation/operator combination, repair operators, fitness function, and initial solution.

Part II of the book starts with three contributions on vehicle and arc routing. The contribution "GRASP with Path Relinking for the Capacitated Arc Routing Problem with Time Windows" by N. Labadi, C. Prins, and M. Reghioui combines several heuristic search approaches such as greedy randomized search, a tour-splitting algorithm which diversifies the search process, a local search, and an optional path relinking process to solve the undirected capacitated arc routing problem with time windows. The proposed approach finds new optimal solutions for the problem and is as effective as state-of-the-art algorithms, while being significantly faster.

The second contribution "A Scatter Search Algorithm for the Split Delivery Vehicle Routing Problem" by V. Campos, A. Corberán, and E. Mota deals with a similar problem and presents a scatter search approach for vehicle routing problems where the demands of clients can be split, this means any client can be served by more than one vehicle. Again, the computational experiments indicate that the proposed heuristic results in similar performance to state-of-the-art methods.

The third contribution on vehicle routing, "Stochastic Local Search Procedures for the Probabilistic Two-Day Vehicle Routing Problem" by K. F. Doerner, W. J. Gutjahr, R. F. Hartl, and G. Lulli, describes a vehicle routing problem where two types of service are provided: an urgent service that delivers within one day and a regular service that needs two days but comes at a lower price. To exploit synergies in building the delivery tours regular orders may be delivered immediately (like urgent services). Assuming a dynamic (online) problem setting with a rolling planning horizon, the problem is formalized as a stochastic problem and solved using an approach based on ant colony optimization.

The contribution "The Oil Drilling Model and Iterative Deepening Genetic Annealing Algorithm for the Traveling Salesman Problem" by H. C. Lau and F. Xiao presents a hybrid approach, which imitates the oil drilling process: Search takes place in many places (a population of candidate solutions) and each place is evaluated by drilling (performing a local search). Search diversification is mainly provided by genetic algorithm concepts, while simulated annealing is used for intensification by means of local search. In the course of the search process the population shrinks and the local search is intensified (drilling deeper). That is, the balance shifts from diversification to intensification. Results show that a proper combination of intensification and diversification elements outperforms a straightforward hybrid algorithm as well as using local or recombination-based search alone.

The contribution "Online Transportation and Logistics Using Computationally Intelligent Anticipation" by P. A. N. Bosman and H. L. Poutré forms the bridge between Part II and Part III of the book. It discusses the importance of anticipation in online decision making and describes how CI can be used to design approaches that perform anticipation. The proposed methods are designed such that they learn from the consequences of previous decisions, which leads to an auto-adaptive design. The authors present two applications: dynamic vehicle routing, which assumes that the loads to be transported are announced while the vehicles are already en route, and inventory management where higher customer satisfaction leads to an increased number of transactions.

The final Part III on supply chain management starts with a contribution on "Supply Chain Inventory Optimisation with Multiple Objectives: An Industrial Case Study" by L. Amodeo, H. Chen, and A. E. Hadji. The authors optimize supply chain inventory policies using a multi-objective optimization approach that combines genetic algorithms with a Petri net-based simulation tool for performance evaluation. In a real-world study, the authors use inventory cost, customer service level, and computation time as optimization goals.

The contribution "Decomposition of Dynamic Single-Product and Multi-Product Lotsizing Problems and Scalability of EDAs" by J. Grahl, S. Minner, and F. Rothlauf shows that certain lotsizing problems are decomposable. Thus, such problems can be solved efficiently by evolutionary algorithms such as estimation of distribution algorithms (EDA). A scalability analysis for EDAs on the dynamic single-product and multi-product lotsizing problem confirms existing scalability theory and shows that solution effort grows with a low-order polynomial depending on the problem size.

The final contribution "Hybrid Genetic Algorithms for the Lot Production and Delivery Scheduling Problem in a Two-Echelon Supply Chain" by S. A. Torabi, M. Jenabi, and S. A. Mansouri considers a single supplier who produces items on a flexible flow line under a cyclic policy and delivers them directly to an assembly facility. The authors formulate the problem as a mixed zero-one nonlinear program and minimize the average setup, inventory-holding, and delivery costs. For solving the problem, a hybrid approach is proposed which combines genetic algorithms and neighborhood search.

In closing we wish to thank all authors who contributed to this book. Without their work and passion, this book would not have been possible. In addition, we thank all reviewers for their help in improving the quality of the contributions in the book. We also want to thank Inka Lölfer for proofreading parts of the book, as well as Jens Czogalla for helping with the technical preparation of some chapters. Last, but not least, we want to thank the editor of this book series, Janusz Kacprzyk, for making this book possible and Thomas Ditzinger who fully supported the project.

Hamburg and Mainz (Germany)
March 2008

Andreas Fink
Franz Rothlauf

Contents

Part I

Traffic and Transport Networks

Combined Genetic Computation of Microscopic Trip Demand in Urban Networks

Theodore Tsekeris[1], Loukas Dimitriou[2], and Antony Stathopoulos[2]

[1] Centre for Planning and Economic Research, Amerikis 11, 10672 Athens, Greece
tsek@kepe.gr

[2] Department of Transportation Planning and Engineering, School of Civil Engineering, National Technical University of Athens, Iroon Polytechniou 5, 15773 Athens, Greece
lucdimit@central.ntua.gr, astath@transport.ntua.gr

Summary. This chapter describes a combined genetic computation approach for estimating time-varying Origin-Destination (O-D) trip demand matrices from traffic counts in urban networks. The estimation procedure combines a microscopic model simulating traffic flow conditions with a genetic algorithm to synthesize the network O-D trip matrix, through determining the turning flow proportions at each intersection. The proposed approach avoids the restrictions involved in employing a user-optimal Dynamic Traffic Assignment (DTA) procedure and carries out a stochastic global search of the optimal O-D trip and turning flow distributions. The multi-objective, single-level optimization formulation of the problem provides a mutually consistent solution between the resulting O-D matrix and path/link flow pattern, which minimizes the difference between estimated and observed link flows. The model implementation into a real arterial sub-network demonstrates its ability to microscopically estimate trip demand with satisfactory accuracy and fast computing speeds which allow its usage in dynamic urban traffic operations.

Keywords: Genetic computation, Trip demand, Urban networks, Traffic flows, Microscopic simulation.

1 Introduction

The time-varying (dynamic) Origin-Destination (O-D) trip demand matrices provide a crucial input for the simulation, management and control of urban road transportation networks. A dynamic O-D matrix specifies the aggregate demand for trip interchange between specific traffic zones of the network over a series of time intervals. The dynamic O-D matrix estimation is commonly based on two sources of information, i.e. measured traffic flow time series (counts) at selected network links and a prior O-D matrix to guide the solution procedure. The resulting O-D matrices are mostly used as input to a Dynamic Traffic Assignment (DTA) procedure for mapping the estimated trip demand into a set of path and link traffic flows. In the DTA procedure, a number of additional constraints can be imposed in order to improve the observability of the dynamic

A. Fink and F. Rothlauf (Eds.): Advances in Computational Intelligence, SCI 144, pp. 3–21, 2008.
springerlink.com

O-D matrices, such as constraints on the turning flow proportions at network intersections. However, the use of a DTA-based solution is typically subject to strong assumptions concerning the quality of input demand information and route choice behavior of users, while it induces severe computational burden for the real-time estimation of dynamic O-D matrices in realistic-size urban networks. Moreover, existing methods for dynamic O-D matrix estimation provide the aggregate demand for interchange between O-D pairs. Such an aggregate treatment does not permit the complete monitoring of the network traffic state, in terms of the knowledge of the spatio-temporal trajectory of each vehicle in the network.

The present study addresses the aforementioned problems through employing a combined genetic computation method, which combines a microscopic traffic simulation model with a genetic algorithm (GA) for the non-DTA-based solution of the dynamic O-D matrix estimation from traffic counts in urban networks. The proposed method, which was originally formulated and briefly analyzed in [25], exploits the hierarchical structure of the network, which is composed of a set of signalized intersections (nodes) connected each other with arterial links. The current approach substitutes the DTA procedure with a series of O-D flow estimation sub-problems at the level of each intersection for the network-wide O-D matrix estimation. The simulation model used here enables to microscopically represent the intricate features of traffic flow and, hence, it allows tracking network O-D trip flows at the basic demand unit of an individual vehicle. The GA-based process carries out a stochastic global search for simultaneously estimating the optimal distribution of turning flows at each intersection and O-D trip flows at the whole network. The GA provides a suitable approach for handling the uncertainty of input information in the estimation procedure and helps reach an optimal (or satisfactory sub-optimal) solution.

The following section provides a general description of the issues and existing approaches involved in estimating turning flow proportions at intersection level and O-D trip flows at network level, and the formulation of the problem as a multi-objective, single-level estimation procedure. Section 3 describes the microscopic model of traffic simulation and the combined genetic computation method. Section 4 provides information about the experimental setup of the study. Section 5 presents numerical results which demonstrate the accuracy and computational efficiency of the model through its implementation into a real urban arterial sub-network, and Sect. 6 summarizes and concludes.

2 Description of the Dynamic Network O-D Matrix Estimation

2.1 The Problem of Dynamic Network O-D Matrix Estimation

The estimation of dynamic network O-D matrices typically employs optimization procedures which consider the relationship between travel demand and path/link flow loading pattern as fixed, based on the equilibrium assignment of the prior O-D matrix onto the network. This relationship is defined by the fractions, known as link use proportions, of each O-D trip population traversing the observed links, and constitute the elements of the assignment matrix. In this case, the problem of inconsistency appears between the initial assignment matrix, on which the estimation process relies, and the final assignment matrix resulting from the estimated demand pattern. Methods that have been hitherto considered for solving the dynamic O-D matrix estimation problem with fixed (multi-path) assignment matrix include the Generalized Least Squares (GLS) [4], the Kalman filtering [1], maximum likelihood [14] and entropy maximization [26].

The development of bi-level programming approaches aimed at treating the inconsistency between the resulting O-D trip matrix and the assignment matrix, by estimating these two separate problems in a common optimization framework [29]. Such approaches include the *iterative* bi-level optimization approach, where the O-D matrix and DTA problems are solved in a sequential manner, and the *simultaneous* optimization approach, where both the O-D matrix and DTA problems are formulated by using a single (composite) objective function and simultaneously solved in each updating iteration as a single-level optimization problem (see Sect. 2.3). The latter approach is concisely referred to here as simultaneous dynamic O-D matrix estimation. The iterative and simultaneous optimization approaches were originally developed for the case of static O-D matrix estimation (see, respectively, [10] and [29]) and later extended through a GLS framework for the estimation of dynamic O-D matrices in simplified test networks (see, respectively, [24] and [27]).

The simultaneous optimization approach can ensure the convergence of the estimation procedure and the mutual consistency of the resulting O-D matrix with the path and link loading pattern, as reflects the time-varying assignment matrix, in comparison to the iterative optimization approach. In addition, it can inherently express through a single objective function the relative contribution of different errors in the resulting O-D trips and traffic flows to the performance of the simultaneous estimation procedure. Nevertheless, the solution of the DTA problem mostly involves increased complexity and heavy computational burden in realistic urban networks. Besides, the simultaneous optimization requires the estimation of partial derivatives, the Hessian matrix of the single objective function and the evaluation of this function in each updating iteration of the solution procedure ([11], [30]). These problems restrict the applicability of the results of the simultaneous O-D matrix estimation for practical traffic operation purposes. Such purposes can primarily involve the online deployment of real-time route guidance information and area-wide traffic signal control systems.

Furthermore, the intractability of DTA-based approaches for the dynamic O-D matrix estimation can be attributed to their inability to replicate actual urban traffic flow conditions. Particularly for the complex case of congested arterial networks, the actual traffic conditions can significantly depart from the user-optimal traffic conditions that are typically sought to be estimated by the DTA models. The inaccuracy of the results of the dynamic assignment procedure, in combination with the uncertainty of the prior demand information and missing flow information, tend to significantly increase the discrepancy between estimated (assigned) and measured link flows.

Stochastic global search techniques, such as Genetic Algorithms (GAs), can suitably address problems of uncertain and missing information in O-D matrix estimation. However, their usage is particularly limited in the literature and restricted to the case of static traffic networks ([16], [23]). In addition, the use of microscopic simulation models could enhance the accuracy of the resulting O-D matrices. Nonetheless, the micro-simulation models are typically used only for the estimation of disaggregate O-D flows at intersections, while their application in network-wide problems involves the use of DTA-based procedures for the estimation of aggregate O-D trip flows [3].

In order to address the problems mentioned previously, the current study describes a non-DTA-based solution of the simultaneous dynamic O-D matrix estimation with traffic counts in urban networks. The DTA procedure is substituted here with a series of intersection O-D matrix estimation sub-problems, which are solved with a micro-simulation model augmented with a genetic algorithm. This combined genetic approach carries out a stochastic global search and simultaneous estimation of the optimal distribution of turning movement flows at each intersection and O-D trip flows at the whole network. The following subsection presents the problem of intersection O-D movement estimation.

2.2 The Problem of Intersection O-D Movement Estimation

The estimation of intersection O-D matrices of turning movements, in terms of the proportions of traffic from each entrance going to a given exit, at signalized intersections plays an increasingly important role in a number of traffic simulation and control procedures in urban networks. These proportions have been historically estimated by manual counting. However, the advent of the third generation of Urban Traffic Control Systems (UTCS), through the on-line, traffic-responsive signal control systems operating on a cycle-by-cycle basis, such as the SCAT, SCOOT [18] and TUC [7], and the real-time traffic adaptive control systems, such as the OPAC [12], indicated the importance of automatically estimating intersection turning proportions from traffic count data. In addition to the UTCS, the estimation of intersection O-D turning flows can support a range of other operations involved in the deployment of the Advanced Traffic Management Systems (ATMS) and the Advanced Traveler Information Systems (ATIS) in urban areas. Such operations include the estimation of O-D trip matrices and the dynamic assignment of demand onto the network, and, subsequently, the deployment of individual (in-vehicle) and collective (e.g., Variable Message Sign or web-based) route guidance systems. Moreover, data on turning flow proportions at signalized intersections can provide valuable information for traffic simulation studies, such as those based on the well-known TSIS platform [9], which are used for generating and evaluating alternative traffic scenarios.

The collection and processing of detector data for estimating turning proportions at signalized intersections of realistic urban networks tend to be highly expensive and time-consuming. For instance, the modeling effort involved in the

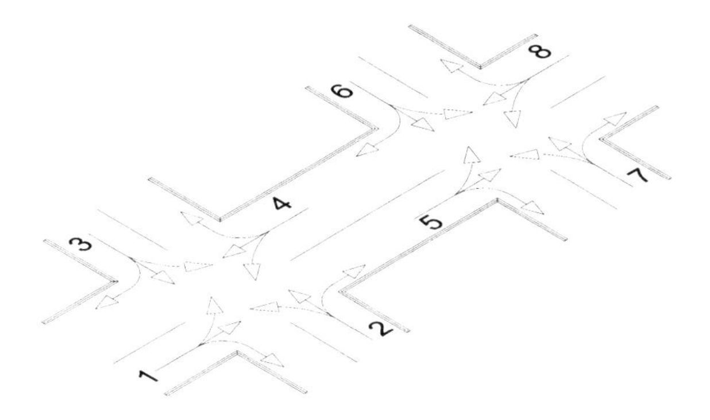

Fig. 1. Turning movements on consecutive intersections of two-way streets

case of a small part of an urban network, composed of a set of two intersections and two-way streets (see Fig. 1), requires six entry flows and twenty four turning flow proportions, apart from the data needed to represent the geometric characteristics, the signal control strategy and other special features, including the characteristics of vehicles and the behavior of drivers. Also, the amount of detector data actually available is typically much less than the amount of data required for measuring the complete set of turning proportions at all signalized intersections, even for the case of an isolated urban arterial.

This fact implies that the problem is underspecified and strongly dependent on the availability, quality and variability of the partial aggregate information about the entry flows and turning proportions. The estimation of turning proportions at signalized intersections with partial traffic counts can be expressed through different types of optimization problems. Such problems include non-recursive (constrained) Least Squares [6], (discounted) recursive Least Squares, Kalman filtering [20], iterative maximum likelihood technique [21], linear programming [19] and prediction error minimization [17]. Nonetheless, the performance of such heuristic optimization methods cannot adequately address the aforementioned problems. Specifically, it has been found to decrease when observed traffic flows are time-dependent and of increased variability, as it typically occurs for the case of real urban arterial networks. Furthermore, it is typically affected by several complicated but crucial modeling tasks, including those corresponding to the representation of queues and platoon dispersion and the appropriate selection of time intervals involved in the estimation process [2].

The adoption of intelligent computational approaches is promising for addressing the above limitations, since they can reduce the level of dependency of the resulting turning proportions on the size and quality of input data. This study employs a stochastic meta-heuristic estimation model of global optimization, which refers to a GA for estimating intersection O-D turning movements, by using as input a partial set of readily available aggregate traffic flows, as obtained from automatic detector measurements (see Sect. 4). The objective function of the turning movement estimation sub-problem at each intersection is incorporated here in the overall (composite) objective function of the global time-varying O-D matrix estimation problem at network level. This objective function is expressed by a metric of the difference between estimated and observed flows, which is sought to be minimized subject to a set of constraints. These constraints are related to the nature of the different types of input data and the natural constraints on the permissible values of the turning proportions (see Sect. 2.3).

The estimation of each intersection turning movement sub-problem is carried out through employing a microscopic traffic simulation approach (see Sect. 3.1), which is combined with the GA. This combination, which was originally developed in [8], enables a better treatment of data requirements associated with more detailed and disaggregated information about vehicle characteristics and drivers' behavior in each arterial link and signalized intersection. In addition, the micro-simulation model allows representing the interaction between individual vehicles and the modeling of the driving behavior of users when moving in the network. In this way, it offers a sound behavioral underpinning to the solution of each intersection O-D matrix sub-problem, in comparison to the existing methods for estimating turning proportions. The following subsection describes the simultaneous optimization formulation of the network-wide dynamic O-D matrix estimation problem.

2.3 Formulation of the Simultaneous Dynamic O-D Matrix Estimation

The simultaneous optimization approach has been considered to address a number of transportation network problems, which can be provided through a bi-level programming formulation (see [28]), including O-D trip matrix estimation. The present formulation ensures a mutually consistent solution between the resulting O-D matrices, link flows and intersection turning proportions at a one-step process. Consider a network composed of G nodes, E directed links, L and M be the number of origins and destinations from and to which vehicular trips are allocated and W be the amount of O-D pairs traversed by flows. Let $\mathbf{x}$ be the (unknown) network O-D matrix whose elements $\tilde{x}_{lm}^{\tau}$ denote the number of vehicular trips departing from origin $l = 1, \ldots, L$ to destination $m = 1, \ldots, M$ during estimation interval $\tau \in T$ and contributing to the flows traversing links during count interval $t \in \tau$, where T is the study period that typically refers to the (morning or afternoon) peak travel period. The trip demand $\tilde{x}_{lm}^{\tau}$ gives rise to path flows p_{lm}^{τ} between each $l - m$ pair. Also, consider J be the total number of observed links, which are equipped with a traffic counter and are traversed by flows that have exited from the upstream node $g \in G_{lm}^{*}$, with G_{lm}^{*} be the set of nodes traversed by the feasible O-D paths between $l - m$ pair, and h_{lmi}^{t} be the flows between $l - m$ pair traversing a link ending at node entrance i during t, with I be the number of node entrances. Moreover, y_j^t and $\tilde{y}_j^t$ denote respectively the variables of the measured and estimated (assigned)flows traversing the observed link j during count interval t and b_{ij}^{lmt} are the proportions of flow between $l - m$ pair turning from entrance i to link j at interval t.

Then, the simultaneous dynamic O-D matrix estimation at an interval $\tau \in T$ can be expressed as a multi-objective optimization problem, through minimizing a composite function F, i.e., a weighted function composed of the Mean Absolute Relative Error ($MARE$) of the resulting O-D trip matrix and link flow estimates, as follows:

$$\min F(\mathbf{x}, \tilde{\mathbf{y}}, \mathbf{b}) = \left\{ \gamma_1 \left[\frac{1}{W} \sum\nolimits_l^L \sum\nolimits_m^M \frac{|x_{lm}^{\tau} - \tilde{x}_{lm}^{\tau}|}{x_{lm}^{\tau}} \right] + \gamma_2 \left[\frac{1}{Z} \sum_{t \in \tau} \sum\nolimits_j^J \frac{|y_j^t - \tilde{y}_j^t|}{y_j^t} \right] \right\} \tag{1}$$

subject to the constraints:

$$\tilde{x}_{lm}^{\tau} = \sum\nolimits_l^L \sum\nolimits_m^M p_{lm}^{\tau} \tag{2}$$

$$p_{lm}^{\tau} = \sum_{t \in \tau} \sum_{g \in G_{lm}^*} \sum\nolimits_i^I h_{lmi}^t \tag{3}$$

$$h_{lmi}^t = \sum\nolimits_l^L \sum\nolimits_m^M \sum\nolimits_k^K \tilde{y}_k^t \tag{4}$$

$$\tilde{y}_j^t = \sum\nolimits_l^L \sum\nolimits_m^M \sum\nolimits_i^I b_{ij}^{lmt} h_{lmi}^t \tag{5}$$

$$\sum\nolimits_j^J \tilde{y}_j^t \leq \sum\nolimits_k^K \tilde{y}_k^t \tag{6}$$

$$0 \leq b_{ij}^{lmt} \leq 1 \tag{7}$$

$$\sum\nolimits_{k \neq i}^K b_{ik}^{lmt} = 1 \tag{8}$$

The scalar Z in function F is given as $Z = J_T \times V$, with J_T be the sum of all links traversed by observed flows that have exited from an upstream node and V be the traffic flow variables (e.g., volume, occupancy, speed) measured by traffic counters. The index k denotes any (observed or unobserved) link traversed by flows that have exited from a node $g \in G_{lm}^*$, with $K \geq J$ be the total number of links traversed by flows exited from that node. The reliability weights γ_1 and γ_2 express the relative confidence assigned to each of the two sources of information, i.e. the prior network O-D matrix, with elements x_{lm}^{τ} (see Sect. 3), and the traffic counts respectively, on the performance of the estimation process. The objective function (1) incorporates the effects of changes in the amount of O-D pairs and traffic counters, the size of O-D demand and measured traffic flow, and the type of traffic flow variable. The equation constraint (2) imposed on O-D trip flows $\tilde{x}_{lm}^{\tau}$, equation constraints (3)–(6) imposed on link traffic flows $\tilde{y}_j^t$ and physical constraints (7)–(8) imposed on turning flow proportions b_{ij}^{lmt} ensure the simultaneous production of a mutually consistent solution between these three variables in each set of count intervals t wherein an estimation interval τ is partitioned.

The above problem formulation circumvents the need for estimating the dynamic user equilibrium (UE) conditions pertaining to the DTA procedure, which are related to the equalization of the travel costs (times) experienced by travelers in all used paths of each $l-m$ pair. In particular, the estimation of the b_{ij}^{lmt} proportions in each node $g \in G_{lm}^{*}$ allows the endogenous construction of the travel path flows and, hence, the trip demand between each $l-m$ pair, without relying on the degree to which the UE-based paths connecting the various O-D pairs contribute to the observed flow at link j, as occurs in the DTA-based methods. The turning proportion estimation and the endogenous path construction are achieved here through a dynamic non-equilibrium procedure of microscopic traffic simulation augmented with a genetic algorithm, as they are described in the following section.

3 A Combined Genetic Computation Approach

3.1 The Microscopic Traffic Simulation Procedure

The present study employs a microscopic simulation model to enable the detailed representation of the urban traffic flow characteristics at the level of each vehicle. The simulation logic of the model relates to the similarity between traffic flow and moving particles and the use of simplified rules to represent the movement and interaction between vehicles. Such an approach can more efficiently handle the information requirements of dynamic traffic operations in urban networks of realistic size, in comparison to the coarse use of macroscopic and analytical traffic assignment models (see [31]). The current approach involves a well-documented traffic micro-simulation procedure for tracking the spatial and temporal trajectory of each vehicle, based on a car-following (acceleration/deceleration) model of collision avoidance logic, as coded in the NETSIM platform [9]. The model also performs detailed representation of other behavioral features of drivers, such as lane changing and gap acceptance.

The use of such a model can better address complexity pertaining to the operational and geometric characteristics and the underlying traffic dynamics of signalized arterial networks. In particular, it enables the coding of such network characteristics as signalized traffic control strategies, transit operations, pedestrian and parking activities. The loading of the network is carried out through providing information about three different types of input data: (a) network-wide data, including network topology, behavioral characteristics of drivers and composition of traffic, (b) partial set of real-time measurements of flows at the entry nodes, including the origin and destination zones, and turning proportions at each node, and (c) signal control strategies. The resulting path flow pattern can involve the use of alternative feasible routes between an O-D pair, which intrinsically reflects the existence of different cost perceptions of travelers. This stochastic dispersion of path flows can be considered as a more realistic route choice assumption, in comparison to that relying on the deterministic (shortest path-based) behavior.

3.2 Description of the Genetic Computation Procedure

The present optimization problem, as described in Sect. 2.3, is convex and, hence, it may ensure a feasible and unique solution for the network O-D matrix estimation process. Nonetheless, the intricate nature of the solution procedure, principally due to the increased dimensionality, complexity of the search space and existence of many local minima in realistic urban networks, imposes an uncertainty in reaching an optimal (or satisfactory sub-optimal) solution. The use of evolutionary computing techniques, such as the GA presented here, can address this uncertainty by relaxing the solution dependencies on the availability, quality and variability of the prior demand and traffic count information, in contrast to the traditional greedy search algorithms. Moreover, the event-based simulation procedure (see Sect. 3.1) employed for processing the movement trajectories of individual vehicles makes the problem highly complex and non-linear so that favors the application of evolutionary computing techniques, instead of using gradient-based methods which commonly appear in the existing literature (see Sect. 2.1).

The combination of the GA with the micro-simulation model (see Sect. 3.1) can provide a powerful computational tool for the given optimization problem. More specifically, the population-based global search process of the GA can address the influence of uncertainty (or variability) of entry flows, missing information about turning proportions and high non-linearity and complexity involved in simulating traffic flow on the resulting solution. In this way, GA can eliminate (or reduce) the effect of such sources of bias in the estimation process and help reach an optimal (or satisfactory sub-optimal) solution.

The solution procedure is performed here with the repeated execution of two stages at each count interval $t \in \tau$, until achieving a satisfactory level of accuracy, as this is expressed by some performance measure (see below). These two stages, which are illustrated in Fig. 2, correspond to (i) the simulation-based microscopic estimation of the link traffic flows and O-D flows are compared with the measured ones (convergence test), and (ii) the execution of the GA operations. Following the initialization of the algorithm, at the first stage the micro-simulation model utilize an initial population of intersection turning flows and O-D trip flows.

In the case where the convergence (or stopping) criterion is satisfied, then, the currently estimated network O-D matrix is the final one, and the algorithm proceeds to the next interval. Otherwise, a 'genetically improved' population of turning proportions for each intersection is produced by the GA operations in order to feed the micro-simulation model. The fitness function provides a metric of the difference between the observed and estimated (simulated) traffic flows at certain measurement locations (sections) of the network. The objective function of the intersection O-D matrix sub-problem refers to the minimization of the fitness function, which is expressed here with the measure of the Mean Absolute Relative Error (MARE), i.e. the second component of function F in equation

(1). GA structures like the present one have been extensively used for addressing problems with composite objective functions, such as multi-objective problems [13].

Since the existing literature concerning the theory, design and implementation of GAs is extensive (e.g., see [5], [13], [15], [22]), the description of the present GA operators will only focus on the current application. Specifically, the GA utilizes a set (population) of strings called chromosomes, each representing a feasible matrix of b_{ij}^{lmt} proportions and, consequently, a feasible solution to the problem. Every entry in the chromosome is called allelic value. The representation of the chromosomes and the assignment of allelic values are based on a binary $\{0, 1\}$ coding scheme (see Sect. 4) to enhance explorability, although a real-numbered coding scheme could be also used. The members of the initial population are created by random perturbation, across a specific range, of those b_{ij}^{lmt} values obtained through the simulation-based assignment of the prior O-D matrix onto the network.

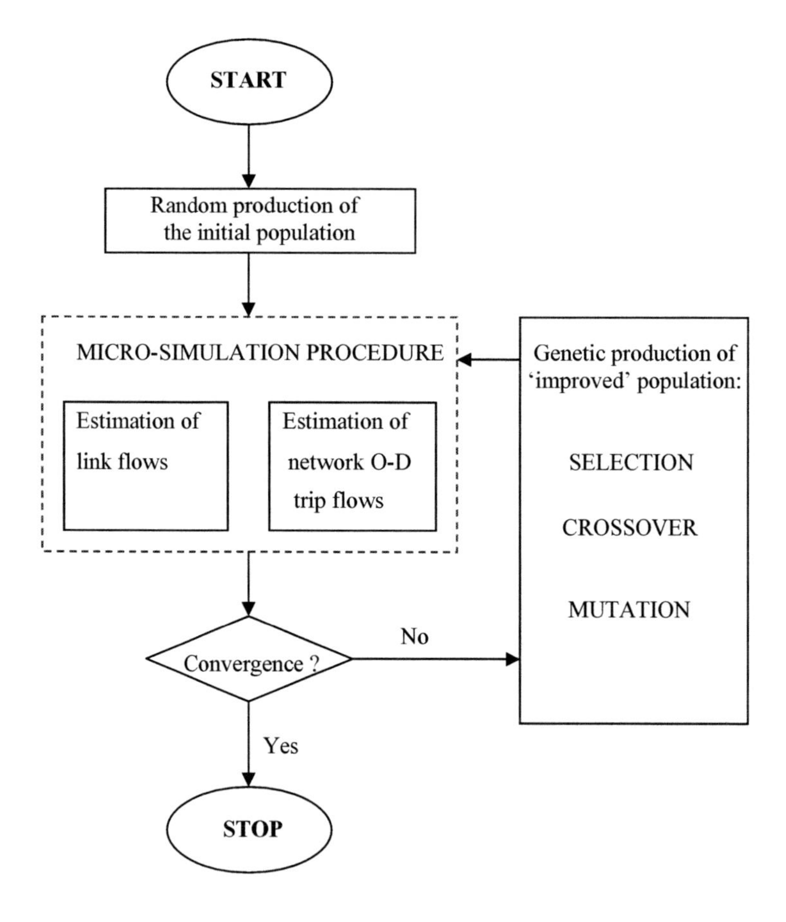

Fig. 2. Flowchart of the combined genetic computation procedure of the model

The algorithm commences with an initial random population of solution sets. Then, the fitness function value f_s is calculated for each string (member) s of the population, which represents a solution set. In the present application, the initial population size is empirically chosen to be rather small (50 individuals). The setup of the GA initialization relies on ensuring a favorable tradeoff between convergence speed and population size for the problem at hand, as it is described in [13]. Namely, the larger the size of the population is, the longer the time required for the calculation of the fitness function in each generation. On the other hand, the probability of convergence increases with the population size. The GA processes carried out in the current application include three phases, i.e. selection, crossover and mutation, as they are analytically described below.

Selection
The survival of a string s according to the value f_s of its fitness function implies that strings with a high value have a higher probability of contribution to one or more offspring in the next generation. The easiest way to implement the reproduction operator is to create a biased roulette wheel, where each string in the current population has a roulette wheel slot sized proportionally to its fitness function value. Given a population with n members, the following formula can be used to calculate the slot size in the roulette wheel with respect to the reproduction probability:

$$\Pi_s = \frac{f_s}{\sum_s^n f_s} \tag{9}$$

where Π_s is the probability of a string s to participate in the production of the new generation.

Crossover
The crossover operation enables the exchange of genetic information between the old population members in order to obtain the new ones. This exchange is carried out by randomly selecting slots among the strings and swapping the characters between the selected positions. Figure 3 illustrates the crossover operator improvement scheme. In relation to the common practice [13], a relatively high rate of crossover is often selected (>50%), which denotes the probability of the selected individuals to exchange genetic information. A crossover rate equal to 70% is chosen here in order to enhance the probability of the selected individuals to exchange genetic information.

Mutation
The mutation operator plays a secondary role with respect to the selection and crossover operators. Nevertheless, mutation is needed to prevent an irrecoverable loss of potentially useful information, which can be occasionally caused by reproduction and crossover. Therefore, it can enhance the possibility of exploring the whole search space, reducing the dependence from the initial population and the probability of finding a false peak. This operator provides an occasional

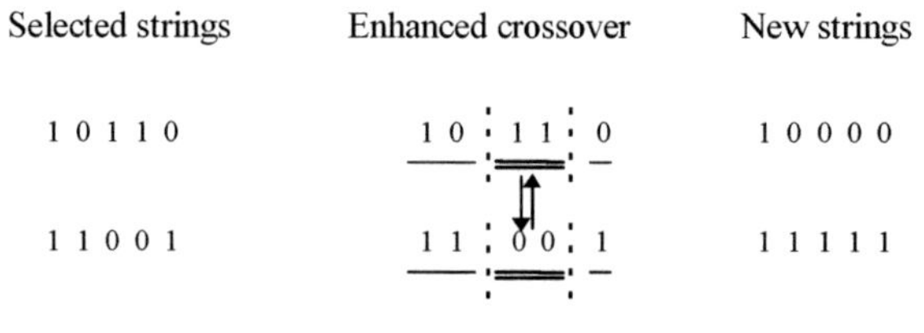

Selected strings | Enhanced crossover | New strings

1 0 1 1 0 | 1 0 : 1 1 : 0 | 1 0 0 0 0

1 1 0 0 1 | 1 1 : 0 0 : 1 | 1 1 1 1 1

Fig. 3. Illustration of the crossover operation

random alteration of the allelic value of a gene with a small probability, which is set equal to 5% in the present application.

4 Experimental Setup

The area of the present application corresponds to a part of the urban road network of Athens, Greece. This area is located in the periphery of the city centre and covers a major arterial sub-network (Alexandras Avenue), which is controlled with a fixed time signal strategy (1.5 Generation). The specific arterial network (see Fig. 4) is composed of 46 links and 32 nodes, 16 of which are entry-exit nodes and 16 are internal nodes. Hence, the dimensions of the O-D matrix are 16×16 and those of the matrix of turning proportions are 46×4, where 4 corresponds to the total number of possible turning movements, i.e. left, through, right and diagonal (see Table 1).

Due to the absence of diagonal movements and prohibitions on several left turning movements, the number of (unknown) turning proportions reduces from $46\times4 = 184$ to 66 (excluding proportions equal to 0 and 100). The GA population is composed of 50 individuals and the coded values for each turning movement lie between 255 (11111111) and 1 (00000001), since a number of 8 alleles is adopted to correspond to each link and turning proportion at each node. Thus, the length of each chromosome is $66 \times 8 = 528$ alleles. The coded values of the turning movements are finally transformed to turning flow proportions.

The estimation process utilizes loop detector measurements corresponding to entry flows at the 16 entry-exit nodes and to link traffic flows traversing 4 selected sections along the major arterial, for both flow directions, i.e. $J_T = 8$. The link flow measurements refer to both smoothed volume and occupancy values, namely, $V = 2$. These four measurement sections, whose location is shown by the markings of Fig. 4, facilitate the constant monitoring of traffic operating conditions along the entire length of the sub-network. The traffic data are collected at the end of every 90-sec signalization cycle and are transmitted to the Traffic Control Center of the city. The traffic measurements of the current data set are aggregated at 3-min count intervals and correspond to a typical peak hour of the morning travel period of the day. A time-partitioned (partial) O-D matrix corresponds here to a typical estimation interval τ of 15 min, since

Table 1. The matrix of turning proportions in the study area

Link No	Upstream Node	Downstream Node	Proportion of Flow Turning: Left	Through	Right	Diagonal
1	801	69	35	65	0	0
2	802	69	0	36	64	0
3	69	58	6	94	0	0
4	58	57	0	100	0	0
5	57	56	0	81	19	0
6	56	55	0	83	17	0
7	55	54	0	100	0	0
8	54	53	6	94	0	0
9	53	52	0	100	0	0
10	52	51	0	90	10	0
11	51	50	0	96	4	0
12	50	49	0	100	0	0
13	49	48	0	85	15	0
14	48	47	0	100	0	0
15	47	46	0	100	0	0
16	46	45	0	95	5	0
17	45	43	37	55	8	0
18	43	45	7	93	0	0
19	45	46	0	100	0	0
20	46	47	0	100	0	0
21	47	48	5	91	4	0
22	48	49	0	100	0	0
23	49	50	20	80	0	0
24	50	51	0	100	0	0
25	51	52	0	100	0	0
26	52	53	0	75	25	0
27	53	54	0	100	0	0
28	54	55	36	64	0	0
29	55	56	0	100	0	0
30	56	57	0	100	0	0
31	57	58	0	93	7	0
32	58	69	44	0	56	0
33	804	58	0	79	21	0
34	805	57	58	0	42	0
35	808	54	51	0	49	0
36	809	52	100	0	0	0
37	810	51	3	65	33	0
38	812	49	32	0	68	0
39	813	50	37	34	29	0
40	814	48	26	27	47	0
41	815	47	0	0	100	0
42	820	45	29	71	0	0
43	819	45	7	65	27	0
44	821	43	6	94	0	0
45	823	43	0	97	3	0
46	822	43	0	69	31	0

such a duration can be considered as adequate for a user to traverse the given part of the network.

The convergence (or termination) of the GA is based on two different empirical criteria. The first criterion refers to the estimation accuracy of the solution procedure, which is set equal to a small value of objective function $F = 0.05$ (or 5%) for the computational purposes of the given problem. The second (stopping) criterion is related to the intended practical usage of the model. The current application concerns the real-time deployment of an area-wide network traffic

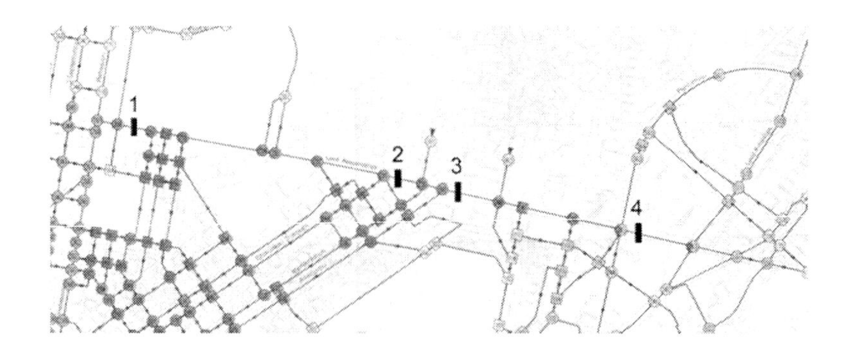

Fig. 4. The layout and coding of Alexandras sub-network

monitoring and control system. For this reason, each partial O-D matrix is regularly updated on a rolling horizon framework, according to the frequency of collecting and processing traffic flow information. In the present study, this frequency refers to a count interval t of 3 min. Thus, a maximum running time of 3 min is set as the stopping criterion for the purposes of the specific application.

The prior O-D matrix was synthesized through the offline implementation of the proposed simultaneous optimization process by using a set of 'typical' (average) traffic flows over the observed links for the given study period. These 'typical' flows were obtained from averaging a series of volumes and occupancies corresponding to the specific period-of-the-day of the past four weeks. This study uses the assumption of the similar reliability and equal contribution of the prior demand and traffic count information on the performance of the estimation process, i.e. $\gamma_1 = \gamma_2 = 0.5$. The O-D trip flows are expressed here in the form of trip rates, as obtained by the ratio of the trip demand across a specific O-D pair to the total demand size in the network. Such a transformation helps relax the dependency of the solution on the scale of demand.

5 Computational Results

The performance of the estimation procedure is investigated with regard to both the solution accuracy and the convergence speed, based on the convergence criteria described in Sect. 4. The micro-simulation of the present network processed a total amount of approximately 2000 vehicle-trips during the given study period. The GA, which was coded in FORTRAN 90/95 workstation, is found to reach a stable solution with the desired level of accuracy, i.e. $F = 5\%$, in an average computing duration of less than 3 min, requiring at most a number of about 2000 runs / fitness evaluations (or 40 generations).

The typical convergence behavior of the GA involves a number of 20–30 generations, while a maximum number of 40 generations is only required for cases of population members of poor quality, i.e. with very small initial fit (less than 5% of the cases). Figure 5 indicates the best value and the mean value of the

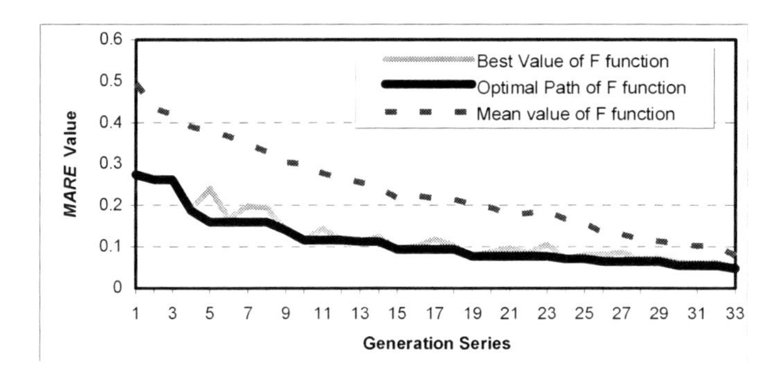

Fig. 5. Results of the objective function F of the GA

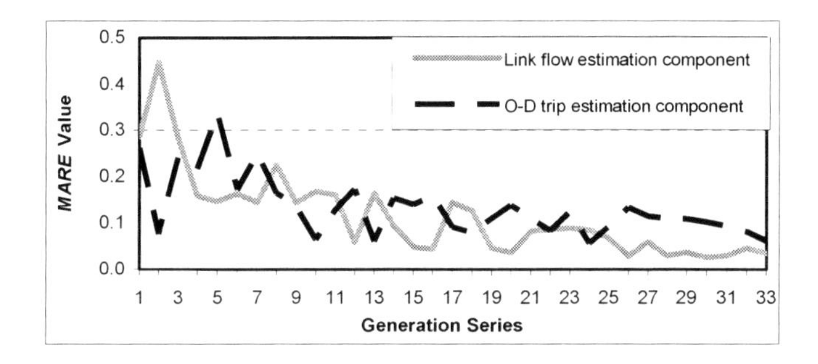

Fig. 6. Results of the O-D matrix and link flow estimation components of the objective function

objective function F in each generation as well as its optimal path across the series of generations for the first (15-min) estimation interval of the study period. The graph shows that the GA performance, in terms of the reduction of the F function, is steadily increasing in the first 20 generations, while convergence is achieved after 30 generations.

Figure 6 illustrates the convergence behavior of the two different components of the objective function, i.e. those of the O-D trip matrix and link flow estimation. The graph indicates a tradeoff mechanism between the $MARE$ of the two components across generations. Nonetheless, the $MARE$ of both components reduces considerably, in the context of the given application, reaching finally a similar level of accuracy, as the GA approaches to convergence. Specifically, the $MARE$ of the O-D matrix estimation component presents an average reduction of 77%, while the $MARE$ of the link flow estimation component presents an average reduction of 88%.

This outcome signifies the fact that the mutually consistent solution achieved through the simultaneous optimization process does not lead to the loss of accuracy of the resulting O-D matrix, in favor of the link flow solution accuracy, as it

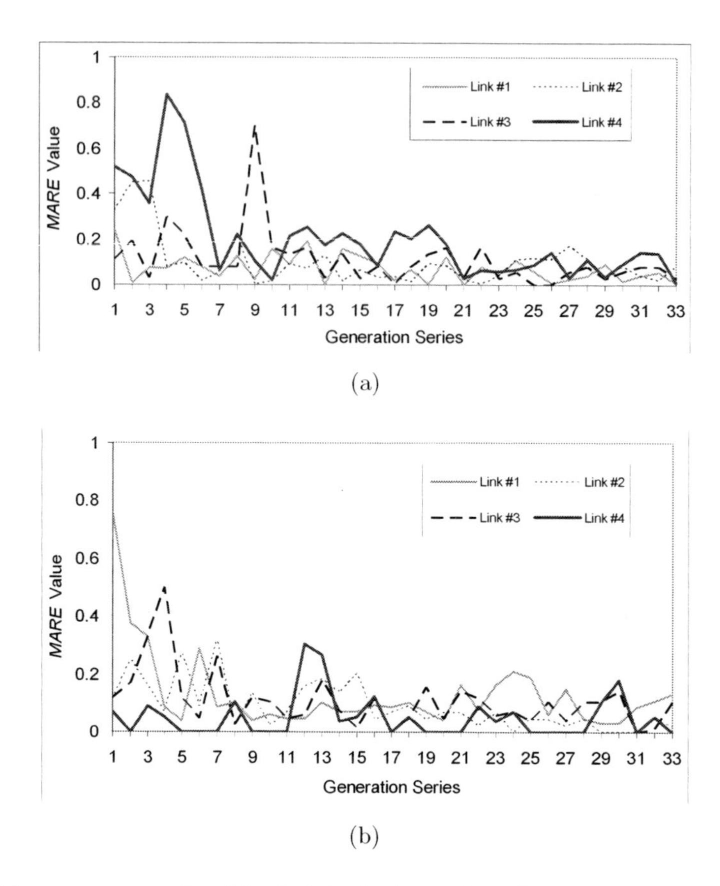

Fig. 7. Convergence results of the link traffic flow estimation at measurement sections of (a) the west-east direction and (b) the east-west direction of the study arterial

typically occurs in the case of the iterative optimization process [24]. In particular, the average improvement of the accuracy of the link flow solution, in terms of the $MARE$, is found to exceed 75% for all measurement sections considered in the study. Figures 7(a) and 7(b) present in detail the GA convergence procedure in terms of the link flow estimation corresponding to the four measurement sections on each direction (west-east and east-west) along the major arterial.

6 Conclusions

This chapter presented a combined genetic computation method for solving the problem of the simultaneous dynamic O-D trip matrix estimation from traffic counts in urban networks. The current approach can ensure the production of a mutually consistent solution between the O-D matrix and the path/link flow loading pattern, circumventing the use of a DTA procedure. In particular, the proposed method relies on the real-time estimation of the turning movements at each intersection, employing data about the prior O-D matrix structure, measurements on entry node flows and link traffic flows, and network loading information obtained from a micro-simulation model. This model is augmented with a genetic algorithm (GA) in order to carry out a stochastic global search and simultaneously estimate the optimal distribution of the intersection O-D turning flows and network O-D trip flows. Such an approach can suitably handle the uncertainty of input information and the complexity of the traffic simulation procedure, and it helps reach an optimal (or satisfactory sub-optimal) solution. The microscopic representation of traffic flows allows tracking the O-D trip demand between each entry-exit node pair at the level of individual vehicle and, hence, it provides the possibility of full monitoring of the network traffic state.

The empirical results from the model implementation into a real urban arterial sub-network showed that the GA is capable of providing substantial gains to both the network O-D matrix and link flow estimation accuracy, in comparison to the initial solution based on the prior demand information. In addition, the computing speed of the algorithm can be considered as satisfactory, taking into account the complexity and processing requirements of the micro-simulation model and the realistic size of the subarea network. The high accuracy of both the resulting O-D trip matrix and link traffic flows, in conjunction with the fast computing speed, indicate the potential of implementing the suggested procedure for purposes of real-time monitoring and control of urban network conditions. In addition, the present approach can be used as a prototype to support the design and evaluation of individual responses to a number of dynamic traffic management strategies. Such strategies can include the deployment of signal plan coordination along urban arterial corridors or subarea networks, collective or personalized route guidance information systems, road pricing and network access control schemes and integrated traffic management measures including combinations of the above strategies.

References

1. Ashok, K., Ben-Akiva, M.E.: Dynamic O-D matrix estimation and prediction for real-time traffic management systems. In: Daganzo, C.F. (ed.) Proceedings of the 12th International Symposium on Transportation and Traffic Theory, pp. 465–484. Elsevier, New York (1993)
2. Bell, M.G.H.: The real time estimation of origin-destination flows in the presence of platoon dispersion. Transportation Research, Part B: Methodological 25(2/3), 115–125 (1991)
3. Ben-Akiva, M., Bierlaire, M., Burton, D., Koutsopoulos, H.N., Mishalani, R.: Network state estimation and prediction for real-time traffic management. Networks and Spatial Economics 1, 293–318 (2001)
4. Cascetta, E., Inaudi, D., Marquis, G.: Dynamic estimators of origin-destination matrices using traffic counts. Transportation Science 27, 363–373 (1993)
5. Coello, C.A., Lamont, G.B., van Veldhuizen, D.A.: Evolutionary Algorithms for Solving Multi-Objective Problems. Springer, New York (2006)

6. Cremer, M., Keller, H.: A new class of dynamic methods for the identification of origin-destination flows. Transportation Research, Part B: Methodological 21(2), 117–132 (1987)
7. Diakaki, C., Papageorgiou, M., Aboudolas, K.: A multivariable regulator approach to traffic-responsive network-wide signal control. Control Engineering Practice 10(2), 183–195 (2002)
8. Dimitriou, L., Tsekeris, T., Stathopoulos, A.: Genetic-algorithm-based microsimulation approach for estimating turning proportions at signalized intersections. In: van Zuylen, H., Middelham, F. (eds.) Proceedings of the 11th IFAC Symposium on Control in Transportation Systems, pp. 159–164. Delft, The Netherlands (2006)
9. Federal Highway Administration. Traffic Network Analysis with NETSIM – A User Guide. Federal Highway Administration, U.S. Dept. of Transportation, Washington (1980)
10. Fisk, C.S.: On combining maximum entropy trip matrix estimation with user optimal assignment. Transportation Research, Part B: Methodological 22, 69–73 (1988)
11. Florian, M., Chen, Y.: A coordinate descent method for the bi-level O-D matrix adjustment problem. International Transactions in Operational Research 2, 165–179 (1995)
12. Gartner, N.H., Pooran, F.J., Andrews, C.M.: Implementation of the OPAC adaptive control strategy in a traffic signal network. In: Proceedings of the 4th IEEE Conference on Intelligent Transportation Systems, pp. 197–202. IEEE, Oakland (2001)
13. Goldberg, D.E.: Genetic Algorithms in Search, Optimization and Machine Learning. Addison-Wesley, Reading (1989)
14. He, R.R., Kornhauser, A.L., Ran, B.: Estimation of time-dependent O-D demand and route choice from link flows. In: Proceedings of the 81st Annual Meeting of the Transportation Research Board, National Research Council, Washington (2002)
15. Holland, J.H.: Adaptation in Natural and Artificial Systems, 2nd edn. MIT Press, Cambridge (1992)
16. Kim, H., Baek, S., Lim, Y.: Origin-destination matrices estimated with a genetic algorithm from link traffic counts. Transportation Research Record 1771, 156–163 (2001)
17. Lan, C.J., Davis, G.A.: Real-time estimation of turning movement proportions from partial counts on urban networks. Transportation Research, Part C 7(5), 305–327 (1999)
18. Luk, J.Y.K.: Two traffic-responsive area traffic control methods: SCAT and SCOOT. Traffic Engineering and Control 25, 14–22 (1984)
19. Martin, P.T.: Turning movement estimation in real time. Journal of Transportation Engineering 123(4), 252–260 (1997)
20. Nihan, N.L., Davis, G.A.: Recursive estimation of origin-destination matrices from input/output counts. Transportation Research, Part B: Methodological 21(2), 149–163 (1987)
21. Nihan, N.L., Davis, G.A.: Application of prediction-error minimization and maximum likelihood to estimate intersection O-D matrices from traffic counts. Transportation Science 23(2), 77–90 (1989)
22. Reeves, C.R., Rowe, J.E.: Genetic Algorithms – Principles and Perspectives: A Guide to GA Theory. Kluwer, Boston (2002)
23. Stathopoulos, A., Tsekeris, T.: Hybrid meta-heuristic algorithm for the simultaneous optimization of the O-D trip matrix estimation. Computer-Aided Civil and Infrastructure Engineering 19, 421–435 (2004)

24. Tavana, H., Mahmassani, H.: Estimation of dynamic origin-destination flows from sensor data using bi-level optimization method. In: Proceedings of the 80th Annual Meeting of the Transportation Research Board, National Research Council, Washington (2001)
25. Tsekeris, T., Dimitriou, L., Stathopoulos, A.: A simultaneous origin-destination matrix estimation in dynamic traffic networks with evolutionary computing. In: Giacobini, M., et al. (eds.) EvoWorkshops 2007. LNCS, vol. 4448, pp. 668–677. Springer, Berlin (2007)
26. Tsekeris, T., Stathopoulos, A.: A real-time dynamic origin-destination matrices from link traffic counts on congested networks. Transportation Research Record 1857, 117–127 (2003)
27. van der Zijpp, N., Lindveld, C.D.R.: Estimation of origin-destination demand for dynamic assignment with simultaneous route and departure time choice. Transportation Research Record 1771, 75–82 (2001)
28. Yang, H., Bell, M.G.H.: Transport bilevel programming problems: Recent methodological advances. Transportation Research, Part B: Methodological 35, 1–4 (2001)
29. Yang, H., Sasaki, T., Asakura, Y.: Estimation of origin-destination matrices from link traffic counts on congested networks. Transportation Research, Part B: Methodological 26, 417–434 (1992)
30. Zhang, X., Maher, M.: An algorithm for the solution of bi-level programming problems in transport network analysis. In: Griffiths, J.D. (ed.) Mathematics in Transport Planning and Control, pp. 177–186. Pergamon, Oxford (1998)
31. Ziliaskopoulos, A.K., Peeta, S.: Review of dynamic traffic assignment models. Networks and Spatial Economics 1, 233–267 (2001)

Genetically Optimized Infrastructure Design Strategies in Degradable Transport Networks

Loukas Dimitriou[1], Theodore Tsekeris[2], and Antony Stathopoulos[1]

[1] Department of Transportation Planning and Engineering, School of Civil Engineering, National Technical University of Athens, Iroon Polytechniou 5, 15773 Athens, Greece
lucdimit@central.ntua.gr, astath@transport.ntua.gr
[2] Centre for Planning and Economic Research, Amerikis 11, 10672 Athens, Greece
tsek@kepe.gr

Summary. This chapter examines the problem of the resource allocation in degradable road transport networks within a stochastic evolutionary optimization framework. This framework expresses the stochastic equilibrium Network Design Problem (NDP) as a game-theoretic, combinatorial bi-level program. Both the discrete and continuous versions of the reliable NDP are considered in order to address different strategies of network infrastructure investment. The estimation procedure employs a Latin Hypercube sampling method for simulating degradation-inducing variations in users' attributes and system characteristics, and hence evaluates the network travel time reliability which constrains the solution. This simulation-based risk assessment technique is combined with a genetic algorithm to handle the complex, non-convex nature of the NDP adequately. The test implementation of the proposed framework demonstrates the significant role of incorporating the stochasticity and reliability requirements in the design process to facilitate the selection of the optimal investment strategies in degradable road networks.

Keywords: Degradable transport networks, Discrete and continuous system structures, System reliability, Stochastic optimum network design, Game theory, Latin hypercube sampling, Genetic algorithms.

1 Introduction

The design of transport networks involves the optimal allocation of infrastructure investments according to specific social, economic, operational and physical criteria. In the context of contemporary metropolitan areas, the road network design is vital to ensure the efficient and safe mobility of people as well as goods. Such a design process typically encompasses several conflicting objectives, including the minimum possible consumption of available land and public fund resources, and the maximum possible provision of infrastructure resources to satisfy the increased demand for passenger and freight transport. Although the provision of infrastructure resources is commonly regarded solely with respect to road capacity expansion, the requirement to ensure a minimum level of service reliability has received an ongoing attention in the literature of the last decade,

A. Fink and F. Rothlauf (Eds.): Advances in Computational Intelligence, SCI 144, pp. 23–43, 2008.
springerlink.com

particularly since the seminal study of Du and Nicholson [13]. The present study uses the concept of the Total Travel Time (TTT) reliability, which denotes the network's ability to respond to the various states of the system variables. This ability can be manifested both in everyday operating conditions and especially during incidents or unpredictable events, like man-made or physical disasters, which potentially degrade the expected (planned) performance of the network. Particularly in the latter case, a reliable design process is supposed to ensure the provision of the sufficient lifelines to support various emergency (for instance, evacuation) operations.

Nonetheless, existing studies mostly merely consider the reliability of one single component of the transport system, such as the one of capacity or travel time (see Sect. 3). The current study provides a unified modeling framework for the evaluation of the way the degradation-inducing events impact on the design process. Specifically, this framework examines the effects of fluctuations in link capacities and hence travel times on condition of demand uncertainty, on system reliability in terms of the network TTT. Furthermore, this study recognizes that alternative infrastructure investment policies can give rise to different design processes. For this reason, the problem of determining the optimum network design known as the Network Design Problem (NDP) is formulated, estimated and analyzed in two different mathematical forms. These forms, whose general reviews can be found in [24], [15], refer to the Continuous-NDP (C-NDP) and the Discrete-NDP (D-NDP).

The C-NDP represents the capacity of the system as a continuous variable, which can be expressed in terms of vehicles, passengers and unit loads. The D-NDP is formulated in terms of discrete (integer or binary) variables, such as the number of new links in the case of network expansion, or the number of lane additions in the case of network enhancement. The selection of the mathematical form of the NDP can rely on different infrastructure investment decision-making mechanisms and planning needs. In particular, the C-NDP is service-oriented, as it refers to the demand units to be served in the network. On the contrary, the D-NDP is infrastructure oriented, as it refers to the number of added lanes and new links to be constructed. The majority of the existing studies deals with the C-NDP, which can be regarded as a relaxation of the D-NDP. Moreover, although several reliability considerations have been incorporated into the structure of the C-NDP [37], [39], [34], no such attempt has been made for the case of the D-NDP. Therefore, there is a need to jointly consider and comparatively evaluate both the discrete and continuous versions of the reliable NDP, that is, the NDP with reliability requirements.

This study formulates the two versions of the reliable NDP within a game-theoretic, combinatorial bi-level optimization framework, which can help represent conflicting objectives involved in the network design process (see Sect. 2). The evaluation of the TTT reliability is carried out by use of mathematical simulation, i.e. the Latin Hypercube sampling method, which represents the stochastic nature of the system variables (see Sect. 3). A genetic optimization procedure is then suggested (see Sect. 4), which is combined with the simulation-based risk assessment technique, for addressing the increased complexity of the solution of the continuous and discrete NDP. Section 5 includes the results obtained by the application of the method to a test network with typical urban road settings, and Sect. 6 concludes.

2 Description of the Equilibrium NDP with Reliability Constraints

As in many other transport planning problems, the network design process is influenced by decisions made on multiple hierarchical levels, concerning both the demand and supply properties of the system [14]. The design process of a transport network (system) can be considered as a game among two players, namely the system designer and the system users, whose individual decisions affect the other's performance. Here, the structure of this game takes the form of a two-stage leader-follower Stackelberg game with perfect information. Specifically, the system designer is the leader who imposes modifications on the network to optimize the system performance, while the users are the followers reacting to alternative design plans. The formulation of such games receives usually the form of bi-level programming problems. For these problems, optimum strategies are sought by taking a number of constraints into account, including those of physical feasibility and budget availability, while regarding the system's demand and supply attributes as known but not necessarily fixed. This study expands the standard game-theoretic, bi-level programming formulation of both C-NDP and D-NDP, so that they include reliability requirements.

Consider a network composed of L links, R origins and S destinations. The travel demand q_{rs} gives rise to equilibrium flows f_k^{rs} along path $k \in K_{rs}$ connecting $r - s$ pair and to equilibrium flows $x_\alpha(y)$ along link α, with $\delta_{\alpha k}^{rs}$ be the path-link incidence variable and $c_{rs}^k \in C^{rs}$ be the cost of traveling along the k th path between $r - s$ pair. The travel cost at equilibrium state of some link α with capacity y_α is denoted as $c_\alpha(x_\alpha(y))$, with y be the maximum link capacity. The maximum capacity is given as $y = y_\alpha + w_\alpha$, where y_α stands for the existing link capacity and w_α for the capacity addition on link α. In the case of the D-NDP, w_α can be expressed as a function of the link or lane capacity, i.e., $w_\alpha = n_\alpha L$, where n_α is the number of lanes which will be added on link α, and L is the effective capacity of an individual lane. Furthermore, $V_\alpha(w_\alpha)$ denotes the monetary expenditures for adding a link α or a lane on link α, B is the total available construction budget for network capacity improvement, and θ is a factor converting monetary values into travel times. Then, the *Upper-Level Problem*, which comprises the objectives of the optimum NDP, and the *Lower-Level Problem*, which provides the path and link equilibrium flows, can be given as follows:

Upper-Level Problem

$$\min_{y} F(x, y) = \sum_{\alpha \in A} (E[c_\alpha(x_\alpha(y), w_\alpha)x_\alpha(y)] + \theta V_\alpha(w_\alpha)) \tag{1}$$

$$\text{subject to} \quad \begin{array}{lll} n_\alpha \leq N_\alpha, & \forall \alpha \in A \quad \texttt{(for the D-NDP)} & (\boldsymbol{a}) \\ w_\alpha \leq W_\alpha, & \forall \alpha \in A \quad \texttt{(for the C-NDP)} & (\boldsymbol{b}) \end{array} \tag{2}$$

$$\sum_{\alpha \in A} (V_\alpha(w_\alpha)) \leq B, \quad \forall \alpha \in A \tag{3}$$

$$P\left(\sum_{\alpha \in A} (c_\alpha(x_\alpha(y), w_\alpha) x_\alpha(y)) \leq T\right) \leq Z, \quad \forall \alpha \in A \tag{4}$$

Lower-Level Problem

$$\min_x G(x) = -\sum_{rs} q_{rs} E\left[\min_{k \in K_{rs}} \{c_{rs}^k\} \, |C^{rs}(x)\right] + \sum_\alpha x_\alpha c_\alpha(x_\alpha) - \sum_\alpha \int_0^{x_\alpha} c_\alpha(\omega) d\omega \tag{5}$$

subject to

$$f_k^{rs} = P_k^{rs} q_{rs} \quad \forall \quad k, r, s \tag{6}$$

$$x_\alpha = \sum_{rs} \sum_k f_k^{rs} \delta_{\alpha k}^{rs} \quad \forall \alpha \in A \tag{7}$$

$$f_k^{rs}, x_\alpha \geq 0 \quad \forall \alpha \in A \tag{8}$$

In the Upper-Level Problem, $F(x, y)$ represents the objective function of the NDP, wherein the first component refers to the travel cost expressed in terms of the expectation E of network *Total Travel Time (TTT)*, and the second component corresponds to the total expenditures (in time units) for capacity improvements. The choice set defined in the alternative relationships (2a) and (2b) reflects the physical restrictions of the link capacity improvements for each type of NDP, with N_α be the maximum allowable number of lanes to be added on link α and W_α be the maximum allowable capacity to be added on link α, while relationship (3) imposes budgetary restrictions. The reliability requirements are introduced in constraint (4) by restricting the probability of the TTT to be lower than or equal to a pre-specified upper limit T, with Z defining the acceptable confidence interval ($0 \leq Z \leq 1$) for this hypothesis. Essentially, such a condition depicts the system's stability [2].

The Lower-Level Problem, which consists of functions (5) to (8), performs the trip demand assignment process, based on the expected (perceived) value E of the path travel cost c_{rs}^k. Specifically, it estimates the response of users to the capacity improvements made at the Upper-Level Problem, by determining the probability P_k^{rs} that a traveler chooses to use path k between $r - s$ pair. The Stochastic User Equilibrium (SUE) model [30] is used here for the assignment of demand onto the network and the solution Method of Successive Averages (MSA) is employed to calculate equilibrium flows.

3 Stochastic Modeling of Network Reliability

3.1 The Stochastic Nature of System Variables

The reliability of transport networks has been examined in literature with the help of a number of definitions, such as travel time reliability, connectivity reliability, network flexibility, network variability and others (e.g., see [2], [25], [33]). These definitions refer to different aspects of the uncertainty involved in the operation of transport systems. As mentioned above, the current study investigates the optimum network design problem subject to Total Travel Time (TTT) reliability requirements. Such a measure of reliability seeks to ensure the stable operation of the system within a pre-specified Level of Service (LoS), since it is defined as the probability that network total travel time will be less than some predefined bound (see Sect. 2). The methodology suggested here employs a holistic approach for modeling the network reliability, in terms of all possible sources of uncertainty pertaining to the system operation.

The operational performance of transport systems typically relies on variables of uncertain nature, as their values are influenced by random events and human decision-making processes. The risk involved in the operation of transport networks can be mainly attributed to the uncertainty incorporated into four different components: demand, supply (capacity), level of service (link travel time) and the users' characteristics (route choice behavior). By and large, travel demand patterns in urban transport networks can be considered as recurrent at typical operating conditions [32]. Nonetheless, demand patterns may experience several disturbances during the operational life of a network, causing significant fluctuations in link travel times. These disturbances can be caused by spatial and temporal (random or non-random) variations of trip flows between Origin-Destination (O-D) pairs, special events in certain network localities, link closures or failures and the nature of the day-to-day route choice process of users. They are additionally influenced by various behavioral features of travelers, which are mostly related to factors involved in the route choice decision-making process, including the perception of travel cost, value of travel time and driving behavior.

Moreover, the variations of supply (capacity) can be considered as a common phenomenon in transport networks. There are several factors influencing capacity, including the composition of traffic, congestion effects, road works, special events, and random phenomena like incidents and adverse weather conditions. Since the link and path travel times are closely related to link capacities, the travel time reliability depends on the fluctuations of link capacities, which are usually referred to as link capacity degradation. The problem of considering network reliability in terms of the link capacity degradation has been extensively investigated in literature [13], [38], [5], [7]. These studies have shown that fluctuations in link capacities lead to fluctuations in the total network capacity and, subsequently, decrease the total travel time reliability. The following subsection presents a mathematical simulation methodology for evaluating the

TTT reliability in relation to the aforementioned components of uncertainty in the operation of transport networks.

3.2 Latin Hypercube Simulation Methodology for Reliability Assessment

The stochastic properties and, in turn, the fluctuations of the values of system variables (i.e. demand, supply and travel time, which affect the network performance) are represented here by the use of a mathematical simulation framework. In particular, the demand is considered as a random variable following the normal distribution $N(\mu_{rs}, \sigma_{rs}^2)$, with μ_{rs} denoting its mean for each $r-s$ pair and σ_{rs}^2 denoting its variance. Despite the fact that travel time can be also described by the normal distribution [2], an alternative, more explanatory assumption is made here, assuming that link speeds follow a multinomial normal distribution correlated with the speeds of the neighboring links. A similar assumption is adopted for the distribution of link capacities.

The current framework connects link travel time to link speed and capacity fluctuations, and it allows the expression of the interacting travel costs of neighboring links. In this way, link travel time variability (which is the result) is intrinsically modeled in the structure of the C-NDP and D-NDP with regard to its causal phenomenon (which is the link capacity and speed variability). More specifically, the Lower-Level Problem enables the estimation of the statistical properties of the TTT, i.e. its mean value and variance, which are subsequently fed to the Upper-Level problem through iterating the solution of the assignment procedure, comprising the set of link and path equilibrium flows, the values of origin-destination demand, link capacity, and the link free flow travel time, in accordance with the stochastic characteristics assigned to these variables, as described previously.

The estimation of the statistical properties of TTT and, hence, the reliability assessment, are performed through the simulation method of the *Latin Hypercube* sampling. In comparison to other simulation methods, such as that of the Monte Carlo simulation, Latin Hypercube is based on a stratified random procedure which provides an efficient way to capture the properties of the stochastic variables from their distributions, namely, it produces results of higher accuracy without the need for increasing the sampling size, and it allows to model correlations among different variables. In particular, the procedure of Iman and Conover [20] is followed here in order to produce correlated random numbers from the normal distribution, based on the Cholesky decomposition of the correlation matrix. The assumptions concerning the usage of the simulation method in the network design process are:

- The duration of changes in link speeds and capacities allows users to re-estimate route choices, and
- Link speed reduction is due to random events (like accident, physical disaster or other) which affect a locality of the network and, hence, link capacities and speeds are correlated with those of neighboring ones.

4 Genetic Optimization of Reliable Network Resource Allocation

4.1 Current Estimation Procedures for the NDP

Both C-NDP and D-NDP can be generally characterized as problems of increased computational complexity. This complexity arises from the fact that bi-level programming problems, even for simple linear cases, are Non-deterministic Polynomial-time (NP)-hard problems [3]. In particular, the stochastic equilibrium NDP is a NP-hard, non-convex combinatorial problem [27], since its set of constraints involves non-linear formulations, such as those of the SUE assignment of the Lower-Level Problem (see Sect. 2). Several algorithms, appropriate for addressing complex, combinatorial optimization problems, have been hitherto proposed and implemented to solve the continuous and discrete form of the NDP. These algorithms can be generally distinguished into: (i) gradient-based methods, and (ii) derivative-free (meta-)heuristic methods.

In the case of C-NDP, a number of gradient-based methods have been proposed in [6], while derivative-free heuristic approaches include the Hooke and Jeeves technique [1] and equilibrium decomposed optimization procedures [35]. Applications of meta-heuristic techniques for solving the C-NDP include the use of Simulated Annealing [16] and Genetic Algorithms ([9], [40]). All these approaches employ static assignment procedures to calculate the traffic flows after having improved the network. Furthermore, models for traffic flow estimation at dynamic disequilibrium states [17] and dynamic equilibrium algorithms [21] have been proposed and tested. A few models have also been considered for treating the uncertainty in various system attributes, such as the demand [37] and travel time [39]. In addition, the C-NDP with travel time reliability requirements has been studied in [34] by the use of approximation methods in order to estimate the statistical properties of the system attributes and, especially, the Probability Density Function (PDF) of the TTT, which ends up with solving a sequential quadratic programming problem.

In the case of D-NDP, suggested estimation procedures include the branch-and-bound method [23], Lagrange relaxation and dual ascent procedures [24], decomposition quasi-optimization methods [31], and support function approaches [18]. Meta-heuristic approaches for solving the D-NDP refer to a Tabu-based search strategy [26], an ant-system method [28] and genetic algorithms, such as those implemented in ([12], [22], [29]). Furthermore, GAs have been applied for the solution of NDP combining continuous and discrete decision variables in ([4], [36]).

The incorporation of reliability requirements in the continuous and discrete form of the NDP and the explicit representation of the uncertainty of the system attributes, as they are proposed in this chapter, increase the complexity of the estimation procedure. This is because some units of the system are treated as stochastic variables and a simulation-based risk assessment is incorporated in the set of problem constraints. Since the current problem is highly nonlinear and complex, its solution cannot be obtained by using classic derivative-based optimization procedures. Global optimization methods which are meta-heuristic and stochastic, such as Genetic Algorithms (GAs) [19], have proved to be able to tackle NP-hard, combinatorial problems of increased complexity, like the present one.

The wide applicability of GAs can be attributed to their convenience in handling variables of stochastic nature and multiple constraints in a seamless way, without requiring information about the nature of the problem but only about the performance of a 'fitness' function for various candidate states. Particularly, GAs have been widely used in solving various bi-level programming problems [8] and they can reach optimal (or adequately near-optimal) solutions, avoiding the possibility of missing the true optimal alternative investment plan when the Braess paradox occurs [22]. For these reasons, a suitable GA is employed here within the framework of the reliable NDP. The present application constitutes a continuation of previous work undertaken in [10] and [11] for using a GA-based evolutionary optimization methodology to solve the reliable C-NDP and D-NDP respectively. The next subsection provides a description of the solution algorithm used for the reliable NDP.

4.2 Description of the Solution Algorithm

GAs are population-based, stochastic, global search methods following the principle of natural evolution. According to it, solutions with increased performance, in terms of the values of the objective function, have also an increased probability to be selected in order to provide the new improved solution population. Each 'individual' of the population is a coded set of the problem variables and forms a string of the variable values, referred to as chromosome. In the current study, the values of the variables follow the binary arithmetic coding scheme, that is, each chromosome is a string of '0s' and '1s', which are called allelic values. In the case of C-NDP, the individuals of the GA population correspond to alternative codings (candidate solutions) referring to the capacity improvements which should be made to the network links. In the case of D-NDP, the individuals of the population correspond to alternative binary codings of the link and lane additions.

The current iterative procedure for solving each form of the reliable NDP combines the use of the mechanics of natural evolution, as it is based on a GA, with the Latin Hypercube sampling methodology (see Sect. 3.2) for modeling the stochastic variables and performing the risk assessment in relation to the reliability requirements. For every individual of the population, a Latin Hypercube simulation is performed, altering the travel demand, link travel time and capacities in order to estimate TTT reliability. The mechanism of the population evolution is based on three genetic operations, i.e. reproduction, crossover and mutation (see below). The steps of the solution algorithm are as follows:

1. *Initialization:*
   Produce an initial random population of candidate feasible solutions (link capacity improvements for the C-NDP, or link and lane additions for the D-NDP) and define the parameters of the genetic operators;

DO UNTIL CONVERGENCE:

2. *Path Enumeration, applied only on the D-NDP:*
   Perform path enumeration for every candidate solution;
3. *Simulation:*
   Estimate the TTT reliability for every candidate solution by Latin Hypercube simulation;
4. *Genetic Evolution Process:*
   a) Check for the consistency of constraints and estimate the "fitness function" of each candidate solution;
   b) Perform a stochastic selection of the "fittest" solution set and the crossover operation among the selected "individuals";
   c) Perform the mutation of individuals;
   d) Produce a new population of genetically improved candidate solutions.

The reproduction operator performs the reproduction of an intermediate population, referred to as mating population, which will produce a new, genetically improved population. The mating population is generated by a selected subset of the current population (parent population) based on the performance of each individual concerning the fitness function. There are various methods for the selection of the individuals from the parent population, such as the roulette wheel and tournament selection. In this study, the tournament selection method is adopted. In this method, the mating population is formed by choosing the most 'powerful' amongst a number of randomly selected individuals from the parent population. After selecting the mating population, the exchange of genetic material, i.e. the crossover operation, is performed among the member individuals. This is a mechanism which leads to the production of a new, improved population. The crossover is performed by randomly mating the individuals and exchanging parts of their chromosomes, according to a pre-specified pattern and rate (probability of two mates to crossover), which is typically selected to be larger than 50%. The current application uses a scattered crossover pattern, wherein randomly selected parts of each chromosome are exchanged to allow the transmission of genetic information among individuals. Finally, the mutation operation provides a mechanism for preventing local convergence through randomly altering some allelic values according to a pre-specified (typically small, such as <5%) rate. Though the extended use of meta-heuristic techniques such as GAs for solving complex problems, the their solutions have met some skepticism. This is mainly because of their dependence on initial conditions and their incorporated random search processes. For this reason, multiple runs of the GA are performed in this study through altering the starting point (the initial population) of the procedure to obtain evidence that the solutions obtained for both the C-NDP and D-NDP do not depend on the initial state and, hence, it can be expected that they are optimal (or adequately near-optimal) with an increased probability. It should be kept in mind that GAs belong to the stochastic approximation class of optimization methods, which implies that no guarantee of optimality can be provided for the final solution. Instead, the final solution should be viewed as the most probable optimal state obtained from a genetically organized sampling procedure in the search space.

5 Model Implementation and Analysis of Results

5.1 Description of the Numerical Experiments

The proposed methodology for solving the C-NDP and D-NDP with reliability requirements is implemented into two different versions of a test network (see Figs. 1(a) and 1(b) respectively). The specific network layout has been used in [18] and is selected here, since it encompasses geometric and operational settings typically found in urban road transport systems. It is composed of a single origin-destination pair (from node #1 to node #12) and 12 nodes. There are 23 links, 6 of which (figured #18-23) are regarded as candidate new links for the case of the D-NDP. These links give rise to a total of 25 paths, after having made all possible improvements.

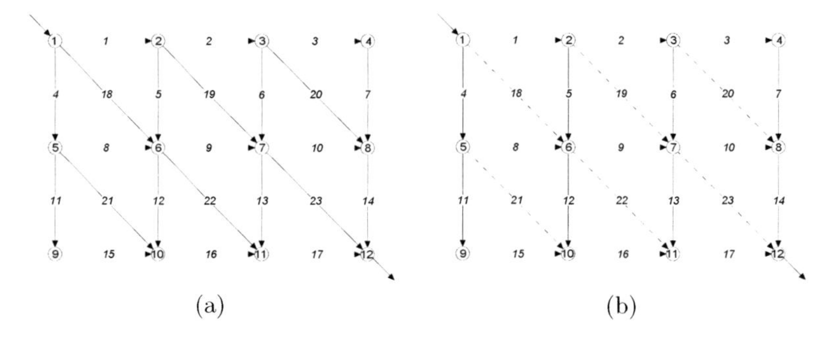

Fig. 1. The test network layout for (a) the C-NDP, and (b) the D-NDP, where existing links available for lane additions are shown in solid lines, and potential new links are shown in dashed lines

The network is considered as fully degradable in the sense that all links exhibit fluctuations in travel speed and capacity. The C-NDP, which is composed of 23 control variables for the given network, can be regarded as complex since it considers possible enhancements in the whole network configuration. Also, the D-NDP forms a complex combinatorial problem, since despite the small scale of the given network it yields $2^L = 2^{23} = 8388608$ possible combinations (alternative construction plans) of a network capacity improvement. In the current study, the link travel time t_α at some link α is expressed as a function of the random free-flow travel time t_α^f, traffic flow x_α and random capacity y_α at this link, and it is calculated by using the standard formulation of the Bureau of Public Roads (BPR), as follows:

$$t_\alpha = t_\alpha^f \left(1 + \beta \left(\frac{x_\alpha}{y_\alpha}\right)^m\right) \tag{9}$$

where β and m are scale parameters depending on the operational characteristics of the network, which have been set here equal to $\beta = 0.15$ and $m = 4$. Although the estimation of the link travel time is based here on the BPR formula, which typically applies to uncongested road networks, other formulations could also be adopted for taking into account the congestion effects, like queues or bottlenecks in the links of the network.

The capacity of each of the existing links is set equal to $y_\alpha = 20$ vehicles per hour (veh/hr). In the D-NDP, the link capacity reaches its maximum level, which is set equal to y=30 veh/hr, after a lane addition, while the maximum capacity of each new link is set equal to 20 veh/hr. In the C-NDP, the maximum link capacity is set equal to y=30 veh/hr. The demand between the origin-destination pair is set equal to q=80 veh/hr. The cost of a lane addition to each of the existing links is set equal to 30 monetary values, while the construction cost of each new link is set equal to 50 monetary values. This study adopts a conversion factor $\theta = 1$. The free-flow travel time, which is proportional to the link length, is set equal to $t_\alpha^f = 1$ min for the existing links, and $t_\alpha^f = 1.4$ min for the new links.

In the D-NDP, the complete reconstruction of the given network, which requires the maximum allowable network capacity improvements in terms of the link and lane additions, amounts to a total of 810 monetary values. Nonetheless, such a scenario may be considered as too expensive and hence impractical in real-world situations. For this reason, half of this amount, i.e. 400 monetary values, is set as the total available construction budget B, which can be regarded as sufficient to enhance the capacity of the existing network. In the C-NDP, the total available construction budget B for network construction plans equals 350 monetary units, which corresponds to the share of the amount needed to expand the capacity of all links to the upper bound.

The convergence of the algorithm to the SUE conditions using the MSA heavily depends on the network configuration and, particularly, the network size, the number of alternative paths connecting the O-D pair and the congestion level. The increase of the number of network links gives rise to more alternative paths between the O-D pair, which, in turn, complicates the route choice problem, since it augments the interaction among the network users. Correspondingly, the increase of the travel demand renders the convergence to the SUE state more difficult, since, particularly under congested conditions, small alterations in the users' path choices can heavily influence the travel cost incurred among alternative paths.

Based on the factors above, various alterative network settings will be employed in terms of the network layouts and demand profiles. During the implementation of the solution algorithm (see Sect. 4.2), an analysis is made in order to ensure that an adequate number of MSA iterations is used for the convergence to the SUE conditions. Figure 2 presents the path choice probabilities and path flows for each MSA iteration, based on a scenario of increased demand (q=120

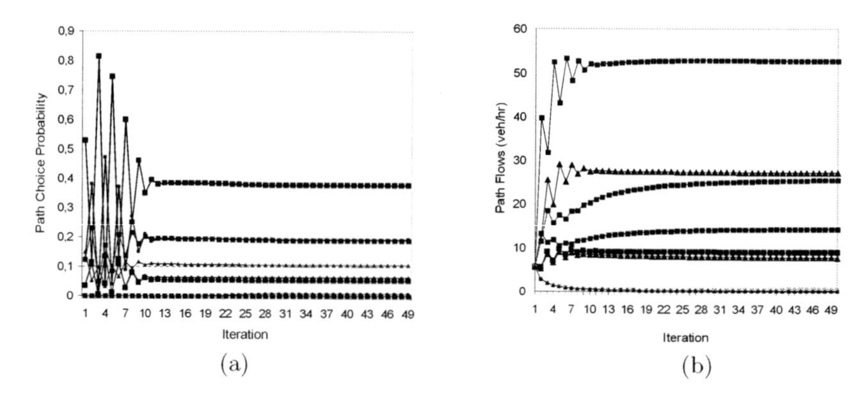

Fig. 2. (a) Path choice probabilities and (b) path flows estimated at each MSA iteration

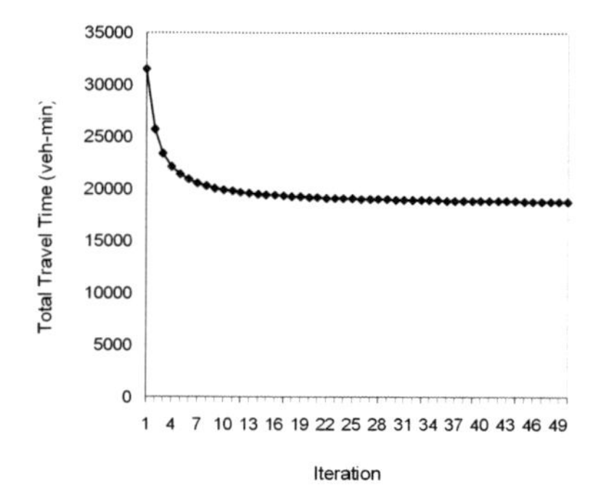

Fig. 3. Convergence of the network traffic assignment problem to a SUE condition

veh/hr), which is the 50% increment of the mean value, for the case where all the 23 links exist in the network in their maximum capacity. Even in this extreme case, wherein the congested network links form 25 alternative paths, the convergence is achieved within almost the first 15 MSA iterations (see Fig. 3). In the current study, a total of 50 MSA iterations is employed to achieve a stable solution to the Lower-Level Problem for the given test network.

Moreover, the distribution of the TTT is analyzed with respect to different levels of link capacity variation. Figure 4 shows that the TTT is significantly influenced by the size of the link capacity variability for the complete network configuration. In particular, when the dispersion of link capacity increases (in terms of the variance) from 10% to 25%, the dispersion of the TTT increases substantially, dropping the reliability of the whole system. Each of the plots below shows the log-normal distribution (and its characteristic values) of the TTT, as it is fitted to the results of the simulation for comparative purposes. The current study adopts the rather conservative assumption that both demand and link capacity variance are equal to 10% of the mean value.

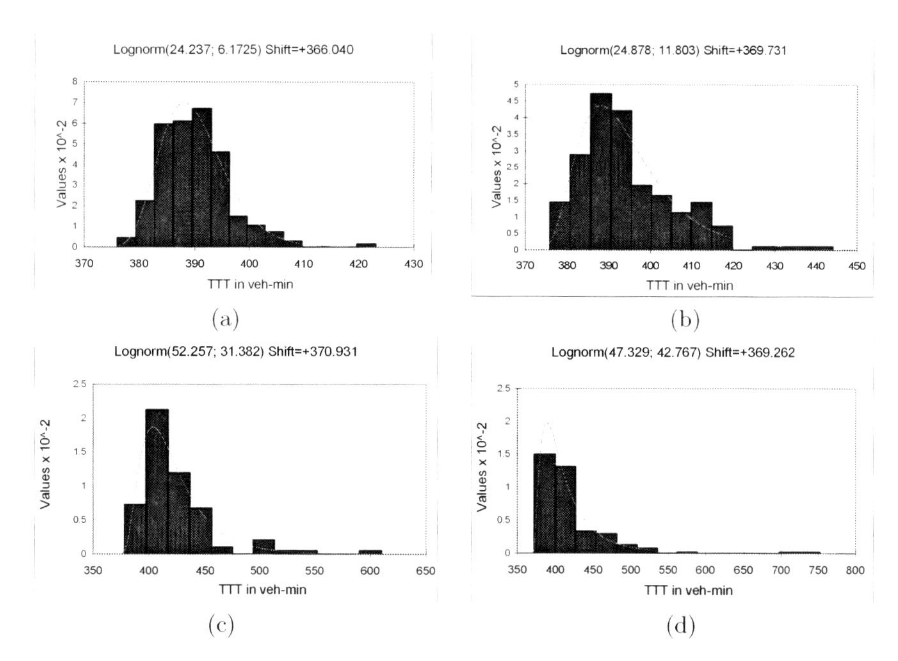

Fig. 4. Log-normal distribution fit of the TTT hypothesizing a link capacity variance equal to (a) 10%, (b) 15%, (c) 20% and (d) 25% of the mean capacity

5.2 Effects of the Design Process on Network Configuration

Before applying the proposed method to the test network, a sensitivity analysis is performed in order to trace those links which mostly influence the TTT. The sensitivity analysis examines the effect that one standard deviation increment of the capacity of each link can have on the standard deviation of TTT. Figure 5 presents the ranking of the links of the complete network configuration with the largest impact on TTT, in terms of the correlation coefficient between the increment of link capacity by one standard deviation and the standard deviation of TTT. As it may be expected, the links, which can be considered as the most critical components of the system, are, in order of significance, #17, #1, #9, #4, #14, #18 and #13, since they are servicing the largest portions of demand between the O-D pair.

An initial solution is first obtained by solving the C-NDP without reliability requirements, as described in constraint (4). The assignment of travel demand onto the initial network (without link improvements) results in a TTT equal to 549 veh-min. The solution refers to a construction plan composed of improvements on links #17, #1, #9 and #4, which results in the reduction of the expected TTT from 549 veh-min to 402 veh-min ($\approx$ 27 % reduction). The construction plan obtained from this solution essentially corresponds to the enhancement of those 4 links which are the most influential to the value of TTT, based on the results of the sensitivity analysis (see Fig. 5). An amount of 40.8

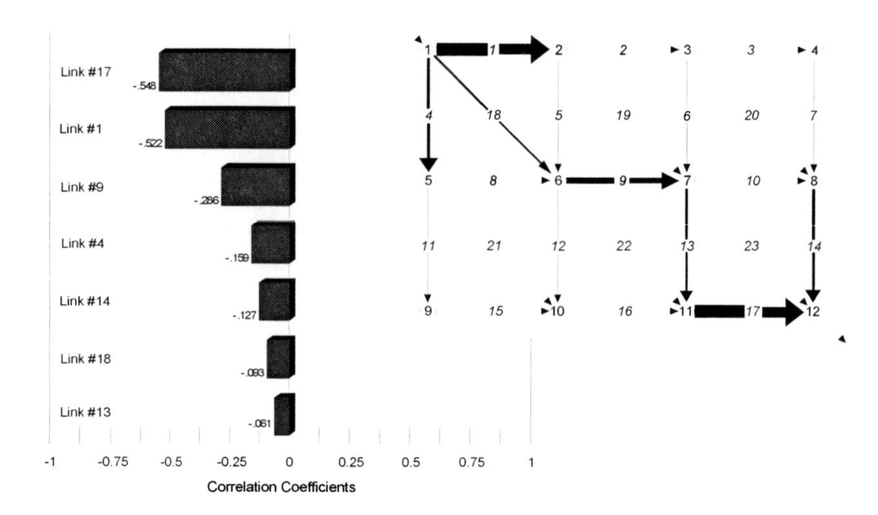

Fig. 5. Sensitivity analysis and classification of the link capacity increment impact on the Total Travel Time, in terms of the correlation coefficients of one standard deviation increment of link capacity with the standard deviation of the TTT

monetary units has been used as construction cost (corresponding to 40.8 veh-min, since θ=1), which in turn leads to a total social cost (TTT + construction cost) equal to 402+40.8=442.8 veh-min.

A new construction plan is then formed by the C-NDP, through adding a reliability requirement, i.e. the probability of the TTT to exceed T=450 veh-min to be less than 5%. This upper limit value expresses approximately the 10% increment of the TTT value (402 veh-min) obtained from the initial problem solution. The new solution refers to a construction plan composed of improvements on links #17, #1, #9, #4, #18, #14, #3, #15, #5, #2, #13, #21, #19, #12, #22. Figures 6(a) and 6(b) indicate the capacity improvements resulting from the solution of the C-NDP with and without reliability requirements respectively. By comparing these two graphs, it can be observed that the capacity improvements corresponding to the reliable C-NDP encompass a much larger number of network links in comparison to the improvements corresponding to the C-NDP without reliability requirements. This outcome verifies the expectation that more capacity is required in order to increase the network reliability and is consistent with the concept of sparse network capacity, or that of increased system redundancy, in engineering terms. More specifically, the total construction cost of the new solution is raised to 67.73 monetary units, while the expected TTT is reduced to 387 veh-min, leading to a total social cost of 387+67.73=454.73 veh-min, which is slightly higher than the total social cost when excluding reliability requirements (442.8 veh-min).

Similar to the C-NDP, an initial solution is first obtained by solving the D-NDP without reliability requirements. The assignment of the travel demand onto

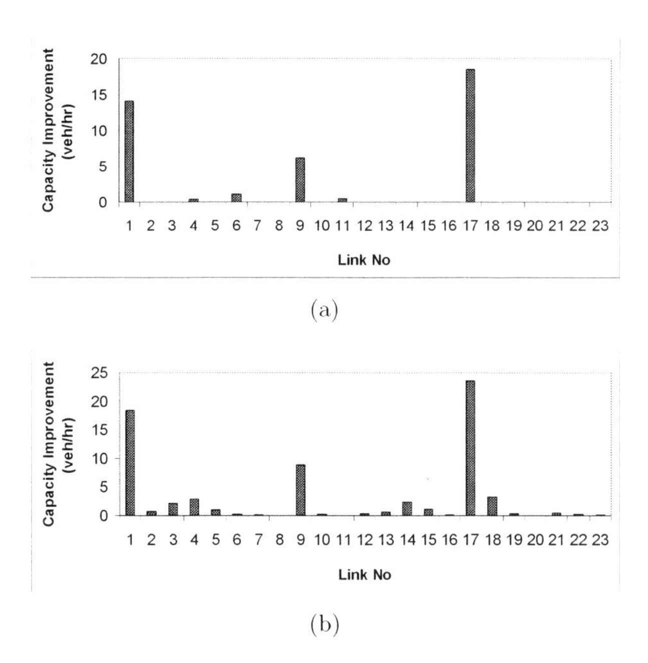

Fig. 6. Construction plans for the case of C-NDP (a) without reliability requirements and (b) with reliability requirements

the initial network (with no link or lane additions) results in a TTT equal to 886 veh-min. The initial solution provides a construction plan which encompasses the addition of the new links #18 and #23 and lane addition to the existing links #1, #9 and #17. The resulting construction cost amounts to 190 veh-min, which is lower than the total available budget (400 veh-min). The TTT reduces from 886 veh-min to 449 veh-min. Hence, the total social cost comes to 449+190=639 veh-min. Figure 7 presents the contribution of each new link to be added, or each existing link with a new lane to be added, in the order of selection, to the TTT and total social cost. As it can be observed, after the 6th sequential addition of a new link or lane, the total social cost increases, although the TTT is reduced, thus leading to inefficient solutions in terms of the social benefit.

The D-NDP provides a new solution by imposing, as reliability requirementm, the probability of the TTT to exceed T =500 veh-min to be less than 10%. This upper limit value represents about the 10% increment of the TTT value (449 veh-min) obtained from the initial solution of the D-NDP. The new solution leads to the construction of two more links, i.e. link #19 and #22, in addition to the improvements resulting from the initial solution. The TTT is further reduced from 449 veh-min to 423 veh-min, while the construction cost is raised to 290 monetary values, which is still less than the total available construction budget (400 monetary values). The new solution results in the increase of the total social cost from 639 veh-min to 423+290=713 veh-min.

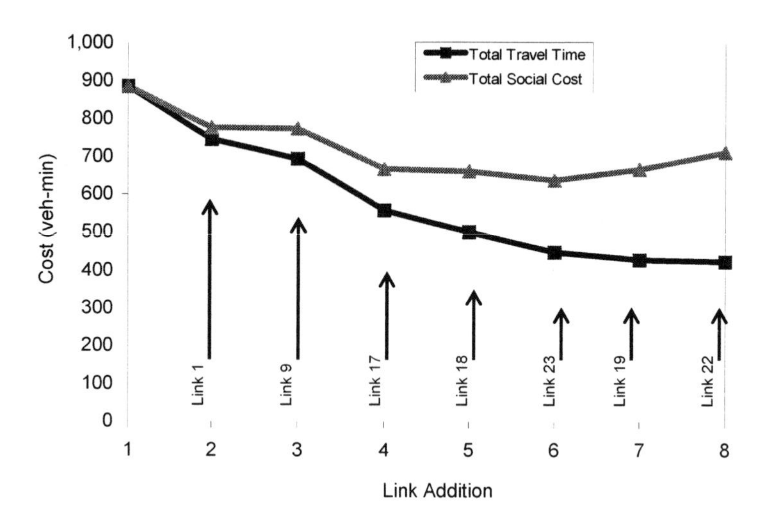

Fig. 7. Impact of capacity expansions on Total Travel Time and total social cost

The above-named results show the existence of a considerable remaining (not allocated) portion of the total available construction budget, which is equal to 350-40.8=309.2 monetary values for the C-NDP and 400-190=210 monetary values for the D-NDP, when ignoring reliability requirements, and 350-67.73=282.27 monetary values for the C-NDP and 400-290=110 monetary values for the D-NDP, when including reliability requirements. This portion can be attributed to the fact that there is a threshold beyond which capacity improvements are regarded as too expensive, with respect to the increase of the total social cost, in comparison to their contribution to the reduction of the expected TTT. Yet, the total available construction budget is much higher than the construction cost obtained from the solution. In terms of the composite objective function, after F value exceeds that specific point (threshold), additional investments in a better link capacity increase the component of the construction cost proportionally more than the reduction caused to the TTT component, independently from the link the investments are made in.

5.3 Effects of the Design Process on Network Reliability

This subsection concentrates on the analysis of the effects of solving the reliable C-NDP and D-NDP on the distribution of the TTT. Figures 8(a) and 8(b) show the resulting TTT distribution, as it is obtained from the solution of the C-NDP, in the cases of ignoring and including reliability requirements respectively. The resulting probability of the TTT to be higher than 450 veh-min was estimated to 9.97%, which is lower than the acceptable upper bound of 10%. As it is shown in the two histograms of Fig. 8, the dispersion of the TTT obtained from the initial solution of the C-NDP without reliability requirements, having $P(TTT \geq 450veh - min) \approx 0.2$ (see left diagram) is significantly wider than

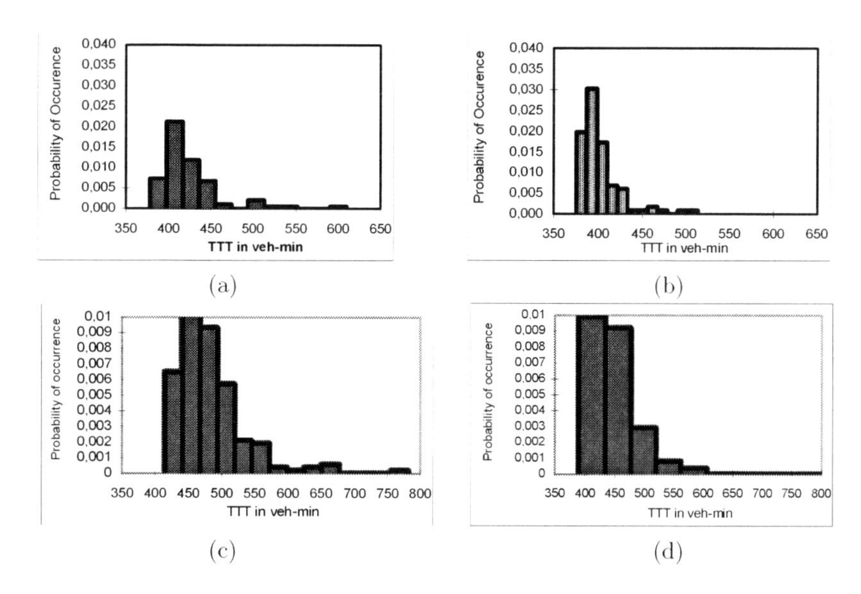

Fig. 8. Distribution of the TTT obtained from the C-NDP (a) without reliability requirements and (b) with reliability requirements, and obtained from the D-NDP (c) without reliability requirements and (d) with reliability requirements

the dispersion of the TTT obtained from solving the reliable C-NDP, having $P(TTT \geq 450veh - min) < 0.1$ (see right diagram).

Correspondingly, Figs. 8(c) and 8(d) indicate the resulting TTT distribution, as obtained from the solution of the D-NDP, in the cases of ignoring and including reliability requirements respectively. The resulting probability of the TTT to be higher than 500 veh-min was estimated to 9.94%, which is lower than the acceptable upper bound of 10%. The two histograms demonstrate that the dispersion of the TTT obtained from the initial solution of the D-NDP without reliability requirements, having $P(TTT \geq 500veh - min) \approx 0.25$ (see left diagram) is considerably wider than the dispersion of the TTT obtained from solving the reliable D-NDP, having $P(TTT \geq 500veh - min) < 0.1$ (see right diagram). The resulting narrower dispersion of the TTT signifies that both the continuous and the discrete forms of the reliable NDP produce a system with increased stability, i.e., a greater ability to respond to the various states of its variables, in comparison to the NDP without reliability requirements.

The proposed optimization framework can be regarded as computational intensive, considering its ability to handle such complex problems. This is because the combination of three procedures, namely the GA combined with the Latin Hypercube random sampling and the traffic assignment, requires a substantial computational effort. In particular, the estimation of the stochastic properties of the system attributes with the Latin Hypercube sampling requires almost 2 min of CPU time for the given networks, using a PC with modest capabilities. Hence, the CPU time required for each generation is equal to 2 min $\times$ 50 individuals =

100 min, and the completion of an experiment utilizing a set of 200 generations (as adopted here) necessitates several days of CPU time. It is straightforward to recognize that the computational effort increases substantially with the problem size. However, this is not an obstacle for adopting the proposed framework into real-size cases, since the NDP relates to strategic decisions of the designer in the planning horizon, where computational burden is irrelevant in relation to the utilization of the results. Nevertheless, advanced methods of computing (e.g. distributed computing) could be further employed to decrease the CPU time in cases where the economy of computational resources is a matter of concern.

6 Summary and Conclusions

This chapter described a game-theoretic, bi-level formulation and a solution algorithm for addressing the reliable Continuous-Network Design Problem (C-NDP) and the reliable Discrete-Network Design Problem (D-NDP) within a stochastic evolutionary optimization framework. The problem formulation seeks the optimal capacity enhancements (for the C-NDP) or the optimal selection of link and lane additions (for the D-NDP) in a degradable transport network, subject to budgetary and physical restrictions, and, additionally, reliability requirements, in terms of the probability of the Total Travel Time (TTT) to be less than a pre-specified value. The model enables the inclusion of four different sources of uncertainty (i.e. demand, capacity, link travel time and route choice) into the reliability assessment, through applying the Latin Hypercube sampling simulation method. The estimation procedure employs a Genetic Algorithm, since the incorporation of simulation techniques into such complex optimization problems renders the use of traditional approaches inappropriate. The implementation of the model into a test network with typical urban road settings demonstrated the ability of the solution algorithm to converge to a stable solution, which, at the lower-level problem, corresponds to the achievement of Stochastic User Equilibrium (SUE) conditions through employing the Method of Successive Averages (MSA).

The results signify the beneficial impact of incorporating reliability requirements into the standard bi-level programming formulation of the continuous and discrete NDP. In particular, the solutions of the reliable C-NDP and D-NDP diminish the expected TTT as well as the dispersion of the TTT, which implies their favorable effect on the stability of the system, in comparison to the solutions of the C-NDP and D-NDP without reliability requirements. These benefits are attained while keeping the required construction costs well below the total available construction budget for both forms of the NDP. The solution of the reliable D-NDP was found to provide larger travel time savings but also higher construction costs than the reliable C-NDP for the current test network settings. The proposed modeling framework facilitates the resource allocation decision-making involved in alternative infrastructure investment strategies, which give rise to different design processes.

The current study can be extended into several research directions. This includes the consideration of various classes of users, e.g. with different values of time and information acquisition mechanisms, in multi-modal networks. The extension above could allow the examination of the road space allocation according to such criteria as social equity, energy conservation and environmental sustainability. This examination could be favored by the adoption of models for capturing more detailed features of the traffic flow process at the microscopic level. Moreover, the current framework could be enriched through incorporating other types of uncertainty which affect network reliability, including those concerning the departure time and schedule delays, and day-to-day (or period-to-period) learning and behavioral adjustments of users.

References

1. Abdulaal, M., LeBlanc, L.J.: Continuous equilibrium network design models. Transportation Research, Part B: Methodological 13, 19–32 (1979)
2. Bell, M.G.H., Iida, Y.: Transportation Network Analysis. Wiley, Chichester (1997)
3. Ben-Ayed, O., Boyce, D.E., Blair, C.E.: A general bi-level programming formulation of the network design problem. Transportation Research, Part B: Methodological 22, 311–318 (1988)
4. Cantarella, G.E., Vitetta, A.: The multi-criteria road network design problem in an urban area. Transportation 33, 567–588 (2006)
5. Chen, A., Yang, H., Lo, H.K., Tang, W.H.: Capacity reliability of a road network: An assessment methodology and numerical results. Transportation Research, Part B: Methodological 36, 225–252 (2002)
6. Chiou, S.-W.: Bilevel programming for the continuous transport network design problem. Transportation Research, Part B: Methodological 39, 361–383 (2005)
7. Cho, D.J.: Three Papers on Measuring the Reliability and Flexibility of Transportation System Capacity. PhD Thesis, University of Pennsylvania, Philadelphia (2002)
8. Colson, B., Marcotte, P., Savard, G.: Bilevel programming: A survey. 4OR: Quarterly Journal of Operations Research 3, 87–107 (2005)
9. Cree, N.D., Maher, M.J., Paechter, B.: The continuous equilibrium optimal network design problem: A genetic approach. In: Selected Proceedings of the 4th EURO Transportation Meeting, Newcastle, pp. 175–193 (1998)
10. Dimitriou, L., Stathopoulos, A., Tsekeris, T.: Reliable stochastic design of road network systems. International Journal of Industrial Systems Engineering 3 (forthcoming, 2008)
11. Dimitriou, L., Tsekeris, T., Stathopoulos, A.: Evolutionary combinatorial programming for discrete road network design with reliability requirements. In: Giacobini, M., et al. (eds.) EvoWorkshops 2007. LNCS, vol. 4448, pp. 678–687. Springer, Heidelberg (2007)
12. Drezner, Z., Wesolowsky, G.O.: Network design: Selection and design of links and facility location. Transportation Research, Part A: Policy and Practice 37, 241–256 (2003)
13. Du, Z.P., Nicholson, A.: Degradable transportation systems: Sensitivity and reliability analysis. Transportation Research, Part B: Methodological 31, 225–237 (1997)

14. Fisk, S.C.: A conceptual framework for optimal transportation systems planning with integrated supply and demand models. Transportation Science 20, 37–47 (1986)
15. Friesz, T.L.: Transportation network equilibrium, design and aggregation: Key developments and research opportunities. Transport Reviews 18, 257–278 (1985)
16. Friesz, T.L., Cho, H.-J., Metha, N.J., Tobin, R.L., Anandalingam, G.: A simulated annealing approach to the network design problem with variational inequality constraints. Transportation Science 26, 18–26 (1992)
17. Friesz, T.L., Shah, S.: An overview of nontraditional formulations of static and dynamic equilibrium network design. Transportation Research, Part B: Methodological 35, 5–21 (2001)
18. Gao, Z., Wu, J., Sun, H.: Solution algorithm for the bi-level discrete network design problem. Transportation Research, Part B: Methodological 39, 479–495 (2005)
19. Goldberg, D.E.: Genetic Algorithms in Search, Optimization and Machine Learning. Addison-Wesley, Reading (1989)
20. Iman, R.L., Conover, W.J.: A distribution-free approach to inducing rank correlation among input variables. Communications in Statistics B11, 311–334 (1982)
21. Karoonsoontawong, A., Waller, T.S.: Dynamic Continuous Network Design Problem: Linear Bilevel Programming and Metaheuristic Approaches. Transportation Research Record 1964, 104–117 (2006)
22. Kim, B., Kim, W.: An equilibrium network design model with a social cost function for multimodal networks. The Annals of Regional Science 40, 473–491 (2006)
23. Leblanc, L.J.: An algorithm for the discrete network design problem. Transportation Science 9, 183–199 (1975)
24. Magnanti, T.L., Wong, R.T.: Network design and transportation planning: Models and algorithms. Transportation Science 18, 1–55 (1984)
25. Morlok, E.K., Chang, D.J.: Measuring capacity flexibility of a transportation system. Transportation Research, Part A: Policy and Practice 38, 405–420 (2004)
26. Mouskos, K.: A Tabu-Based Heuristic Search Strategy to Solve a Discrete Transportation Equilibrium Network Design Problem. Ph.D. Dissertation, The University of Texas at Austin (1992)
27. Papadimitriou, C.H., Steiglitz, K.: Combinatorial Optimization: Algorithms and Complexity. Dover Publications, Mineola (1998)
28. Poorzahedy, H., Abulghasemi, F.: Application of ant system to network design problem. *Transportation* 32, 251–273 (2005)
29. Poorzahedy, H., Rouhani, O.M.: Hybrid meta-heuristic algorithms for solving network design problem. European Journal of Operational Research 182, 578–596 (2007)
30. Sheffi, Y.: Urban Transportation Networks: Equilibrium Analysis with Mathematical Programming Methods. Prentice-Hall, Englewood Cliffs (1985)
31. Solanki, R.S., Gorti, J.K., Southworth, F.: Using decomposition in large-scale highway network design with a quasi-optimization heuristic. Transportation Research, Part B: Methodological 32, 127–140 (1998)
32. Stathopoulos, A., Karlaftis, M.: Temporal and spatial variations of real-time traffic data in urban areas. Transportation Research Record 1768, 135–140 (2001)
33. Stathopoulos, A., Tsekeris, T.: Methodology for processing archived ITS data for reliability analysis in urban networks. IEE Proceedings Intelligent Transport Systems 153, 105–112 (2006)
34. Sumalee, A., Walting, D.P., Nakayama, S.: Reliable network design problem: Case with uncertain demand and total travel time reliability. Transportation Research Record 1964, 81–90 (2006)

35. Suwansirikul, C., Friesz, T.L., Tobin, R.L.: Equilibrium decomposed optimization: A heuristic for the continuous equilibrium network design problem. Transportation Science 21, 254–263 (1987)
36. Ukkusuri, S.V., Mathew, T.V., Waller, S.T.: Robust transportation network design under demand uncertainty. Computer-Aided Civic Infrastructure Engineering 22, 6–18 (2007)
37. Waller, S.T., Ziliaskopoulos, A.K.: Stochastic dynamic network design problem. Transportation Research Record 1771, 106–113 (2001)
38. Yang, H., Bell, M.G.H.: A capacity paradox in network design and how to avoid it. Transportation Research, Part A: Policy and Practice 32, 539–545 (1998)
39. Yin, Y., Iida, H.: Optimal improvement scheme for network reliability. Transportation Research Record 1783, 1–6 (2002)
40. Zhang, G., Lu, J.: Genetic algorithm for continuous network design problem. Journal of Transportation Systems Engineering and Information Technology 7, 101–105 (2007)

Genetic Algorithm for Constraint Optimal Toll Ring Design

Agachai Sumalee

Department of Civil and Structural Engineering, The Hong Kong Polytechnic University, Hung Hom, Kowloon, Hong Kong
ceasumal@polyu.edu.hk

Summary. This chapter considers the optimal toll ring design problem in a general urban traffic network. Several constraints on outcomes of the toll ring scheme are imposed on the design (e.g., equity impact or revenue). In this chapter, the GA based algorithm proposed by [13] is integrated with a penalty based approach to tackle the problem. Three penalty methods including static, dynamic, and self-adaptive penalties are investigated. The algorithm is tested with a realistic traffic network.

Keywords: Road pricing, Optimal toll location, Genetic algorithms, Constraint handling.

1 Introduction

Road pricing is a fiscal policy for managing the demand of road usage. There are several types of road pricing schemes including point-based (which charge for crossing a designated point), area-based (which charge for traveling inside a specific area), distance-based (which charge per distance traveled), time-based (which charge per time spent traveling), and toll ring ones (which charge for crossing a designated cordon line). From these possible schemes, May et al. [6] report that the toll ring based scheme is the most favored option by practitioners thanks to its simplicity. A toll ring is composed of a number of tolled roads surrounding a designated area. Drivers traveling from the outside of the toll ring cannot enter the designated area without paying the toll.

The main objective of road pricing is to apply an appropriate toll scheme to control the travel demand and traffic distribution in order to maximize the system's performance (normally measured by the social welfare index). Nevertheless, an inappropriate design of the scheme may decrease its benefit or even degrade social welfare [6]. In practice, apart from the main objective of social welfare maximization, the design of a toll ring scheme must also take several outcome constraints into account (e.g., equity impact, generated revenue, or congestion reduction) to enhance the practicality and acceptability of the scheme [12].

There are several approaches for solving the optimal toll ring design problem (see, e.g., [13, 14, 18]). However, none of these approaches have explicitly

A. Fink and F. Rothlauf (Eds.): Advances in Computational Intelligence, SCI 144, pp. 45–61, 2008.
springerlink.com

considered the outcome constraints of the design. The main focus of this chapter lies thus on the development of an algorithm to design an optimal toll ring scheme subjected to several outcome constraints. The Genetic Algorithm (GA) based method proposed by [13] is extended to consider different outcome constraints.

There are three main groups of the methods for handling the constraint in GA including (i) penalty based methods, (ii) search restrictions, and (iii) hybrid methods. An excellent review on these methods can be found in [2]. This chapter studies the different methods and focuses on the most flexible and particularly appropriate ones for the considered problem. Constraints are considered by including a penalty in the objective function to penalize infeasible solutions. Three types of penalty functions can be adopted including static, dynamic, and self-adaptive penalty functions. This chapter applies these three approaches and compares their performances. The chapter is structured into five sections. The next section presents the mathematical formulation of the optimal toll ring design problem with constraints, and defines the indicators for different outcomes. Section 3 presents the branch-tree concept proposed by [13] which is used to encode a toll ring using a string-based representation. Section 4 introduces the GA-based method for solving the optimal toll ring design with constraints. Section 5 presents numerical results for a real-world problem. The final section concludes the chapter.

2 Problem Formulation and Outcome Indicators

2.1 Optimal Toll Ring Design Problem with Constraints

The optimal toll ring design problem (OTP) involves selecting roads and toll levels to charge drivers for using roads. The set of tolled roads must form a toll ring. Topologically, a toll ring surrounds a designated area (e.g., city centre) so that all trips from the outside of the toll ring to the area inside the toll ring are charged. Usually, drivers respond to the introduction of tolls by rerouting, switching mode, or giving up their trips. Thus, the responses of the drivers to the toll must be considered in the evaluation of the benefit and outcomes of a toll ring scheme.

The assumption of Wardrop's user equilibrium [16] is normally used to model travelers' responses to the toll. Several off-shelf traffic modeling software for finding equilibrium traffic flows are available; see, e.g., SATURN [15]. We do not discuss the details of the equilibrium assignment here but refer to [11].

For the purpose of this chapter, the equilibrium assignment is viewed as a mapping from a toll vector to the equilibrium demand and route flow vectors. In this chapter, two types of response from the drivers are considered, namely rerouting and elastic demand. The elastic demand represents the drivers' decisions to make or give up their trips (if the travel cost is too high). Let $\mathbf{T}(\tau)$ and $\mathbf{F}(\tau)$ be the vector mapping from a toll vector $\tau \in \Re^J$ to equilibrium origin-destination (OD) demand flows and path flows $\mathbf{T} \in \Re^I$ and $\mathbf{F} \in \Re^P$, respectively. J, I, and P are the total numbers of links, OD pairs, and routes in the network in that order.

Let $\epsilon \in \{1,0\}^J$ be a binary vector, where $\epsilon_j = 1$ indicates that link j is tolled. An additional constraint $\tau_j = toll \times \epsilon_j$ (where $toll$ is the toll for all links included in the toll ring) is imposed on all τ_j to ensure that the toll is only implemented on the tolled links. Thus, the OTP involves choosing ϵ and $toll$ so that the objective function $W(\mathbf{T}(\tau), \mathbf{F}(\tau), \epsilon)$ is maximized subject to several outcome constraints $h_e(\mathbf{T}(\tau), \mathbf{F}(\tau), \epsilon) \leq 0$:

$$\begin{aligned} \max_{(toll,\epsilon)} \ & W(\mathbf{T}(\tau), \mathbf{F}(\tau), \epsilon) \\ s.t. \ & \epsilon \in \Psi \\ & \tau_j = toll \cdot \epsilon_j \qquad \forall j \\ & h_e(\mathbf{T}(\tau), \mathbf{F}(\tau), \epsilon) \leq 0 \ \forall e \\ & 0 \leq toll \leq M, \end{aligned} \tag{1}$$

where $\Psi \subset \{1,0\}^J$ is an arbitrary set of all feasible combinations of tolled links which form toll rings. M is the upper bound of the toll. The objective function of the OTP maximizes the net social benefit which is the social welfare deducted by the costs of the scheme (i.e., implementation and operation costs). This can be defined as:

$$W = \sum_i \int_0^{T_i(\tau)} D_i^{-1}(x)dx - \sum_j \sum_p \delta_{jp} \cdot F_p(\tau) \cdot t_j(v_j) - \sum_j \epsilon_j s_j, \tag{2}$$

where D_i^{-1} is the inverse demand function for OD pair i, $t_j(v_j)$ is the link travel time function of link j (as a monotone function of the link flow v_j), s_j is the cost for implementing the toll on link j, and δ_{jp} is a binary parameter in which $\delta_{jp} = 1$ if link j is on route p, and $\delta_{jp} = 0$ otherwise. Note that $v_j = \sum_p \delta_{jp} F_p(\tau)$, i.e., link flow is equal to the sum of all route flows using that link. The next section defines of the outcome constraint, $h_e(\mathbf{T}(\tau), \mathbf{F}(\tau), \epsilon) \leq 0$.

2.2 Outcome Indicators and Constraints

This section defines two particular outcome constraints, h_1 and h_2. First, h_1 lies on the revenue of the toll scheme. In some cases, one needs to guarantee a certain level of revenue generated from the toll scheme to ensure that the scheme is self-financed (for its implementation and operation) and can provide sufficient funds for other infrastructure projects. The revenue of a particular toll scheme is defined as $\sum_j \tau_j v_j(\tau)$. Thus, $h_1 \equiv Rev - \sum_j \tau_j v_j(\tau)$, where Rev is the minimum required revenue. $h_1 \leq 0$ implies that $Rev \leq \sum_j \tau_j v_j(\tau)$.

Second, h_2 addresses negative effects of the road pricing scheme which is the *equity impact*. The equity impact describes unequally distributed social benefits over different groups of the population from different areas. If a toll scheme favors some groups of people from selected areas, the toll ring scheme is unfair and can be viewed as an inequitable policy.

In economics, the *Gini coefficient* is commonly used to measure income inequality. Figure 1 depicts the *Lorenz curve* or *empirical distribution* which represents the actual income distribution over the population and the equality curve or

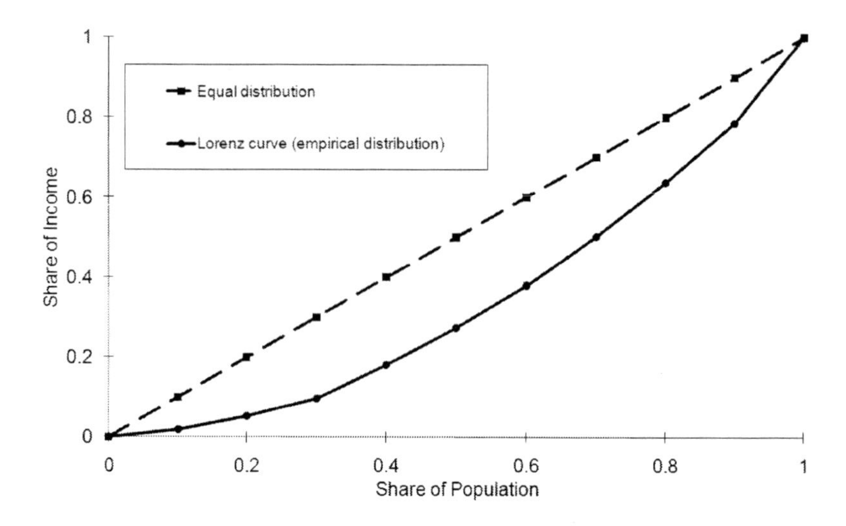

Fig. 1. Lorenz distribution and Gini coefficient

uniform distribution line (representing equal income distribution). The Gini coefficient is the ratio between the area between the Lorenz curve and the uniform distribution curve and the area below the uniform distribution curve. Therefore, the value of the Gini coefficient lies between 0 and 1. Low Gini coefficients lead to a more equal income or wealth distribution.

In this chapter, the Gini coefficient is adopted to measure the spatial distribution of the social benefit caused by the toll scheme. Let $W_i = \int_0^{T_i} D_i^{-1}(x)dx - T_i \cdot \pi_i$ be the social benefit for OD pair i, where π_i denotes the minimum travel time for OD pair i (calculated by the traffic assignment). The Gini coefficient G can then be calculated as

$$G(\tau) = \frac{1}{2 \cdot T^2 \cdot \overline{W}} \cdot \sum_{i=1}^{I} \sum_{r=1}^{I} T_i \cdot T_r |W_i - W_r|, \tag{3}$$

where $0 \leq G(\tau) \leq 1$, T is the total demand in the no-toll scenario and $\overline{W}$ is the average value of W_i. Thus, the second outcome constraint can be defined as a constraint on the equity impact, i.e., $h_2 \equiv G(\tau) - Eqi$, in which $h_2 \leq 0$ implies $G(\tau) \leq Eqi$. Eqi denotes the acceptable level of equity impact.

3 Branch Tree Framework for Toll Ring Representation

A Genetic Algorithm (GA) is used to solve the above-described OTP. The OTP's phenotype is the toll ring and its corresponding *toll*. The genotype is based on the branch-tree framework proposed by [13]. The problem-specific crossover and mutation operators (see Sects. 4.6 and 4.7) are applied to branch-trees to produce new branch-trees. The branch-tree structure and the design of the GA operators ensure that the constraints are always satisfied during the GA optimization process.

A toll ring surrounds a designated area and imposes a toll on all drivers traveling from the outside of the toll ring to its insides. From the network structure, all routes originating from all zones outside the toll ring passing through one of the nodes inside the toll ring must be tolled *at least once*.

3.1 Notation

Let $G(A,N)$ be a directed graph representing an urban traffic network where A and N are sets of links and nodes, respectively. A link connects two nodes, i and j, where $i \neq j$. The link direction is from i to j. i is a *tail node* of j and j is a *head node* of i. $\Xi_j = \{i|\, i \text{ is a tail node of } j\}$ is defined as a set of all tail nodes of j, where $\|\Xi_j\|$ is the size of set Ξ_j (total number of tail nodes of j).

Let $\beta_s = \{(j, d_j)\}$ be a set whose members are the pairs of nodes, j, and its degree, d_j. The nodes included in the set β_s are included in the branch-tree rooted from node s. The degree of j is the number of children nodes of j included in the branch-tree. For node j, only node $i \in \Xi_j$ can be included as a child node of node j in the branch-tree. In the branch-tree framework d_j can only be either 0 or $\|\Xi_j\|$. This implies that in the branch-tree framework each node j can either have no children node or include all of its tail nodes ($\forall i \in \Xi_j$) as its children nodes in the branch-tree. If $d_j = 0$, then j is denoted as *leaf node*. Figure 2 shows an example of a branch-tree. This branch-tree consists of five nodes. The root node of this branch is node A. Nodes B and C are children of node A. This implies that links (B, A) and (C, A) exist in the full traffic network. Similarly, nodes D and E are children of node B. Thus, the degrees of nodes A and B are two. Nodes C, D, and E have no children and hence have degree zero since they are leaf nodes. This branch-tree can be defined as $\beta_A = \{(A, 2), (B, 2), (D, 0), (E, 0), (C, 0)\}$.

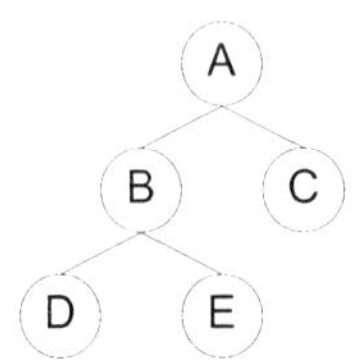

Fig. 2. Example of a branch tree

A branch-tree can be expanded by including all tail nodes of a selected leaf node into the branch-tree. This process is named *branching process*. At node $(j, 0)$ in the branch-tree (which is one of the leaf nodes) the branch-tree can be expanded by changing the degree of node j from 0 to $\|\Xi_j\|$ and adding nodes $\forall i \in \Xi_j$ to the branch-tree. Figure 3 illustrates the branching process. The whole traffic network for this example is shown in Fig. 3a. In

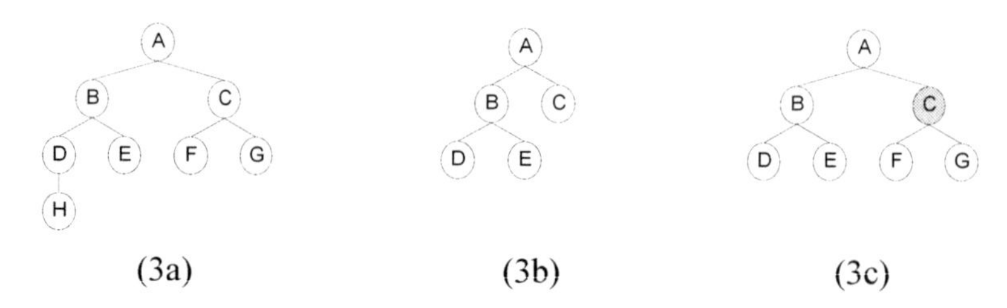

Fig. 3. Example of the branching process

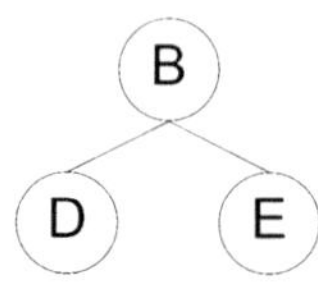

Fig. 4. Example of a sub-branch

this example, the current branch-tree is shown in Fig. 3b. A new branch-tree (associated with a new toll ring) is created by applying the branching process to node C. Thus, the new branch-tree can be constructed by setting $d_C = 2$ and including $(F, 0)$ and $(G, 0)$ into β_A. The new branch-tree is then $\beta_A = \{(A, 2), (B, 2), (D, 0), (E, 0), (C, 2), (F, 0), (G, 0)\}$ as shown in Fig. 3c.

The last notation to be introduced is the *sub-branch*. For a branch-tree β_s, a sub-branch $\overline{\beta}_j \subset \beta_s$ is defined as a branch-tree rooted from node $j \in \beta_s$. Figure 4 shows $\overline{\beta}_B$ which is a sub-branch rooted from node B of the branch β_A shown in Fig. 2.

3.2 Branch-Trees and Toll Rings

For a given $\beta_s = \{(j, d_j)\}$, the tolled links for the toll ring represented by this branch-tree are defined by the set of leaf nodes and their head nodes in the branch-tree. For instance, from the branch-tree in Fig. 2, the tolled links from this branch-tree are (C, A), (D, B), and (E, B). The set of tolled links form a toll ring. This is based on the branching rule in which $d_j = 0$ or $d_j = \|\Xi_j\|$. In the beginning, an initial toll ring should be provided and then a new toll ring can be created based on this initial toll ring by using the branching process as shown in Fig. 3. With the initial toll ring, all routes passing through the nodes in the branch-tree are tolled at the tolled links defined by the leaf nodes and their associated head nodes. By branching out at node C, additional tolled links are (F, C) and (G, C) and all routes passing through node C are still tolled. This guarantees that all routes passing the nodes in the branch-tree are still tolled, and hence the new branch-tree represents a new valid toll ring.

Figure 5a illustrates the relationship between the branch-tree and toll ring. Cordon 1 in Fig. 5a is defined as the initial toll ring. From this initial toll ring, a virtual root node (named $C1$), representing nodes A, B, C, and D in the

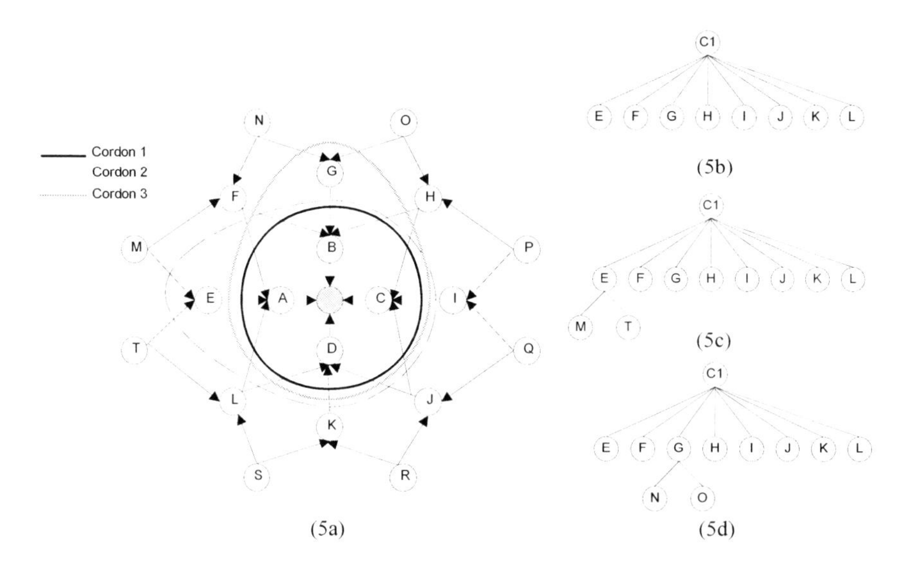

Fig. 5. Demonstration of the relationship between branch trees and closed cordons

network, is defined as root node for the branch-tree. Nodes A, B, C, and D are predefined by the user and are named *target nodes*. The nodes at the next level of $C1$ are all tail nodes of A, B, C, and D (as shown in Fig. 5b). The branch-tree in Fig. 5b is then expanded at nodes E and G to create two new branch-trees as shown in Figs. 5c and 5d, respectively. These two new branch-tress represent Cordon 2 and Cordon 3 shown in Fig. 5a. Those interested in further details of the branch-tree framework should consult [13]. In addition, an extension of the branch-tree framework for multiple toll rings can be found in [14].

4 Applying Genetic Algorithms to Solve the OTP

Each GA chromosome encodes a toll ring (as a branch-tree) and its uniform toll. The fitness of each chromosome can be found by finding the user equilibrium (UE) demand and link flows (by running any traffic assignment software with the toll scheme) and then evaluating (Equation 2). The GA iteratively applies selection and search operators (crossover and mutation) to evolve the population. In this section, the chromosome encoding as well as the search operators are described. The GA based algorithm is based on GA-AS proposed by [13]. Figure 6 shows the overall procedure of GA-AS.

4.1 Chromosome Design

A solution is encoded as a string. Each solution consists of three parts: the *node string*, *degree string*, and *toll string*. The node and degree strings represent a branch-tree (and the toll ring). The toll string represents the toll level of the toll ring. The node string contains all nodes included in a branch-tree, i.e., $\forall j \in \beta_s$.

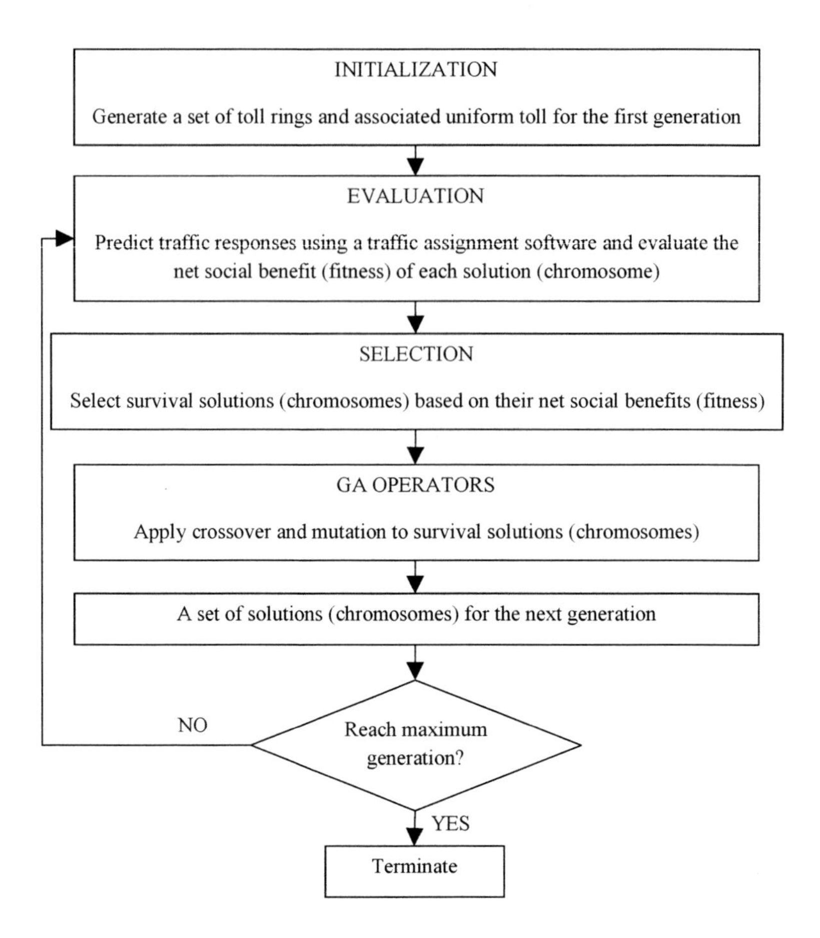

Fig. 6. Overall process of GA-AS

Let $\eta(s) = [n_{col}]$ denote the node string (row vector) of the branch-tree s where n_{col} is the node number in column *col* of the string. The degree string contains the degree of the node in the corresponding column in the node string. Let $\Lambda_s = [\theta_{col}]$ be the degree string for the branch-tree s, where θ_{col} is the value in column *col* of the string, i.e., $\theta_{col} = d_{n_{col}}$. Figure 7 exemplifies the node and degree strings of a branch-tree. In this example, the branch-tree comprises of nodes A, B, D, E, and C with the degrees of 2, 2, 0, 0, and 0 in that order. A recursive program is used to encode and decode the branch-tree from the genotype.

Recursive program for encoding the branch-tree

1. Consider a branch-tree $\beta_A = \{(j, d_j)\}$. Set $col = 1$. Set $j = 1$. Let A be the root node and $d_A = \|\Xi_A\|$. Set $n_{col} = A$ and $\theta_{col} = d_A$.
2. If $d_{n_{col}} = \|\Xi_{n_{col}}\|$ then set $P_j = \Xi_{n_{col}}$; otherwise set $P_j = \emptyset$. Set $\bar{P}_j = \emptyset$, and go to step 3.

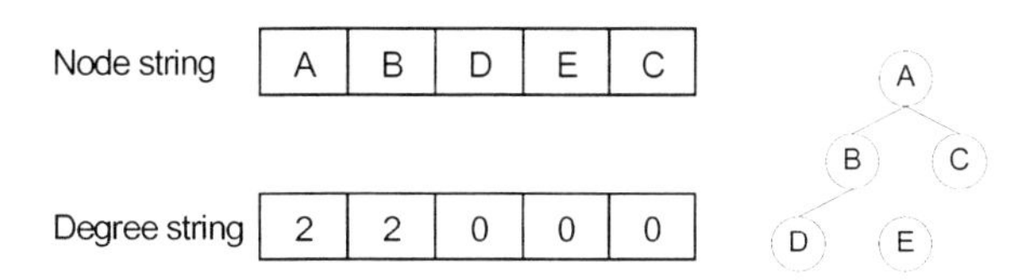

Fig. 7. Chromosome structure of a branch tree

3. If $\bar{P}_j = P_j$, then set $j = j - 1$ and then go to step 7. Otherwise, go to step 4.
4. Set $col = col + 1$.
5. Set $o = n \in \{P_j - \bar{P}_j\}$ (i.e., pick one node which is in P_j but not in $\bar{P}_j$). Set $n_{col} = o$. Set $\theta_{col} = d_o$. $\bar{P}_j = \bar{P}_j + o$.
6. If $d_o = 0$ then go to Step 3. Otherwise, set $j = j + 1$ and go to step 2.
7. If $j = 0$, then terminate. Otherwise, go to step 3.

The toll string is binary encoded and indicates the toll level. A set of possible tolls is defined *a priori*. Suppose that there are seven different toll levels, namely (i) £0.50, (ii) £0.75, (iii) £1.00, (iv) £1.25, (v) £1.50, (vi) £2.00, and (vii) £3.00. A toll string of 101 represents the fifth toll level which is equivalent to £1.50.

4.2 Initialization

The recursive program explained in the previous section is used to randomly generate a set of toll rings. The variable *Prop* is defined as the probability of a node to be expanded, i.e., by applying the branching process to this node. As discussed, a set of tolled links must be defined to set up an initial toll ring. The tail nodes of these tolled links become the target nodes (see Fig. 5). A random number x, is generated for each leaf node in turn. If $x \leq Prop$, the branching process is applied to this leaf node. For the toll level, a random binary toll string is generated for each toll ring.

4.3 Fitness Evaluation

The fitness of each chromosome is measured by the social benefit of the toll ring as defined in Equation 2. A traffic assignment model is used to compute the equilibrium demands and link flows. The travel demands and link flows are used in Equation 2. In addition, the outcome constraints h_1 and h_2 are considered.

4.4 Penalty-Based Approach for Constraint Handling

The constrained problem is transformed into an unconstrained one by introducing the auxiliary fitness function

$$\hat{W} = W - \sum_{e=1}^{m} \left(\kappa_e \cdot \max\left(h_e, 0\right)\right), \tag{4}$$

where κ_e is the penalty applied to the fitness function if constraint h_e is violated. If $h_e \geq 0$, then $\max(h_e, 0) = h_e$ and W is reduced by $(\kappa_e \times h_e)$. Otherwise, $h_e \leq 0 \Rightarrow \max(h_e, 0) = 0$, and no penalty is applied.

The adjustment of the penalty factor is crucial to the success of the algorithm [9]. Choosing a penalty which is too low may cause the search to waste too much time searching in infeasible regions. On the other hand, using a high penalty restricts the search region and traps the search at a local optimum [3]. Three types of penalty functions are considered in this chapter: (i) the static penalty, (ii) dynamic penalty, and (iii) self-adaptive penalty.

For the static penalty, the penalty term κ_e does not change according to the generation numbers of the GA. In most cases, the user defines κ_e as a constant. In contrast to the static penalty, the dynamic penalty function changes from one generation to another. Joines and Houck [5] proposed a simple dynamic penalty function in which the penalty term increases with the GA generations. A more complex dynamic penalty function was proposed by Michalewicz and Attia [8] based on the idea of simulated annealing. The self-adaptive penalty function is another class of the dynamic penalty function. The self-adaptive penalty function incorporates some characteristics of current population (e.g., average fitness) to adjust the penalty for the next generation [4, 9].

4.5 Selection Process

The selection process is based on "stochastic universal sampling" which uses a single wheel spin [7]. The so called "roulette wheel" is constructed where each chromosome is assigned to one slot. The probability for each chromosome to be selected is defined by the linear ranking approach proposed by Whitley [17]. The chromosomes are first ranked in descending order by their fitness values, i.e., the first rank chromosome possesses the highest fitness value in that generation. Then, the probability of the chromosome with rank r to be selected is defined as

$$p_r = \frac{1}{\|P\|} \cdot \left(2 - c + (2c - 2) \cdot \left(\frac{\|P\| - r}{\|P\| - 1}\right)\right), \tag{5}$$

where p_r denotes the selection probability of the chromosome with the rank r, $\|P\|$ is the population size, and $1 \leq c \leq 2$ is the selection bias. The higher the values of c, the more deterministic is the selection process (i.e., focus on selecting fitter chromosomes). The fittest and weakest chromosomes are thus associated with the selection probabilities of $\frac{c}{\|P\|}$ and $\frac{2-c}{\|P\|}$ respectively. In addition to the stochastic selection, the *elitism selection* is employed to ensure the survival of the fittest chromosomes.

4.6 Crossover Operator

When recombining two chromosomes, a set of common nodes included in both chromosomes is identified. If a random number is less than the probability of crossover Pr_{cross}, the crossover operator is applied to these mated chromosomes.

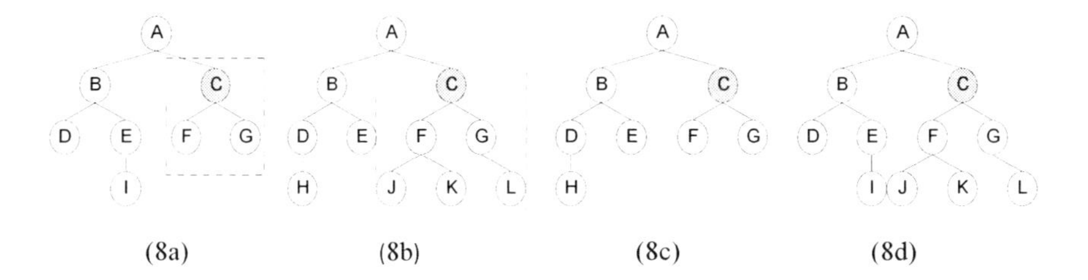

Fig. 8. Illustration of crossover process with branch trees in GA-AS

The crossover operator randomly selects one of these common nodes. The sub-branches (see Sect. 3.1) rooted from the selected common node are defined. The crossover operator swaps these two sub-branches. For the toll string, a simple two-points crossover method is applied. Figure 8 illustrates the crossover process. In this example, there are two mated branch-trees (as shown in Figs. 8a and 8b). The set of common nodes is $\{A, B, C, D, E, F, G\}$. Suppose that node C is randomly selected by the crossover operator. The sub-branches in both branch-trees rooted from node C are shown in the dash-line boxes in Figs. 8a and 8b. These two sub-branched are then swapped, creating two new branch-trees as shown in Figs. 8c and 8d.

4.7 Mutation Operator

The mutation operator randomly selects one node from the branch-tree. If a random number is less than the predefined probability of mutation Pr_{mut}, mutation is applied to this node. If the selected node is a leaf node, the branching process is applied to this node to expand the branch-tree. On the other hand, if the selected node is not a leaf node, then that node is changed to a leaf node by removing the sub-branch rooted from that node and changing the degree of that node to 0. Figure 9 illustrates the mutation process. The two branch-trees on the left side of the figure illustrate the mutation at node F (non-leaf node). The two branch-trees on the right hand side demonstrate the mutation at node I (leaf node).

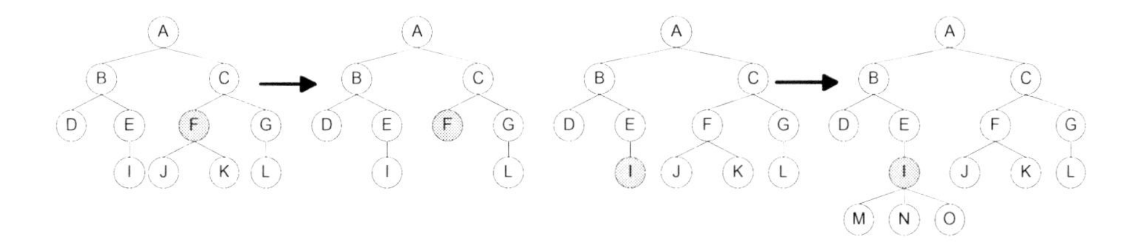

Fig. 9. Illustration of mutation process in GA-AS

5 Numerical Tests with the Edinburgh Network

This section presents the tests of the GA-AS with a realistic urban traffic network. Two sets of tests are presented. The first set in Sect. 5.1 compares the behaviors of different penalty methods (i.e., static, dynamic, and self-adaptive). The second set in Sect. 5.2 applies the modified GA-AS with the self-adaptive penalty to the OTP. The tests are carried out with the network of Edinburgh (see Fig. 13) which was also used in [13]. The GA parameters are set as follows [12]: population number = 50, generation number = 200, probability of crossover = 0.35, probability of mutation = 0.05, and number of elitist individuals = 5.

5.1 Comparison between Static, Dynamic, and Self-adaptive Penalty Methods

The tests involve solving the OTP using three different penalty methods to deal with a constraint on the revenue, i.e., $h_1 \leq 0$ where Rev =£50k. This implies the revenue should be greater than or equal to £50k. The static penalty method is adopted in this test with the penalty function

$$\hat{W} = W - \sum_{e=1}^{m} \kappa_e(h_e) \cdot \min(h_e, 0) \tag{6}$$

where $\kappa_e(h_e)$ are the penalty coefficients for the constraint e. Note that the level of the penalty coefficient depends on the level of violation of that constraint. In this test, three levels of violation are defined and the penalty coefficients for different violation levels are given in Table 1.

Table 1. Levels of violations and penalty coefficients for static penalty function

Level of violation	Penalty coefficient ($\kappa_e(h_e)$)
net revenue < £19,000	0.8
£19,000 ≤ net revenue < £38,000	0.1
£38,000 ≤ net revenue < £50,000	0.05

This chapter adopts the dynamic penalty function proposed by Joines and Houck [5]:

$$\hat{W} = W - \left((C \times gen)^{\alpha} \cdot \sum_{e=1}^{m} \min(h_e, 0)^{\beta} \right) \tag{7}$$

where C, α, and β are parameters given by the user ($C = 0.005$, $\alpha = 1$, and $\beta = 1$ in this test), and gen is the generation number. With $\beta = 1$, the tested dynamic penalty is a linear function of gen.

The self-adaptive penalty function is

$$\hat{W} = W - \left(\mu \cdot \left(\frac{gen}{Gen} \right)^{\rho} \cdot \overline{W}_{gen-1} \cdot \sum_{e} \min(h_e, 0) \right), \tag{8}$$

where *Gen* is the total number of generations, $\overline{W}_{gen-1}$ is the average fitness (without being penalized) in the generation *gen*-1, and ρ and μ are parameters given by the user. In this test, ρ and μ are set to be 0.95 and 0.0000005 respectively.

Figures 10–12 show the ***average** net benefits* and ***average** net revenues* in different generations (not the best found) for the tests with static, dynamic, and self-adaptive penalties respectively. All three methods converge to the same best found solution with the net benefit of around £7k per peak hour and the net revenue of around £57k per peak hour which satisfies the constraint. It is noteworthy that the result of this test does not imply any conclusive comparison of these three methods. The main purpose of this test is only to illustrate different behaviors of different penalty functions.

The trends of the average net benefits and net revenues in Fig. 10 for the static penalty are relatively flat. With the static penalty, the chromosomes violating the constraint are immediately penalized in early generations and hence have low probabilities to survive. On the other hand, as shown in Figs. 11 and 12, the dynamic and self-adaptive penalties gradually increase the penalties, which provides higher chances for infeasible chromosomes to survive in early generations.

The dynamic penalty increases linearly with the generation number (regardless of the change in the average net benefit) which gradually forces the average

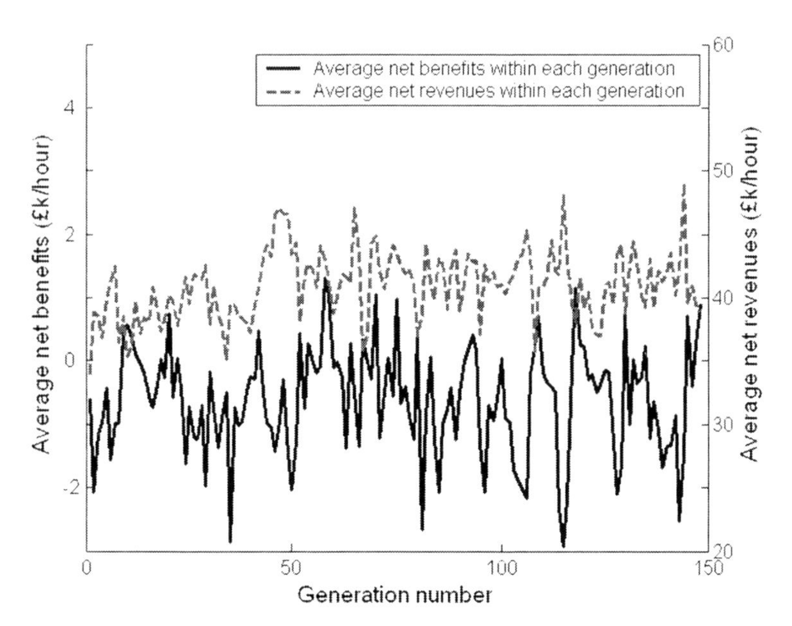

Fig. 10. Average net benefits (dash line) and net revenues (static penalty method)

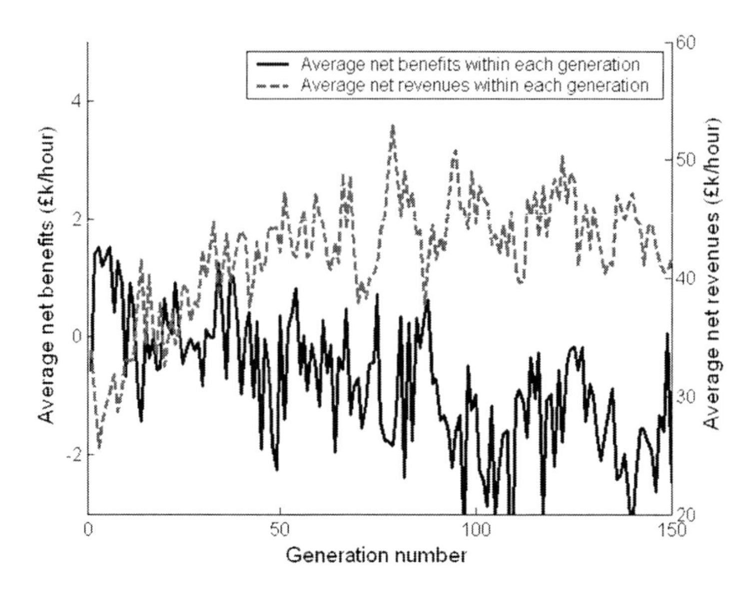

Fig. 11. Average net benefits (dash line) and net revenues (dynamic penalty method)

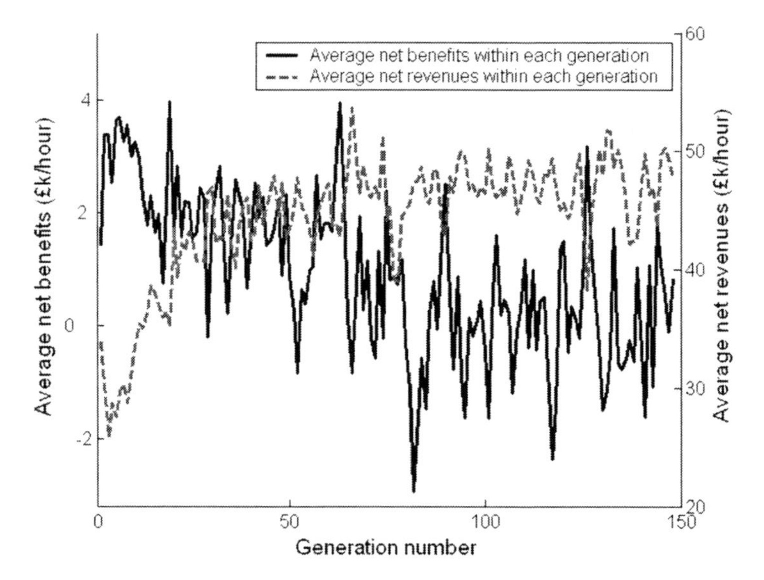

Fig. 12. Average net benefits (dash line) and net revenues (self-adaptive penalty)

net revenue to around the level of the constraint set (i.e., only the feasible chromosomes can survive). On the contrary, the self-adaptive penalty responds to the change in the average net benefit. In early generations, the low average net benefit signals the GA to apply a lower penalty. Once the average fitness of the population increases, the self-adaptive penalty increases to penalize the infeasible solutions to ensure the feasibility of the solutions.

5.2 Results of the OTP

Two different outcome constraints are included in the tests: $h_1 \leq 0$ and $h_2 \leq 0$. Three different tests are set up:

1. Maximize net benefit with revenue constraint (£50k $\leq$ Net revenue)
2. Maximize net benefit with equity constraint ($G \leq 0.30$)
3. Maximize net benefit with equity and revenue constraints ($G \leq 0.30$ and Net revenue$\leq$£45k)

Table 2 shows the results of all tests. The optimal toll rings found for the tests 1–3 are shown in Fig. 13 (named CON-REV, CON-GINI, and CON-REV-GINI respectively). OPC1 is the optimal toll ring without any outcome constraint as reported in [13].

Table 2. Overall results for constrained cordon designs

Cordon	Optimal toll (£)	No of toll points	Net benefit (£k/hr)	Net revenues (£k/hr)	Gini
OPC1	£1.50	13	7.21	43.70	0.41
CON-REV	£2.00	13	6.99	56.44	0.40
CON-GINI	£0.75	14	5.79	27.16	0.28
CON-REV-GINI	£1.50	17	4.38	48.55	0.29

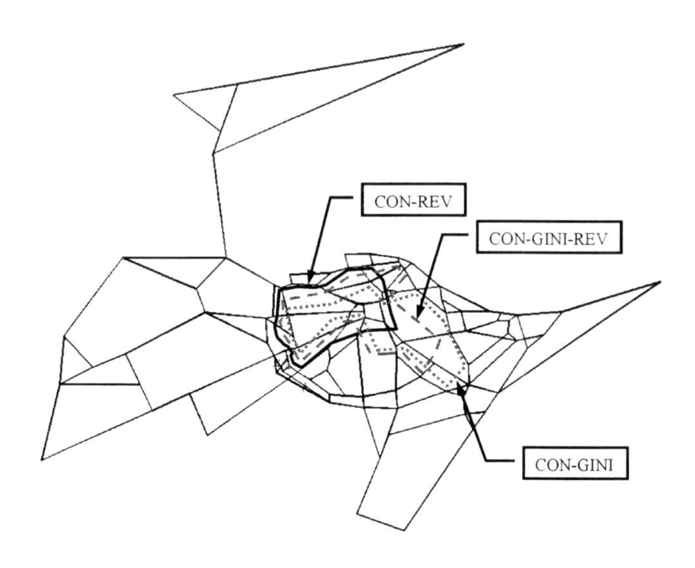

Fig. 13. Optimal cordon locations

The results show significant trade-offs between the scheme benefit and the introduction of the outcome constraints. From the tests, the revenue constraint does not seem to reduce the benefit of the scheme substantially. On the other hand, the benefit drops significantly when the equity constraint is introduced. Introducing both constraints further reduces the scheme benefit. The optimal uniform toll of CON-GINI is also substantially lower than the optimal toll of the unconstrained case (OPC1). In contrast, the optimal toll for CON-REV is higher than the toll of OPC1.

6 Conclusion

This chapter presents a GA based algorithm for solving the optimal toll ring design with outcome constraints. The problem involves the selection of a combination of tolled roads forming a toll ring or cordon line so that the social benefit is maximized. Furthermore, additional outcome constraints are considered. A branch-tree framework [13] is used to encode toll rings. This encoding, together with the problem-specific GA operators, guarantees that new chromosomes represent valid toll rings. Three penalty methods (static, dynamic, and self-adaptive) are used to handle the outcome constraints of the design. An auxiliary fitness function is introduced in which the fitness is penalized if some of the constraints are violated.

The GA based method is tested for a real-world problem, the Edinburgh network. The test results illustrate different behaviors of the three penalty functions. Further tests show a satisfactory performance of the proposed method in finding the optimal design given additional outcome constraints. Future studies should look into the performance and effect of different penalty functions and the chromosome structure. In particular, the issue of the chromosome structure's locality (see, e.g., Sect. 3.3 in [10]) should be investigated, in which a small perturbation of the chromosome structure may result in a significant change in the shape of the toll ring.

References

1. Bard, J.F.: Practical Bilevel Optimization: Algorithms and Applications. Kluwer Academic Publishers, Boston (1998)
2. Coello, C.A.C.: Theoretical and numerical constraint-handling techniques used with evolutionary algorithms: A survey of the state of the art. Computer Methods in Applied Mechanics and Engineering 191(11-12), 1245–1287 (2002)
3. Coello, C.C.: Use of a self-adaptive penalty approach for engineering optimization problems. Computers in Industry 41(2), 113–127 (2000)
4. Coit, D.W., Smith, A.E., Tate, D.M.: Adaptive penalty methods for genetic optimization of constrained combinatorial problems. INFORMS Journal on Computing 8(2), 173–182 (1996)
5. Joines, J., Houck, C.: On the use of non-stationary penalty functions to solve nonlinear constrained optimization problems with GAs. In: The First IEEE Conference on Evolutionary Computations, pp. 579–584. IEEE, Los Alamitos (1994)

6. May, A.D., Liu, R., Shepherd, S.P., Sumalee, A.: The impact of cordon design on the performance of road pricing schemes. Transport Policy 9, 209–220 (2002)
7. Michalewicz, Z.: Genetic Algorithms + Data Structures = Evolution Programs. Springer, New York (1992)
8. Michalewicz, Z., Attia, N.F.: Evolutionary optimization of constrained problems. In: The 3rd Annual Conference on Evolutionary Programming, pp. 98–108 (1994)
9. Richardson, J.T., Palmer, M.R., Liepins, G., Hilliard, M.: Some guidelines for genetic algorithms with penalty functions. In: Schaffer, J. (ed.) Proceedings of the 3rd International Conference on Genetic Algorithms. Morgan Kaufmann, San Francisco (1989)
10. Rothlauf, F.: Representations for Genetic and Evolutionary Algorithms. Springer, New York (2006)
11. Sheffi, Y.: Urban Transportation Networks: Equilibrium Analysis with Mathematical Programming Methods. Prentice-Hall, Englewood Cliffs (1985)
12. Sumalee, A.: Optimal Road Pricing Scheme Design. PhD Thesis, University of Leeds, Leeds (2004)
13. Sumalee, A.: Optimal road user charging cordon design: A heuristic optimisation approach. Computer-Aided Civil and Infrastructure Engineering 19, 377–392 (2004)
14. Sumalee, A.: Multi-concentric optimal charging cordon design. Transportmetrica 3(1), 41–71 (2004)
15. Van Vliet, D.: SATURN – A modern assignment model. Traffic Engineering and Control 23(12), 578–581 (1982)
16. Wardrop, J.G.: Some theoretical aspects of road traffic research. In: Proceedings of the Institute of Civil Engineers, PART II, vol. 1, pp. 325–378 (1952)
17. Whitley, D.: The GENITOR algorithm and selection pressure: Why rank based allocation of reproductive trial is best. In: Schaffer, J. (ed.) Proceedings of the 3rd International Conference on Genetic Algorithms. Morgan Kaufmann, San Francisco (1989)
18. Zhang, X., Yang, H.: The optimal cordon-based network congestion pricing problem. Transportation Research 38B(6), 517–537 (2004)

Real Time Identification of Road Traffic Control Measures

Khaled Almejalli, Keshav Dahal, and M. Alamgir Hossain

MOSAIC Research Group, School of Informatics, University of Bradford,
Great Horton Road Bradford, BD7 1DP, United Kingdom
{k.a.al-mejalli,k.p.dahal,m.a.hossain1}@bradford.ac.uk

Summary. The operator of a traffic control centre has to select the most appropriate traffic control action or combination of actions in a short time to manage the traffic network when non-recurrent road traffic congestion happens. This is a complex task, which requires expert knowledge, much experience and fast reaction. There are a large number of factors related to a traffic state as well as a large number of possible control actions that need to be considered during the decision making process. The identification of suitable control actions for a given non-recurrent traffic congestion can be tough even for experienced operators. Therefore, simulation models are used in many cases. However, simulating different traffic actions for a number of control measures in a complicated situation is very time-consuming. This chapter presents an intelligent method for the real-time identification of road traffic actions which assists the human operator of the traffic control centre in managing the current traffic state. The proposed system combines three soft-computing approaches, namely fuzzy logic, neural networks, and genetic algorithms. The system employs a fuzzy-neural network tool with self-organization algorithm for initializing the membership functions, a genetic algorithm (GA) for identifying fuzzy rules, and the back-propagation neural network algorithm for fine tuning the system parameters. The proposed system has been tested for a case-study of a small section of the ring-road around Riyadh city in Saudi Arabia. The results obtained for the case study are promising and demonstrate that the proposed approach can provide an effective support for real-time traffic control.

Keywords: Road traffic control, Fuzzy logic, Neural networks, Genetic algorithms.

1 Introduction

The traffic congestion problem becomes alarming as the number of vehicles and the need for transportation grow. Traffic congestions do not only cause considerable costs due to unproductive time losses, but they also increase the probability of accidents and have a negative impact on the environment (air pollution, lost fuel) and on the quality of life (health problems, noise, stress) [9]. Therefore, traffic management and control have been a major problem in developing as well as in developed countries. Governments have been spending hefty amounts to develop the traffic control centres using different methodologies using the benefits of advanced information technology.

A. Fink and F. Rothlauf (Eds.): Advances in Computational Intelligence, SCI 144, pp. 63–80, 2008.
springerlink.com

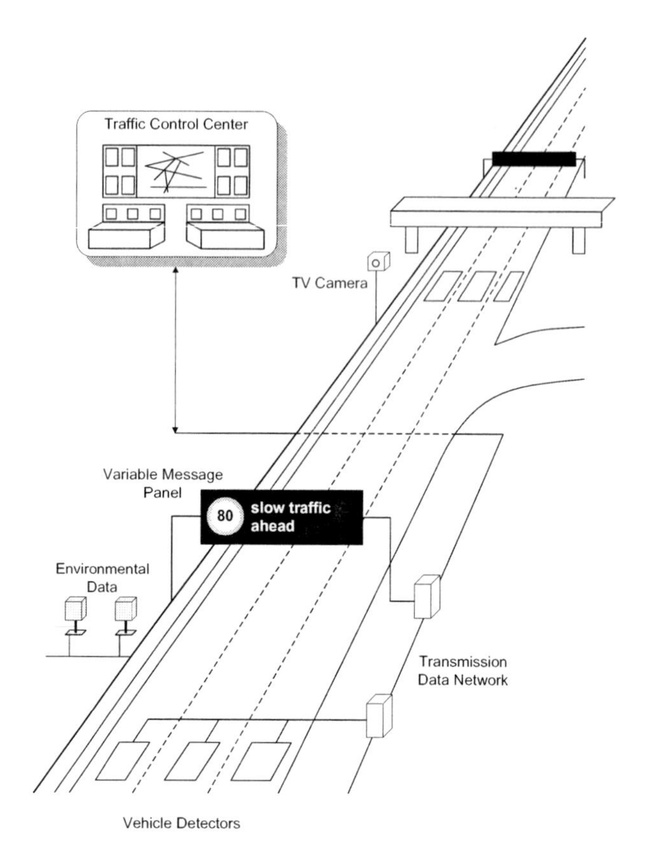

Fig. 1. Typical information infrastructure for real-time traffic control [26]

Modern traffic control centres are connected to monitoring devices such as detectors, weather sensors, and cameras to record data related to the traffic state on-line, e.g., speed, flow, demand, environmental conditions, etc. Moreover, the control centres use advanced dynamic control devices such as ramp metering, dynamic route information panels (DRIPs) and/or variable message signs (VMSs). Figure 1 shows a typical form of infrastructure for real-time traffic control which can be found in different cities [26].

When non-recurrent congestion happens, the operator of the traffic control centre has to assess the severity of the situation, predict the most probable evolution of the state of the network, and quickly select the most appropriate control actions [12]. This complex task requires expert knowledge and much experience which can often only be obtained after an extensive training. There is a large number of factors related to a traffic state and a large number of possible control measures that need to be considered during the decision-making process. The identification of suitable control actions for a given non-recurrent traffic congestion can be tough even for experienced operators [8]. Therefore, one needs an advanced traffic control system that integrates the traffic state data with traffic monitoring and control software to help operators in their decision-making. Road traffic simulation models are used in many cases. However, simulating different traffic actions for a number of control measures in a complicated situation can be very time-consuming [8].

Our previous work [1] reported the preliminary research on the development of an intelligent support system which controls road traffic using a number of machine-learning techniques. A learning algorithm was developed in [2] for the machine-learning techniques and tested for a number of different applications. This investigation deepens their works by combining the learning algorithm with the machine-learning techniques for road-traffic management.

This chapter presents an intelligent real-time traffic control identification system to assist the human operator of the traffic control centre to manage traffic on highways and urban ring roads. The inputs of the system are the current traffic state and the available control actions, and the output is a ranked list of all possible of control actions. The proposed system is based on a fuzzy logic approach in combination with other soft-computing approaches: neural networks and genetic algorithms. These approaches within the identification system provide the initialization and self-organization of the membership functions for fuzzy input variables, identification of fuzzy rules, and fine tuning of the system parameters. The learning algorithm presented in [2] has been used to generate the fuzzy rules to the fuzzy neural network tool. We have tested the system for a case-study of a small section of the ring-road around Riyadh city in Saudi-Arabia.

The chapter is organized as follows. The next section reviews the related work published in the literature. Section 3 describes the proposed Road-Traffic Decision Support System (RTDSS), its design and functions. This is followed by the application of the developed identification system for a case study in Sect. 4. The conclusions and future works of this research work are given in Sect. 5.

2 Related Work

Different intelligent approaches have employed to develop decision support systems for road traffic control [26, 12, 8, 31, 41, 7, 35, 4, 42]. Some of these applications used the fuzzy logic technique in their decision process [12, 8], while others used neural networks [35, 42]. The TRYS system described in [26, 7] is an agent-based environment for building intelligent traffic management systems applications for urban, interurban and mixed traffic areas. TRYS system is based on knowledge frames, and some of these frames use fuzzy logic. Other knowledge-based real-time decision support systems for road traffic management are described in [31, 41]. Hegyi et al. [12] have presented a fuzzy logic based traffic control system for efficiently managing non-recurrent congestions, which later has been extended by De Schutter et al. in [8] as a multi-agent traffic control system. The presented fuzzy traffic control system is part of a larger traffic support system which uses a case base and fuzzy logic to generate a ranked listing of combinations of traffic control measures and their estimated performance for a given traffic situation. Since the proposed system does not use any knowledge from experts or heuristic rules and it is only based on a case based reasoning system, therefore, the quality of its result depends basically on the quality of the case base.

Ramp controlling or ramp metering is a traffic control technique used to control the traffic inflow into a freeway by limiting the number of vehicles entering the freeway. Usually, the main goal of the ramp metering system is to avoid congestion and reduce vehicle's total travel time. For solving freeway ramp-metering control problems, Wei [35] has developed artificial neural network models. Inputs to neural network models are traffic states in each time period on the freeway segments while outputs are the desired metering rate at each entrance ramp. Also Zhang et al. [42] have used the neural network technique for ramp control, while Bogenberger et al. [4] have applied adaptive fuzzy logic system for ramp controlling problem. The adaptive fuzzy logic algorithm has been used to determine the traffic responsive metering rate.

Traffic signal control is another traffic control problem with a number of complex and sometimes conflicting variables and objectives. Therefore, different kinds of traffic signal control methods have been presented. For example, Wei et al. [36] have presented a fuzzy logic adaptive traffic signal controller for an isolated four-approaches intersection with through and left-turning movements. The controller has the ability to make adjustments to signal timing in response to observed changes. Other traffic signal control systems are described in [34, 3].

Fuzzy Neuron Network has been employed in the traffic management field in several papers. For example, Henry et al. [13] have developed a neuro-fuzzy control method for controlling of traffic lights of an intersection. The system offered good results for simple and medium-complexity intersections but poor performance on a complex intersection. Another fuzzy neuron network system has been proposed in [27] to the analysis and prediction of traffic flow. The system has been fully trained and subsequently used for short-term traffic flow prediction. The prediction results are shown to be promising.

3 The Proposed Decision Support System

3.1 Overall Framework

As mentioned earlier a large number of factors that determine the current traffic state need to be considered in the decision making process. These input factors are usually measured by the on-line monitoring system using sensors, detectors and cameras (alternatively, the traffic state can be forecasted by a traffic flow simulation model). These input factors include traffic densities, average speeds, traffic demand, etc. Similarly, there are many possible control actions that can be employed to control the road network depending on the nature of traffic problems and available road control facilities. The proposed system receives the current values of the input factors (e.g., from monitoring system) and all possible control actions. The RTDSS then outputs a ranked list of the control actions to assist

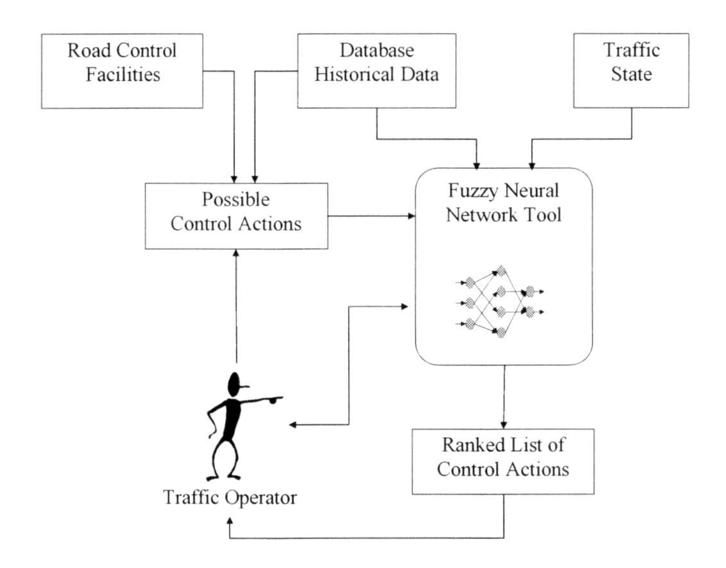

Fig. 2. Structure of the Road-Traffic Decision Support System (RTDSS)

the human operator of the traffic control centre to manage the traffic network in real-time. The overall structure of the proposed RTDSS is depicted in Fig. 2.

In general, the proposed intelligent system works as follows. Let S be the set of all possible control actions, which can be used to control the considered road network. S is created for a given road network off-line using the available road control facilities, the traffic operator's experience, and historical data. This also considers the interrelations between the traffic control measures at different locations in the network. Control action $(c_i \in S)$ can be one control measure such as lane closure, ramp, etc., or a combination of the control measures.

Given a current traffic state from the on-line monitoring system and set of all possible control actions (S) for the given road network, the RTDSS employs a pre-trained fuzzy-neural network tool (FNN_Tool) (see details in the next section) to predict the performance of each c_i for the current traffic state. Then the system provides the operator with a ranked list of the control actions in real-time. The following pseudo-code summarizes the main functions of the proposed RTDSS:

```
Identify S = {c1, c2, c3, ..., cn};
For ( i=1; i<=n; i++) do
    {
        Set Control_Input = ci;
        Run FNN_Tool(Traffic_State_Input, Control_Input);
        Calculate the Aggregated Performance of ci;
    }
Rank S;
Show ranked list;
```

3.2 Performance Prediction

There is a range of traffic criteria such as queue lengths, total travel times, the number of vehicles entering the network, number of vehicles leaving the network, etc., which can be considered to assess the performance of a control action. These performance criteria can be calculated using a traffic simulation model such as METANET [25]. The aggregated performance of each control action c_i can be calculated by considering one or more of the performance criteria, or by using a weighted sum [8], which is defined as:

$$P_{c_i} = \sum_{k=1}^{N} (w_k * E_{k,c_i}) \tag{1}$$

where P_{c_i} is the aggregated performance of control action c_i for the given traffic state ; E_{k,c_i} is the evaluation of control action c_i over performance criterion k for the given traffic state; w_k is the weight of the performance criterion k; and N is the number of considered performance criteria. These weights (w_k) are usually selected by the operators based on current traffic management policies and other considerations.

3.3 Fuzzy Neural Network Tool (FNN_Tool) Structure

In principle a simple neural network can be used as a decision support tool within the proposed framework. It can give an accurate output provided it is trained on all possible cases. However, given the high-dimensionality of the prediction problem addressed here, training a neural network on all possible traffic cases is impossible. For example, if the conditions in a network are described by the time of day, densities of its links, traffic demands on the network boundaries, control actions that have been applied, and the incident status, then the description of the conditions on a 25 link network will yield approximately 10^{24} cases. Clearly, it is unfeasible to consider such a number of traffic cases in the training process [15]. Therefore, fuzzy neural network is used to address this problem.

Fuzzy neural networks are hybrid intelligent systems which combine the advantages of both neural networks and fuzzy logic. The neural fuzzy system is a fuzzy system that uses the learning ability of the neural networks to determine fuzzy sets, fuzzy memberships and fuzzy rules. The neural network provides the fuzzy systems a self-adaptive capability to elicit membership function, map fuzzy sets to fuzzy rules, and implement defuzzification [22, 33].

The structure of the neural fuzzy network tool (FNN_Tool) used in our proposed system is similar to the structure proposed in [27]. It is a five-layer structure, as shown in Fig. 3, where each layer performs an operation to build the fuzzy system. The inputs of our FNN_Tool are the current traffic state which is characterized by input factors (e.g., densities, average speeds, traffic demand) and all possible control actions. The outputs of the FNN_Tool are the evaluation of those control actions for the current traffic state according to a number

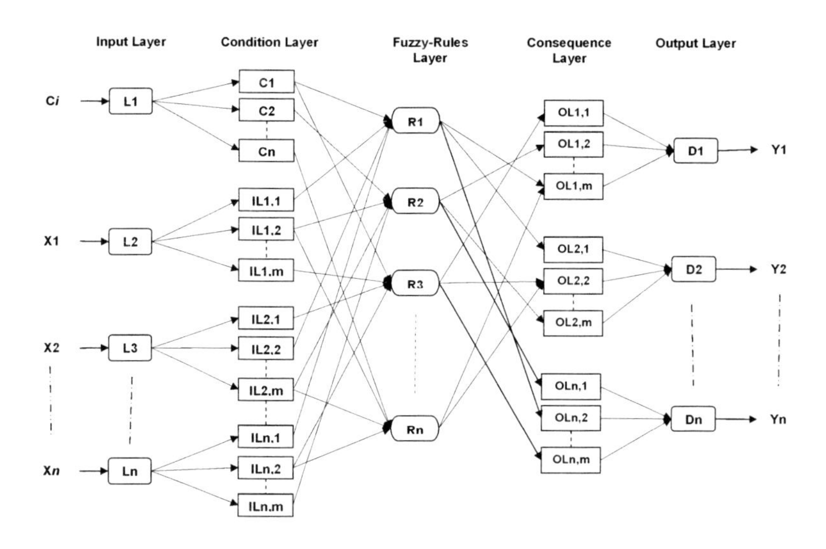

Fig. 3. Structure of the fuzzy neural tool (FNN_Tool)

of performance criteria such as queue lengths, total travel times, waiting times, etc. The process of each layer is described below (see Fig. 3):

Layer 1: is the input layer. Neurons in this layer represent input linguistic variables such as "speed", "density", and "control action" and directly transmit non-fuzzy input values to the next layer. In our case, the neuron inputs represent the *Traffic_input* of the current traffic state, which are represented as vectors $X^T = [x_1, x_1, x_2, ..., x_n]$, and the *Control_input* of the road network, which is represented as c_i. The link weight w_i between this layer and the next layer is 1. The input and the output of this layer are given as follows:

$$o_i^{(1)} = i_i^{(1)} \tag{2}$$

where $i_i^{(1)}$ is the input and $o_i^{(1)}$ is the output of input neuron i in layer 1.

Layer 2: is the fuzzification layer, which defines the fuzzy sets and membership for each of the input factors. Neurons in this layer act as a membership function and represent the terms of the respective linguistic variable, such as "low", "high", and "action 1". In our model the neurons of this layer are modeled as a common bell-shaped membership function [32], so that the input $i_{i,j}^{(2)}$ and the output $o_{i,j}^{(2)}$ of fuzzification neuron i in the layer 2 are given as follows:

$$o_{i,j}^{(2)} = e^{-\frac{\left(i_{i,j}^{(2)} - m_{i,j}\right)^2}{\sigma_{i,j}}} \tag{3}$$

where $m_{i,j}$ and $\sigma_{i,j}$ are the centres and the widths of the membership function for the input-label neuron $LI_{i,j}$ respectively.

Layer 3: is the fuzzy rule layer, which defines all possible fuzzy rules to specify qualitatively how the output parameter is determined for various instances of the input parameters. Each neuron in this layer represents a fuzzy rule, for example, "if the time is morning, the average speed high, the density low and the control action is "action 1", then the total travel time is medium and the waiting time is low". Different approaches are used to identify the fuzzy rules in the fuzzy neural networks, such as using experts' linguistic information [40, 20], using unsupervised learning algorithms [24, 29], and using supervised leaning algorithm (such as the backpropagation technique) [21]. In this research we have proposed a GA based learning algorithm to identify fuzzy rules (see Sect. 3.5 for details). The input and the output of a rule neuron in the layer 3 are given as follows:

$$y_i^{(3)} = min\left(x_{k,i}^{(3)}\right) \tag{4}$$

where $x_{k,i}^{(3)}$ are the inputs, and $y_i^{(3)}$ is the output of fuzzy rule i in layer 3.

Layer 4: is the consequence layer (or the output membership layer). Neurons in the consequence layer represent fuzzy sets (such as "low", "medium", and "high") used in the consequent part of a fuzzy rule. The input and output of a consequence neuron in layer 4 are given as follows:

$$y_i^{(4)} = min\left(1, \sum_l x_{l,i}^{(4)}\right) \tag{5}$$

where $x_{l,i}^{(4)}$ is the input (the output of neuron l in the fuzzy rule layer), and $y_i^{(4)}$ is the output of membership neuron i in the layer 4.

Layer 5: is the output layer (or the defuzzification layer). Each neuron in the output layer represents a single output variable such as "average speed" and "travel time". The input and the output of an output neuron in layer 5 are given as follows:

$$x_i^{(5)} = \Sigma\left(a_{c,i}/b_{c,i}\right) y_i^{(4)} \tag{6}$$

$$y_i^{(5)} = \frac{x_i^{(5)}}{\sum_{i=1}\left(y_i^{(4)}/b_{c,i}\right)} \tag{7}$$

where $x_i^{(5)}$ is the input and $y_i^{(5)}$ is the output of neuron i in layer 5, $a_{c,i}$ and $b_{c,i}$ are the centre and width of the fuzzy set respectively.

3.4 Learning Process of the Proposed Intelligent System

There is a wide variety of fuzzy neural systems. In general, we can subdivide the existing fuzzy neural networks based on their construction and learning techniques into two types, as follows:

Class I: fuzzy neural networks which are constructed using linguistic information provided by human beings and whose learning process is achieved by using neural network techniques [28].

In this case, experts are required to provide a clear description of the membership functions and fuzzy rules used in the system. Although this type of fuzzy neural networks can converge faster during the learning process and performs better in decision making, the design of such a fuzzy neural network is rather subjective, since linguistic information from experts may vary depending on the person and the time. In addition, linguistic information is too general and wide-ranging to focus on certain situations [28]. Several researchers' efforts concentrated on this class of fuzzy neural systems such as [40, 20, 30].

Class II: fuzzy neural networks which are constructed using numerical information and whose learning process is achieved by using neural network techniques [28].

This class of fuzzy neural networks is similar to the previous one as regards the tuning and configuration of the parameters and structures. However, the initial set of parameters and the structure of such fuzzy neural networks are not derived from linguistic information. Instead, they are constructed using an unsupervised learning algorithm from a set of training data and are fine-tuned on the basis of the numerical information. This type of fuzzy neural networks is suitable for the applications where one may have direct observations from the system but is unable to find experts who can provide an organized description of the system. However, since the set of training data is the only source of information employed in this type of fuzzy neural networks, it has to be representative of the system's behavior [28]. Several authors have presented this type of fuzzy neural networks, such as [39, 6, 5].

Our proposed FNN_Tool belongs to the second class of fuzzy neural network. We suggest three stages of learning. The first stage is the initialization of the membership functions of both input and output variables by determining their centres and widths. To perform this stage, we have employed a self-organizing algorithm [19] as in other works [27, 24, 37]. A GA-based learning algorithm is performed in the second stage to identify the fuzzy rules that are supported by the set of training data. In the last stage, the derived structure and parameters are fine tuned by using the back-propagation learning algorithm [32]. During this stage, all adjustable parameters (i.e., centroids and widths of input-label membership functions, centroids and widths of output-label membership functions, and weights of fuzzy rules nodes) are fine tuned. All those three stages of learning process are done off-line.

3.5 GA-Based Learning Algorithm

An important topic of a fuzzy neural network design is the identification of appropriate fuzzy rules. However, there is no systematic design procedure at present [27]. The recent research direction in the identification of the fuzzy rules in fuzzy neural networks is to learn and modify the rules from past experience [29]. In this section we propose a GA-based learning algorithm to make use of the known membership function to identify the fuzzy rules. This GA-based learning algorithm belongs to the unsupervised learning algorithms.

GA is a class of evolutionary algorithms which applies operators inspired by the mechanics of natural selection to a population of solutions encoding the parameter space at each generation [14, 10]. Several authors have proposed genetic algorithms (GAs) for fuzzy neural parameters optimization to adjust the control points of membership functions or to tune the weightings such as [18, 16, 17, 23]. The pioneer was Karr [18], who used GAs to adjust membership functions. Ishibuchi et al. [16] proposed a genetic-based method for selecting a small number of significant fuzzy rules to construct a compact fuzzy classification system with high classification power. Ishibuchi and Yamamoto further developed this idea by using mult-objective genetic local search algorithms in [17]. Lin [23] proposed a GA-based hybrid learning algorithm for parameter learning. The hybrid algorithm used GA to tune membership functions at the precondition part of fuzzy rules, while the least-squares estimate method was used to tune parameters at the consequent part.

To explain the design of the proposed GA-based learning algorithm for generating fuzzy rules, we consider a simple example of FNN_Tool with two input linguistic variables x_1 and x_2, and one output linguistic variable y. After performing the self-organization learning algorithm, each linguistic variable has a number of fuzzy sets, say we have three fuzzy sets low (L), medium (M), high (H). The proposed GA learning algorithm considers all possible rules as shown in Fig. 4. In our simple example there are a total of twenty seven possible rules. In fact these rules are made of nine possible antecedents (preconditions). These antecedents of fuzzy rules are represented by neurons $R_1 ... R_9$ of the Fuzzy-Rules Layer. Each antecedent is linked to three possible decision fuzzy sets (neurons in Consequence Layer: L, M and H). For example, the three possible fuzzy rules associated with neuron R1 are:

If x1 is L and x2 is L, then y is L.
If x1 is L and x2 is L, then y is M.
If x1 is L and x2 is L, then y is H.

In this way the total number of fuzzy rules includes all possible fuzzy rules associated with all neurons. However, some of these rules can be redundant and are not used for making decisions. Filtering out these redundant rules will reduce the computational time in the decision- making process. We use a GA-based learning approach to identify the appropriate and relevant rules.

A number of decisions must be made in order to implement the GA to generate appropriate fuzzy rules. There are problem specific decisions which are concerned

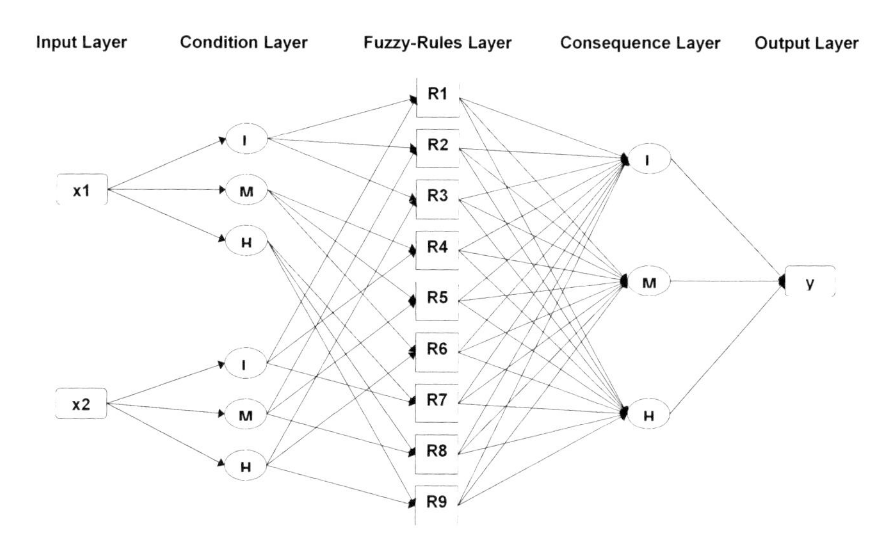

Fig. 4. FNN_Tool with two inputs variable and one output variable with all possible fuzzy rules

with the search space (and thus the representation) and the form of the fitness function.

The encoding of the problem using an appropriate representation is a crucial aspect of the implementation of the GA technique. The encoding used to represent chromosomes (solutions) defines the size and the structure of the search space. Here we propose integer strings as chromosomes to represent candidate solutions of the problem. The string is given by $t_1, t_2, t_i, ..., t_n$, where t_i is an integer $0 \leq t_i \leq m$ which indicates the link of neuron R_i (i.e., neurons in Fuzzy-Rules Layer) with output neurons (i.e., neurons in Consequence Layer). n is the number of neurons in the Fuzzy-Rules Layer and m is the number of neurons in the consequence Layer. For our example, the chromosome has nine integers, and $0 \leq t_i \leq 3$. $t_i = 0$ indicates that there is no link of R_i with output neuron; $t_i = 1$ indicates that there is a link with L neuron in the consequence Layer and so on. (An example of decoding a chromosome is shown in Fig. 5).

The correctness of every chromosome is evaluated by using a fitness function. The fitness function can be any nonlinear, nondifferentiable, or discontinuous positive function, because the GA only needs a fitness value assigned to each chromosome. We use a set of training data to calculate the fitness of each chromosome based on the following fitness function:

$$FIT\,(i) = \frac{1}{RMS\_ERROR\,(i)} \tag{8}$$

where $RMS\_ERROR\,(i)$ represents the root-mean-square error between the fuzzy-neural network outputs and the desired outputs for the the fuzzy rule set represented by ith GA string. The GA aims to maximize the fitness function

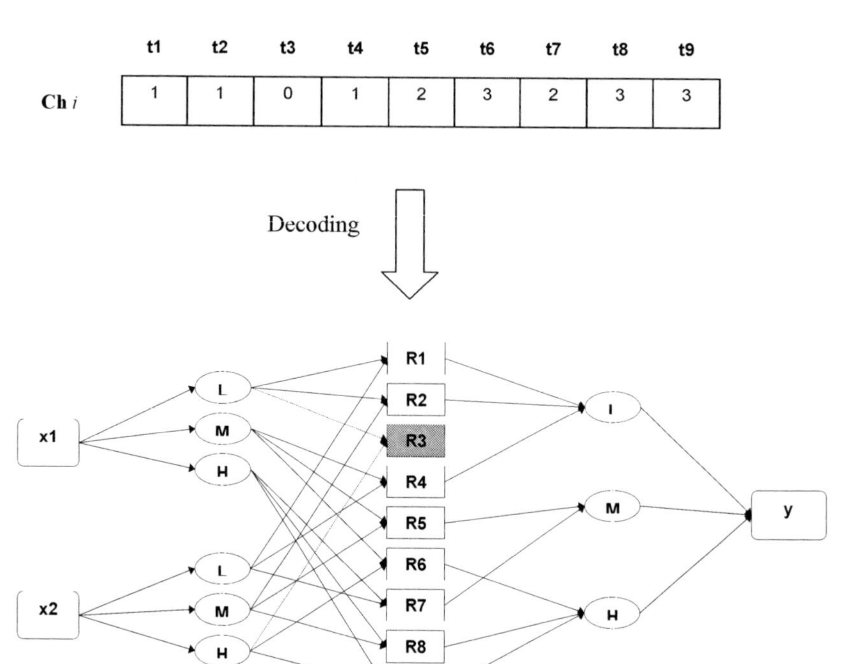

Fig. 5. An example of decoding a chromosome

(8) to minimize the error value (e). This error value depends on the selected fuzzy rules. A gene in a GA string with $t_i \neq 0$ represents a fuzzy rule to be considered by the tool and with $t_i = 0$ suggests a fuzzy rule to be ignored. The weight for all rules is assumed to be 1 at this stage. However, our experiment showed that the inclusions of some of the rules suggested for ignoring (i.e., $t_i \neq 0$) with low weightings can still improve the error value.

In order to correctly identify the minimum number of the appropriate fuzzy rules without ignoring any relevant rule that might improve the error value, the fitness value of a chromosome is calculated in two stages: Firstly, a chromosome is evaluated as given by GA (fit_1), i.e., the fitness of the chromosome is calculated considering all rules represented by ($t_i \neq 0$) (taking weight 1), and rules represented by ($t_i = 0$) are ignored. Secondly, for each rule (R) represented by ($t_i = 0$), the fitness of the chromosome is calculated again considering a low weight LW (e.g., 0.01) for each possible rule associated with that rule (Ri) (in our example there are three possible rules) and then the best one is selected (fit_2). Then these two fitness values (fit_1, fit_2) are compared and the best fitness value is taken. The chromosome is adjusted if the second fitness (fit_2) appeared to be the better one.

Based on our previous experience with GA and a number of experiments, we have selected GA operators and their parameters to be used in this application.

The population size is 40 and the GA operators used are the steady state replacement approach [38], tournament selection [11], standard two-point crossover (with 0.7 probability) and a mutation operator. We use a higher mutation priority of 0.1 in our case in order to maintain the diversity of the population. The steady state approach directly inserts a new solution into the population pool replacing a less fit solution. The tournament selection method picks a subset of solutions randomly from the population to form a tournament selection pool, from which two solutions are selected with the probability based upon the fitness values of the solutions. The two-point crossover operator splits the selected solutions at two randomly chosen positions and exchanges the centre sections with a crossover p robability. The mutation operator changes the integer at each position in the solution within the allowed range with a defined mutation probability. The elitist approach, which ensures that the best solution in the population pool is always retained, has been applied. The initial population of chromosomes is created at random. The stopping criterion for a GA run is to achieve the pre-specified error level (e).

When the GA learning process is completed (i.e., when pre-specified error level is achieved) after running the GA over a large number of runs, we choose the best GA chromosome. This best chromosome is decoded to get the structure of the FNN_Tool by keeping only the links that are indicated by the chromosome. The GA approach can take a lengthy computation time for the optimization process. However, this computation time is not considered problematic for offline training.

4 Case Study

In order to evaluate and test the proposed system discussed in the previous section, we derived a test case study using a small section of the ring-road around Riyadh, the capital city of Saudi-Arabia. The selected section is one of the busiest parts of the Riyadh ring-road, because it is used mostly for traffic approaching the city centre as shown in Fig. 6. This section includes 10-km of the main road with three lanes in each way and a service road with a limited capacity. The service road parts with the main road at point B and runs parallel to the main road and gives access to Mather Street and then joins the main road again at point C (see Fig. 6). A, B, and C are join points between service roads and the main road and they are controlled by ramp metering devices. Before point B there is a DRIP which can display queue information or give some alternative routes to drivers. In this case study, we only consider the traffic going from the south to the north.

Since the aim of this stage is assessing the technical feasibility of the proposed system, only a limited number of inputs, control actions, and training data have been considered. However, the increase in the number of inputs, possible control actions, and training data should not affect the validity of the proposed system. We have considered the following variables in our case study:

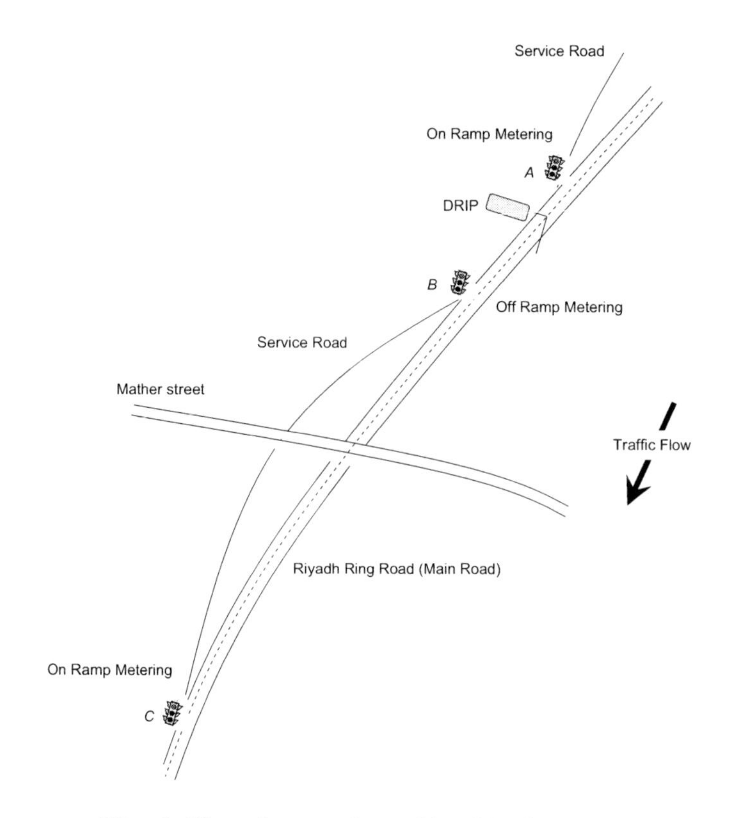

Fig. 6. The sub network considered in the prototype

Table 1. The performance evaluation of the control actions C1–C5 on a selected traffic state

Traffic State:

Traffic Flow : Main Road = 2600 Service Road =1000

Traffic Demand : Main Road = 1780 Service Road = 900

	Riyadh-Traffic Simulation Model			Proposed System		
Control Actions	TTT	TFC	AQL	TTT	TFC	AQL
C1	1243	3589	15	1240	3581	15
C2	987	2989	19	984	2988	19
C3	1056	3345	22	1050	3345	21
C4	890	2879	15	888	2872	15
C5	1267	3108	17	1266	3103	16

Two traffic factors to represent the current traffic state:

- Traffic flow (vehicle/hour): on both the main road and the service road.
- Traffic demand (number of vehicles inflow into the network): for both the main road and the service road.

Five traffic control actions:

- C1: metering the on-ramp at point A, and using DRIP.
- C2: metering the off-ramp at point B, and using DRIP.
- C3: metering the on-ramp at points A and C, and off-ramp at points B.
- C4: using DRIP to display the queue information.
- C5: doing nothing.

We use three evaluation criteria for calculating the performance of the control actions:

- Total travel time (TTT) (hours).
- Total fuel consumption (TFC) (liters).
- Average queue length (AQL) (number of vehicles).

The data needed for the training process has been generated using a traffic simulation model (METANET macroscopic flow model [25]). Real data is not available, because obtaining real data which represents the performances of different control actions for each real traffic state is impossible. All our variables have been considered and simulated for the period from 4pm to 8pm.

In order to test the performance of the proposed system, we compared the proposed system and the traffic simulation model. The results obtained for the performance of the five control actions by the proposed system and the simulation model on a selected traffic state are shown in Table 1. We found that the results obtained by the proposed system are very close to the results given by the traffic simulation model, which confirms the validity of the proposed model in obtaining results. However, the time the proposed system needs to calculate the performance of a given control action is much less than the time needed by the simulation model. Since the proposed system has been developed using trained FNN_Tool, the main advantage of the proposed system is its execution speed. Furthermore, the proposed system allows the user to evaluate a set of control actions in one process instead of evaluating one by one.

5 Conclusion and Future Work

This chapter has described a real-time decision support system for a traffic control centre. The proposed system uses a decision support tool with FL, NN and GA techniques to assist the human operator of the traffic control center to manage the current traffic state. When a non-recurrent congestion takes place, the operator can use the proposed system in real-time to assess the approximate performance of several control actions in real-time. For constructing and training the fuzzy neural network tool used in the proposed system we performed three different algorithms: the self-organization algorithm for initializing the fuzzy sets, a GA-based learning algorithm for identifying appropriate fuzzy rules, and the back-propagation algorithm for fine tuning the system's parameters. A case study of a section of the ring-road around Riyadh is presented and discussed in order to evaluate and test the proposed system. The results of the proposed system clearly demonstrate its merits and capabilities in terms of processing speed and flexibility.

This research investigation has demonstrated the technical feasibility of the proposed system with a traffic network and limited inputs (traffic variables and control actions). It is noted that the proposed system can provide an effective tool to assist traffic control operators in real-time decision making for a traffic network with a large number of traffic variables and control measures.

References

1. Almejalli, K., Dahal, K., Hossain, A.: Intelligent traffic control decision support system. In: Giacobini, M., et al. (eds.) EvoWorkshops 2007. LNCS, vol. 4448, pp. 688–701. Springer, Heidelberg (2007)
2. Almejalli, K., Dahal, K., Hossain, A.: GA-based learning algorithms to identify fuzzy rules for fuzzy neural networks. In: de Janeiro, R. (ed.) The 7th International Conference on Intelligent Systems Design and Applications (ISDA), pp. 289–296. IEEE Computer Society Press, Los Alamitos (2007)
3. Bingham, E.: Reinforcement learning in neurofuzzy traffic signal control. European Journal of Operational Research 131(2), 232–241 (2001)
4. Bogenberger, K., Keller, H.: An evolutionary fuzzy system for coordinated and traffic responsive ramp metering. In: The 34th Annual Hawaii International Conference on System Sciences, p. 3038. IEEE, Los Alamitos (2001)
5. Chen, M., Linkens, D.: Rule-base self-generation and simplification for data-driven fuzzy models. In: The 10th IEEE International Conference on Fuzzy Systems, Melbourne, pp. 424–427 (2001)
6. Chunshien, L., Chun-Yi, L.: Self-organizing neuro-fuzzy system for control of unknown plants. IEEE Transactions on Fuzzy Systems 11(1), 135–150 (2003)
7. Cuena, J., Hernandez, J., Molina, M.: Knowledge-based models for adaptive traffic management systems. Transportation Research Part C 3(5), 311–337 (1995)
8. De Schutter, B., Hoogendoorn, S., Schuurman, H., Stramigioli, S.: A multi-agent case-based traffic control scenario evaluation system. In: Proceedings of Intelligent Transportation Systems, pp. 678–683. IEEE, Los Alamitos (2003)
9. FHWA. FHWA Administrator Testifies That Growing Traffic Congestion Threatens Nation's Economy, Quality of Life (2002) (October 15, 2006), `http://www.fhwa.dot.gov/pressroom/fhwa0220.htm`
10. Goldberg, D.: Genetic Algorithms in Search, Optimization and Machine Learning. Addison Wesely, Reading (1989)
11. Goldberg, D., Deb, K.: A comparative analysis of selection schemes used in genetic algorithms. In: The First Workshop on Foundations of Genetic Algorithms, pp. 69–93. Morgan Kaufmann, San Francisco (1991)
12. Hegyi, A., De Schutter, B., Hoogendoorn, S., Babuska, R., van Zuylen, H.: A fuzzy decision support system for traffic control centers. In: Proceedings of Intelligent Transportation Systems, pp. 358–363. IEEE, Los Alamitos (2001)

13. Henry, J., Farges, J., Gallego, J.: Neuro-fuzzy techniques for traffic control. Control Engineering Practice 6(6), 755–761 (1998)
14. Holland, J.: Adaptation in Natural and Artificial Systems. MIT Press, Cambridge (1992)
15. Hoogendoorn, S., Schuurman, H., De Schutter, B.: Real-time traffic management scenario evaluation. In: The 10th IFAC Symposium on Control in Transportation Systems (CTS 2003), Tokyo, pp. 343–348 (2003)
16. Ishibuchi, H., Nozaki, K., Yamamoto, N., Tanaka, H.: Selecting fuzzy if-then rules for classification problems using genetic algorithms. IEEE Transactions on Fuzzy Systems 3(3), 260–270 (1995)
17. Ishibuchi, H., Yamamoto, T.: Fuzzy rule selection by multi-objective genetic local search algorithms and rule evaluation measures in data mining. Fuzzy Sets and Systems 141(1), 59–88 (2004)
18. Karr, C.: Genetic algorithms for fuzzy controllers. AI Expert 6(2), 26–33 (1991)
19. Kohonen, T.: Self-Organization and Associative Memory. Springer, New York (1984)
20. Krause, B., von Altrock, C., Limper, K., Schafers, W.: A neuro-fuzzy adaptive control strategy for refuse incineration plants. Fuzzy Sets and Systems 63(3), 329–338 (1994)
21. Lee, M., Lee, S., Park, C.: A new neuro-fuzzy identification model of nonlinear dynamic systems. International Journal of Approximate Reasoning 10(1), 29–44 (1994)
22. Lin, T.: Neural Fuzzy Control Systems With Structure and Parameter Learning. World Scientific, Singapore (1994)
23. Lin, C.: A GA-based neural fuzzy system for temperature control. Fuzzy Sets and Systems 143(2), 311–333 (2004)
24. Lin, C., Lee, C.: Neural-network-based fuzzy logic control and decision system. IEEE Transactions on Computers 40(12), 1320–1336 (1991)
25. Messmer, A.: METANET – A Simulation Program for Motorway Networks. Technical University of Crete, Dynamic Systems and Simulation Laboratory (2000)
26. Molina, M., Hern, J., Cuena, E.: A structure of problem-solving methods for real-time decision support in traffic control. International Journal of Human-Computer Studies 49(4), 577–600 (1998)
27. Quek, C., Pasquier, M., Lim, B.: POP-TRAFFIC: A novel fuzzy neural approach to road traffic analysis and prediction. IEEE Transactions on Intelligent Transportation Systems 7(2), 133–146 (2006)
28. Quek, C., Zhou, R.: POPFNN: A pseudo outer-product based fuzzy neural network. Neural Networks 9(9), 1569–1581 (1996)
29. Quek, C., Zhou, R.: The POP learning algorithms: Reducing work in identifying fuzzy rules. Neural Networks 14(10), 1431–1445 (2001)
30. Petrovic-Lazarevic, S., Coghill, K., Abraham, A.: Neuro-fuzzy modelling in support of knowledge management in social regulation of access to cigarettes by minors. Knowledge-Based Systems 17(1), 57–60 (2004)
31. Ritchie, S.: A knowledge-based decision support architecture for advanced traffic management. Transportation Research Part A: General 24(1), 27–37 (1990)
32. Rumelhart, D.E., Hinton, G.E., Williams, R.J.: Learning internal representations by error propagation. In: Rumelhart, D.E., McClelland, J.A. (eds.) Parallel Distributed Processing: Explorations in the Microstructure of Cognition, vol. 1: Foundations, pp. 318–362. MIT Press, Cambridge (1986)
33. Tay, J., Zhang, X.: Neural fuzzy modeling of anaerobic biological wastewater treatment systems. Journal of Environmental Engineering 125(12), 1149–1159 (1999)

34. Trabia, M., Kaseko, M., Ande, M.: A two-stage fuzzy logic controller for traffic signals. Transportation Research Part C: Emerging Technologies 7(6), 353–367 (1999)
35. Wei, C.: Analysis of artificial neural network models for freeway ramp metering control. Artificial Intelligence in Engineering 15(3), 241–252 (2001)
36. Wei, W., Zhang, Y., Mbede, J., Zhang, Z., Song, J.: Traffic signal control using fuzzy logic and MOGA. In: 2001 IEEE International Conference on Systems, Man and Cybernetics, pp. 1335–1340 (2001)
37. Werbos, P.: Neurocontrol and fuzzy logic: Connections and designs. International Journal of Approximate Reasoning 6(2), 185–219 (1992)
38. Whitley, D.: Using reproductive evaluation to improve genetic search and heuristic discovery. In: The Second International Conference on Genetic Algorithms and Their Application, pp. 108–115. Lawrence Erlbaum Associates, Mahwah (1987)
39. Yager, R.: Modeling and formulating fuzzy knowledge bases using neural networks. Neural Networks 7(8), 1273–1283 (1994)
40. Yager, R.: Implementing fuzzy logic controllers using a neural network framework. Fuzzy Sets and Systems 100(1), 133–144 (1999)
41. Zhang, H., Ritchie, S.: Real-time decision-support system for freeway management and control. Journal of Computing in Civil Engineering 8(1), 35–51 (1994)
42. Zhang, H., Ritchie, S., Jayakrishnan, R.: Coordinated traffic-responsive ramp control via nonlinear state feedback. Transportation Research Part C: Emerging Technologies 9(5), 337–352 (2001)

Simultaneous Airline Scheduling

Tobias Grosche[1] and Franz Rothlauf[2]

[1] airconomy GmbH & Co. KG, Frankfurt Airport Center 1, Hugo-Eckener-Ring, 60549 Frankfurt, Germany
grosche@airconomy.com
[2] Department of Information Systems and Business Administration, University of Mainz, Jakob-Welder-Weg 9, 55128 Mainz, Germany
rothlauf@uni.mainz.de

Summary. Currently, there are no solution approaches available to construct and optimize airline schedules within a single model. All existing approaches decompose the problem into smaller and less complex subproblems and solve those subproblems separately. This chapter presents a metaheuristic for simultaneous airline scheduling where several different subproblems are integrated into one single optimization model, except for crew scheduling. The problem-specific metaheuristic uses an adaptive procedure for operator selection to allow an efficient choice between a variety of different operators. Experiments are conducted as proof-of-concept and to calibrate free parameters. Comparing different search strategies and studying operator probabilities show that efficiently solving the airline scheduling problem requires the application of both, local and recombination-based search operators.

Keywords: Airline scheduling, Fleet assignment, Genetic algorithms, Threshold accepting.

1 Introduction

In 2005, for the first time the number of passengers of the worlds scheduled airlines exceeded two billions [24]. For the future, demand for airline travel is expected to grow at an average rate of about 5% per year until 2025 [24]. However, despite these positive market trends, an airline's individual profit margin is considerably small and highly dependent on general economical performance [9]. To be successful, an airline has to make the most efficient and effective use of its resources to match passengers' demand. Its major instrument is the airline schedule. It includes the flights an airline carries out and the assignment of resources (aircrafts and crews) to these flights.

The objective of airline schedule optimization ("airline scheduling") is to construct an airline schedule with high operational profit. This planning task is not only the most important but also the most complex task an airline is confronted with. Until now, a single optimization approach of the complete airline scheduling problem is believed to be computational intractable and even its formulation impossible [3]. Instead, the problem is usually decomposed into several less complex subproblems that are solved in a sequential manner. The solution of one

A. Fink and F. Rothlauf (Eds.): Advances in Computational Intelligence, SCI 144, pp. 81–108, 2008.
springerlink.com

subproblem is used as input for the next subproblem. However, to account for any interdependencies between the subproblems, feedback loops and iterations are necessary. Therefore, it remains questionable if the decomposition of the overall airline scheduling problem reduces the quality of the resulting schedules compared to schedules that would result from an integrated airline scheduling approach.

The goal of this chapter is to overcome the artificial decomposition of the airline scheduling problem into subproblems and, thus, to make a step from the status-quo of airline scheduling towards the researcher's ultimate goal of a fully integrated airline scheduling approach. It presents an optimization approach for an airline scheduling problem integrating the choice of flights and the aircraft assignment. It includes more decision variables in one optimization model than existing airline scheduling models while representing airline operations on a higher level of detail without simplifying assumptions. This integrated model is solved using metaheuristic optimization.

The chapter is structured as follow. In the next section, a short introduction to airline scheduling is given. Then, Sect. 3 presents the conceptual design for a metaheuristic approach on airline scheduling addressing elements like representation, search operators, fitness function, initialization, and search strategy. Section 4 includes experimental results from the application on test scenarios. They focus on calibration of free parameters, the choice of the best search strategy, and the analysis of the search process. Finally, Sect. 5 concludes this contribution and suggests directions for future work.

2 Airline Scheduling

An airline schedule represents the central element within an airline's corporate planning system since it affects almost every operational decision and has the largest impact on profitability [36, 34, 5, 3]. Thus, many factors such as demands in various markets, competition, and available resources have to be considered to achieve optimal solutions [16, 3]. The high number of variables and their interdependencies make this problem too complex to be believed solvable in one single optimization model. Instead, an airline schedule usually is constructed in several steps emerging from a decomposition of the overall airline scheduling problem into smaller subproblems [34, 27, 18, 3].[1] The subproblems are less complex and can be solved independently, each with its individual objective function [12]. The major subproblems that have to be addressed when solving an airline scheduling problem are:

- *Flight Schedule Generation:* In this step, flights including their departure and arrival times and airports are chosen to be offered to potential passengers. These flights determine the airline's route network and the frequency of connections. The selection of flights is influenced by traffic forecasts, tactical and strategic initiatives, seasonal demand variations, and passengers' connection

[1] For approaches see for example [34, 5, 27, 33, 18, 1, 17, 25, 11, 3, 26, 29].

possibilities. Major constraints in this planning step are the size and composition of the aircraft fleet and other resources, and legal factors like traffic rights [13, 5, 18, 2]. The flight schedule generation affects all subsequent planning steps and has a high impact on passenger demand [5, 2, 11].

- *Fleet Assignment:* Airlines usually posses several fleets of aircrafts having some characteristics like cruising speed, capacity, conformance to noise restrictions, or crew requirements in common [4, 17]. As a result, different fleets produce different revenues if assigned to the same flight. The objective of the fleet assignment problem is to find an assignment of aircraft types to the flights that minimizes operating costs and opportunity costs of lost revenue due to too low capacities. Major constraints in the fleet assignment problem include the available number of aircrafts per type, the coverage of each scheduled flight, the flow balance at each airport (the numbers of arrivals and departures of each type have to be the same), and operating restrictions (e.g. noise limitations, runway lengths, or endurance).
- *Aircraft Routing:* Given the fleet assignment, the objective of the aircraft routing is to find a feasible and profit-maximizing assignment of physical aircrafts to the flight legs [4]. An aircraft routing consists of a number of flight legs that can be carried out by the same aircraft. A sequence of aircraft routings that starts and ends at the same location and that can be flown by one aircraft is denoted as rotation. A rotation can be flown by more than one aircraft (in parallel), and many airlines construct one rotation per fleet to have an equal utilization of the aircrafts and easier operational planning [22, 6]. The major constraints in this solution step are maintenance restrictions, since the aviation authorities and the airlines usually require each aircraft to undergo regular maintenance every three or four days at an appropriate maintenance station [16, 35]. Thus, the rotations need to incorporate intermediate stops of sufficient length to allow regular maintenance opportunities.
- *Crew Scheduling:* The objective of the crew scheduling problem is to find an optimal assignment of crew members to the flights given by the previous planning steps. This problem represents a very complex optimization problem because of many constraints defined by work-rules given by legal regulations, union agreements, and company policies [19]. Therefore, the crew scheduling problem is usually solved in a two-step process, in which at first generic crew schedules (pairings) are constructed that are feasible and minimize crew costs, which then are assigned to individual crew members by either bidding of crew members or direct assignment [3].

To further reduce the complexity of the overall scheduling problem, airlines usually consider a pattern or base schedule for a shorter time interval (a day or a week), so that the cyclical extension of this schedule represents the schedule over the whole planning horizon [13, 18].

To solve these subproblems, many different (optimization) models have been developed. Not surprisingly, the capability of these models has constantly grown. For example, the number of decision variables was increased and more practical requirements included. However, none of the existing models solves the complete flight schedule generation, fleet assignment, and aircraft routing in an integrated or simultaneous approach.[2]

3 Conceptual Design

Metaheuristics have been proven to allow the solution of complex real-world problems that are intractable for traditional exact optimization techniques. Thus, they represent a promising approach to heuristically solve the integrated airline scheduling problem. The metaheuristic presented in this chapter integrates the steps flight schedule generation, fleet assignment, and aircraft routing. This section focuses on conceptual design of the metaheuristic. It presents its four basic design elements: representation and corresponding search operators, fitness function, initialization, and search strategy.

3.1 Representation and Search Operators

A representation determines the mapping between a phenotype and a genotype. The phenotype is a solution to the given problem, the genotype represents this solution such that the heuristic's search operators can be applied. Thus, the representation and the operators work together and cannot be developed independently [32].

Representation

Because our goal is to solve the integrated airline scheduling problem, a complete airline schedule must be fully determined by the genotype and the genotype-phenotype mapping. Thus, all decision variables that are necessary to construct a complete schedule must be included either into the genotype or in the genotype-phenotype mapping.

A genotype consists of a fixed number S of segments. Since a segment encodes the flight program of one day for one aircraft, S is equal to the number of available aircrafts. All segments of aircrafts of one fleet type are grouped together on succeeding positions in the linear genotype.

Each segment contains a sequence of tuples, where L_s denotes the number of tuples in segment $s \in [1, S]$ and $l_s \in [1, L_s]$ denotes the position of a tuple in segment s. Each tuple consists of an airport number or name a_{l_s} and a waiting time t_{l_s}, indicating the airport and additional time the aircraft is located on the ground. Figure 1 gives an overview of this representation concept with seven aircrafts, two different fleet types, and six tuples in the fourth segment denoting the airports that are visited by an aircraft of fleet type B. The aircraft is required to fly between the encoded airports in the indicated sequence, thus, there is an

[2] Overviews of such solution models can be found for example in [20, 12, 36, 34, 5, 33, 1, 17, 18, 25, 11, 3].

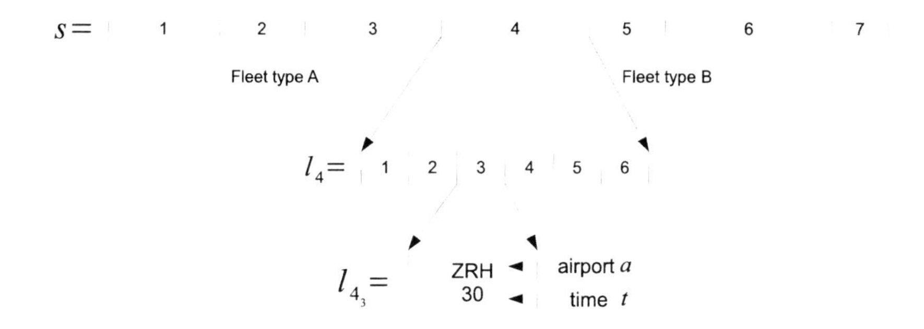

Fig. 1. Representation concept

indirect encoding of the flights depending on the ground activities. Using this indirect and aircraft-based representation avoids infeasible aircraft assignments.[3]

The departure time of a flight depends on the departure times of the preceding flights. The airport a_{l_s} is the departing airport of the flight f_{l_s} which connects the airports a_{l_s} and a_{l_s+1}. The time interval t_{l_s} at each position in the segment indicates the amount of time that the aircraft is scheduled to remain on the ground after the minimum turn time (necessary for refueling, loading, catering etc.) has elapsed and before conducting the next flight. The departure time $t_{l_s}^{dep}$ of flight f_{l_s} can be calculated recursively based on the departure time of the previous flight $t_{l_s-1}^{dep}$, its block-time $t_{a_{l_s-1},a_{l_s}}^{block}$ which specifies the time period between departure from the gate of the departure airport until arrival at destination airport, the minimum required turn time $t_{a_{l_s}}^{turn}$ and the scheduled ground time t_{l_s}:

$$t_{l_s}^{dep} = t_{l_s-1}^{dep} + t_{a_{l_s-1},a_{l_s}}^{block} + t_{a_{l_s}}^{turn} + t_{l_s}. \tag{1}$$

If a departure time is scheduled when night flying restrictions at the corresponding airport are into effect, $t_{l_s}^{dep}$ is set to the earliest time at which flights are allowed again. This is also the procedure to calculate the first departure time t_1^{dep} of an aircraft in a segment. If the airport allows flight operations throughout the night, t_1^{dep} can be set to a pre-specified time point. Using the decision parameters (a and t) and the given data (t^{block}, t^{turn}, and night flying restrictions), a feasible flight schedule can be encoded. If, like in Fig. 1, more than one aircraft of the same fleet type exists, their segments are concatenated and the last flight of each segment is heading towards the starting airport of the next segment. This procedure results into one single rotation per fleet type, and each aircraft of the same fleet type accomplishes the identical routing consisting of a number of days equal to the number of aircrafts in the fleet. Often, airlines

[3] By encoding the flight program of one day, the flights take place on a daily base. Thus, the concept presented at this stage is suitable for example for airlines with a regional scope. On the other hand, intercontinental flights are more likely to be scheduled on a weekly base. Thus, to represent weekly schedules the daily model can easily be extended by constructing 7 flight programs (one per day) per aircraft.

prefer a single rotation per fleet because of evenly distributed use of the aircrafts and ease of scheduling [22, 6].

Search Operators

The search operators can be divided into two groups: local search operators and recombination operators. In this study, for each group of operators different variants are developed.

Local Search Operators

Local search operators construct a new solution by applying small changes to the current solution. Thus, a local search operator should produce a solution that is in the original solution's neighborhood in the search space, keeping most of the original solution's properties. Based on the given problem here and the representation used, a genotype determines several elements of an airline schedule (like visited airports, times, aircraft assignments). A neighboring solution is obtained by modifying one element of the genotype. Different local operators were developed, each resulting in a different neighborhood.

- Delete ground time (*locDelGT*):
  This local search operator chooses a random segment s of the genotype and a random position l_s within this segment and sets the corresponding ground time $t_{l_s} = 0$.
- Insert ground time (*locInsGT*):
  The ground time t_{l_s} of a random position l_s in a randomly chosen segment s is increased by a time value between 0 and a parameter t^{init}. Therefore, all flights following the encoded position are displaced by t^{init}.
- Change airport (*locChgApt*):
  This operator randomly changes the airport a_{l_s} at a random position l_s in a randomly chosen segment s. The operator results in two different flights, since a_{l_s} represents an arrival and a departing airport of two succeeding flights. In addition, if the overall block time of the new flights changes, the departure times of succeeding flights are different. Figure 2 illustrates the application of the operator (airport DUS is changed into airport HAM) and the corresponding changes in the phenotype.
- Delete airport (*locDelApt*):
  The tuple at a random position l_s in a randomly chosen segment s is removed from the genotype. This operator replaces the two flights connected via a_{l_s} with one flight from a_{l_s-1} to a_{l_s+1}. In addition, because the new flight usually has a lower block time than the cumulated block time of the two original flights, all flights following l_s in the segment will depart earlier.
- Insert airport (*locInsApt*):
  This operator inserts a new random tuple at a random position l_s in a randomly chosen segment s, replacing one flight with two new flights connecting via a_{l_s}. The new ground time is chosen randomly between 0 and t^{init}. Usually, succeeding flights have later departure times.

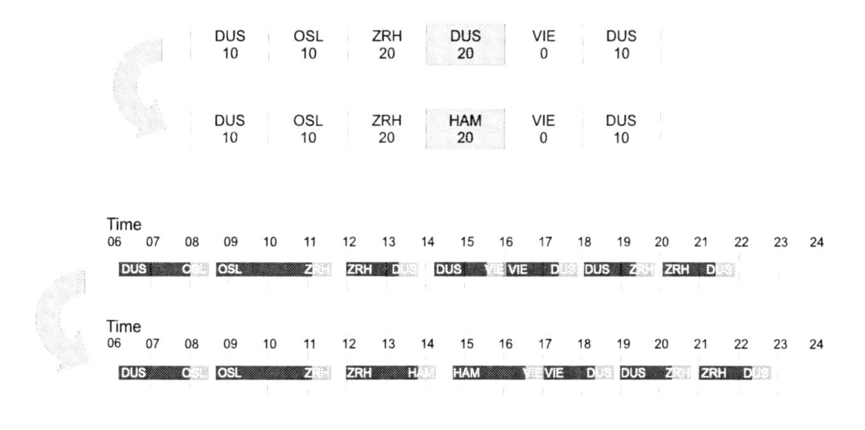

Fig. 2. Application of locChgApt search operator

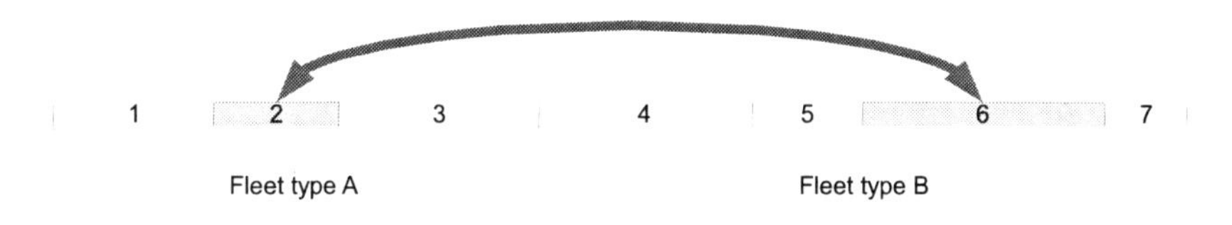

Fig. 3. Application of locChgFA

- Change fleet assignment (*locChgFA*):
  This operator changes the order of the segments in the genotype. Because the segments of a genotype are ordered by the fleet types, exchanging two segments s_1 and s_2 changes the fleet assignment if the two segments belong to different fleet types. In addition to a different fleet assignment, the flights at the connecting points between the segments might be changed if the segment is part of a multi-segment rotation. Figure 3 illustrates the functionality of this operator.
- Change airports with similar market size (*locChgAptMS*):
  This operator randomly changes the airport at a random position l_s of a randomly chosen segment s. To increase the locality of the search space with this type of operator, similarities between different airports are considered. Therefore, the operator uses the total market size (passenger demand) of an airport as a characteristic to measure similarity. The smaller the difference between the two market sizes of two airports, the higher their similarity. Given the market size ms_{od} for every market of an originating airport $o \in A$ and a destination airport $d \in A$, the total market size ms_a of airport $a \in A$ is calculated as:

$$ms_a = \sum_{d \in A} ms_{ad} + \sum_{o \in A} ms_{oa}. \qquad (2)$$

The probability p_{ab} of an airport b to replace an airport a in the genotype increases with decreasing difference in their market sizes:

$$p_{ab} = \frac{\min(|ms_a - ms_b|^{-1}, 1)}{\sum\limits_{c \in A} \min(|ms_a - ms_c|)^{-1}, 1)}. \tag{3}$$

- Change airports with similar distance (*locChgAptDist*):
  This operator works in analogy to *locChgAptMS*, instead it uses the geographical distance as indication of similarity of airports. The closer two airports, the higher the probability that they are exchanged in the genotype.

Recombination Operators

The purpose of recombination operators is to create new solutions by combining meaningful elements of different solutions. To apply recombination operators, a population of solutions is necessary because each new solution ("offspring") emerges from at least two preceding solutions ("parents"). In the following, different recombination operators are presented.

- one-point crossover (*1xover*):
  The traditional one-point crossover splits two genotype strings at a random position and exchanges the partial strings. When using the absolute position l as crossover point, a new offspring with a large time displaced can be generated. If, for example, one parental solution contains many short flights and the second parent only long flights, the same absolute position would indicate a flight of the first parent departing much earlier than the corresponding flight in the second parent. Therefore, *1xover* does not use as crossover points absolute positions in a segment but relative positions. In a first step, it randomly chooses a

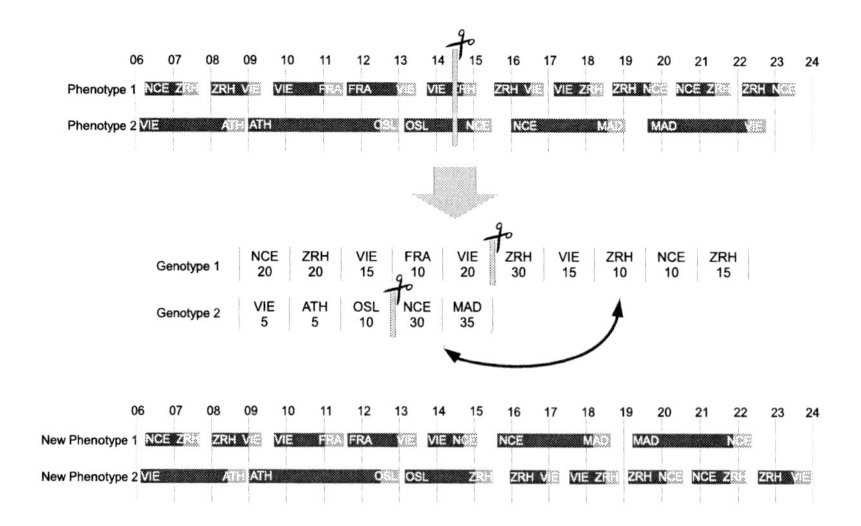

Fig. 4. Application of 1xover

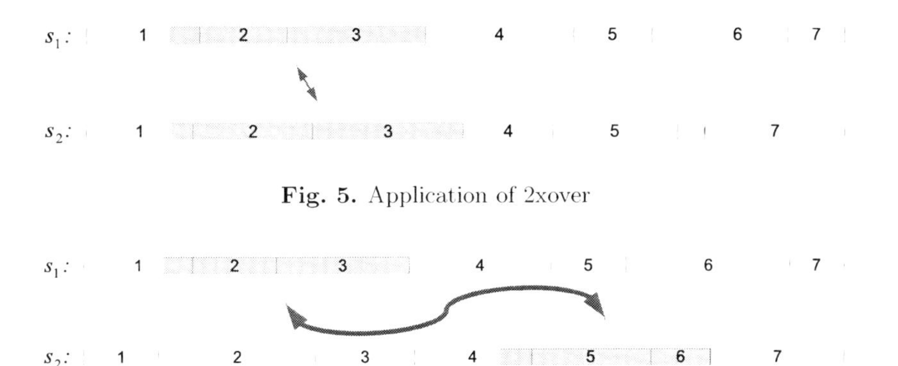

Fig. 5. Application of 2xover

Fig. 6. Application of stringxover

segment s and a time t^{cross}. Then, the corresponding tuple l_s in the two parents is indicated and the tuples following after the crossover point are exchanged. Therefore, displacement of flights after the crossover position is low. Figure 4 illustrates the functionality of the operator ($t^{cross} = 14{:}30$).

- two-point crossover (*2xover*):
  The operator *2xover* works analogously to *1xover* besides randomly choosing an additional time t^{cross2} defining a second crossover position ($t^{cross} + t^{cross2}$). Thus, each genotype is divided into three substrings, and the inner string is exchanged between the parents. An example is given in Fig. 5.
- string crossover (*stringxover*):
  When using *stringxover* elements of the genotype are exchanged between approximately the same positions. The operator works analogously to *2xover*, however, the segment s_1 at one parent where the substring of the other parent is inserted does not need to correspond to this segment s_2 ($s_1 \neq s_2$). This operator allows to change the fleet assignment. The functionality of the operator is illustrated in Fig. 6.

Repair Operators

The representation and the corresponding search operators ensure that only feasible aircraft assignments exist. Because of the aircraft-based relative encoding, each flight is covered by exactly one aircraft, the number of aircrafts per fleet is not exceeded, and the aircraft flow at each airport is balanced. In addition, the search operators consider operational restrictions of the fleets, for example they do not insert an airport in a segment where the corresponding fleet type is not able to operate or where the insertion would result in a flight being too long for an aircraft's endurance.

However, there are other restrictions that might still be violated when applying the search operators. If there are either too many stops in one segment or the operators replace flights by flights with longer block times, the subsequent flights are postponed because of the relative encoding of their departure times. This could lead to the situation that the last flights of the segment would be scheduled to depart at the next day. If the end of day is exceeded by a segment, a repair operator searches for ground times that can be minimized. If there is not sufficient ground time available, the repair operator randomly deletes stops until the resulting flight program can be completed in one day.

Night flight restrictions are always met at the beginning of each day, since the departure times are encoded relatively to either the ready time of the previous flight or the time an airport allows flight operations. However, the arrival of a flight might violate some restrictions, if for example a flight departs at an airport without night restrictions heading to an airport with restrictions. To deal with such problems, a repair operator tries to minimize the scheduled ground times if the last flight arrives too late in the evening. If minimizing is not sufficient stops are randomly deleted from the segment.

3.2 Fitness Function

The objective of the airline scheduling problem is to construct airline schedules with maximum operating profit. Thus, the fitness function has to estimate the estimated operational profit of an airline. The operational profit is defined as the profit directly related to and dependent of the flights in the schedule, or, to be more specific, the yield of all passengers of all flights minus the costs for operating the flights. The prediction of operating costs is straightforward, since in each solution all information on the flights is given, allowing to easily determine the corresponding costs of these flights. However, estimating the number of passengers demands more sophisticated approaches which are discussed in the following paragraphs.

Passenger Estimation

To estimate the number of passengers on individual flights, many different (commercial) applications exist. Such tools imitate passenger behavior on a very detailed level and take into account many factors that influence passenger demand, leading to computation times varying from minutes to many hours for evaluating one schedule. Using such tools (and corresponding models) for the fitness evaluation in a metaheuristic would lead to excessive overall computation times. In addition, their acquisition costs, the lack of parameters and data required by those tools, and the general difficulty to obtain detailed information due to their proprietary nature [26] make it impossible to use them for this study. Thus, a custom model for passenger estimation is developed. Because of its complexity, a detailed presentation would be beyond the scope of this contribution, thus, in the following only the main steps are presented.[4] In each step, some parameters exist that have to be calibrated. For the calibration, real-world data was available consisting of MIDT (Market Information Data Tapes) bookings of 2004 for airline travels between Germany and European countries with non-stop or one-stop service. This data set consists of 1,365,497 records including a total of

[4] See [20] for a more detailed description of this evaluation procedure.

7,808,041 passenger bookings. Each record contains detailed flight information like departure time, operating airlines, route of flight, and others.

Given the total airline passenger demand for each market (city-pair), the flight schedules of other airlines, and the schedule subject to evaluation the number of passengers is estimated in three steps:

1. *Itinerary construction*
2. *Itinerary market share estimation*
3. *Passenger allocation*

Itinerary Construction

An itinerary is a travel alternative between two cities, which could be either a direct flight or a sequence of connecting flights [8]. Since itineraries are constructed by concatenating single flights, the number of city-pair connections exceeds the number of flights.

Because passengers choose among competing itineraries (and not among flights), the demand has to be estimated on the itinerary level. Therefore, in a first step it is necessary to construct valid itineraries from the set of individual flights in the complete schedule including flights and flights offered by other airlines. Because each flight represents a valid nonstop itinerary, the focus is on the construction of connecting flights (or simply *connections*). Since the number of possible connections grows exponentially with the number of flights, airlines limit the number of connections to those itineraries that are reasonable and are likely to be chosen by passengers.

Only few articles discuss under which conditions a connection forms a valid itinerary. The following conditions for determining whether a connection represents a valid itinerary are used in this study. Some conditions include parameters which have been calibrated using the MIDT data set described above.

- The geographical distance $dist_{cnx}$ of a connecting flight must not exceed the direct distance $dist_{dir}$ between the origin and departure airport by a maximum detour factor d_{max}. Thus, $dist_{cnx} \leq d_{max} dist_{dir}$.
- Connecting flights must not depart before the flight before has arrived and a minimum connecting time t^{cnx}_{min} has elapsed.
- The travel time t^{cnx} of a connecting flight must not exceed the travel time $t^{shortest}$ of the shortest itinerary in the market by a certain factor t_{delay}: $t^{cnx} \leq t_{delay} t^{shortest}$. Since the time delay is believed to be perceived differently depending on the type of the shortest itinerary (direct or connection) $t^{shortest}$ differentiates between both types, resulting in t^{cnx}_{delay} and t^{dir}_{delay}.
- The second flight leg has to depart before a maximum connecting time t^{cnx}_{max} has elapsed.
- Connections of different airlines (interline connections) are only allowed if in the given data at least $n_{interline}$ connections could be observed between these airlines.

The parameters need to be calibrated to obtain a complete connection building procedure that reproduces the travel behavior observed in reality in a proper way.

The MIDT data describes the behavior in reality because each record represents a chosen itinerary and the number of passengers that have chosen it. Given all single flights in the observed data, the calibrated connection building sequence should result in a set of connections that were chosen by the passengers in the past. Because this number could easily be maximized by constructing all possible connections (leading also to a vast number of non-chosen connections and to excessive computation times), a second (conflicting) objective is to minimize the total number of connections constructed.

The calibration process represents an optimization problem for which a simple hill-climbing metaheuristic using the parameters of the rules as decision variables has been applied. Table 1 summarizes the outcome of the optimization procedure (details of the optimization procedure are omitted for brevity, details can be found in [20]) and lists the mean and standard deviation σ of the parameters. Setting the parameters to the listed values results in a connection builder that produces a set of connections similar to the connections that can be observed in the MIDT booking data.

Table 1. Optimal parameters for itinerary construction

parameter	mean	σ
d_{max}	1.26	0.41
t^{dir}_{delay}	1.39	0.36
t^{cnx}_{delay}	1.74	1.16
t^{cnx}_{max}	113.1	44.8
t^{cnx}_{min}	55.1	24.7
$n_{interline}$	81.3	11.3

Itinerary Market Share Estimation

The direct and connecting flights represent the set of itineraries a passenger can choose from. The objective of itinerary market share estimation is to forecast the attraction of each itinerary for a single passenger. The attraction of an itinerary depends on various attributes of the itinerary such as convenience of travel, travel time, departure and arrival time, average fare, aircraft type, and airline preferences. The attraction of an itinerary is measured as the market share of the itinerary, thus, if multiplied with market sizes, the total passenger demand for each itinerary can be calculated.

Multinomial logit (MNL) models are commonly used for market share estimation [7]. Because for this study the variables required and used in the published models are not available, the existing models cannot be directly used. In addition, it remains unclear whether the assumptions of MNL models (like logistic function or linear-in-parameter utility) limit the forecasting accuracy in comparison to other estimation models. In [21] MNL models, artificial neural networks, and a custom model for itinerary market share estimation are compared on a real-world data set. The MNL performs worst with respect to the forecasting quality; in contrast, the custom model yield the highest prediction accuracy. Thus, the model of [21] is used for market share estimation.

Passenger Allocation

Given market sizes and the relative share (attraction) of each itinerary, calculating the absolute passenger demand for each itinerary is straightforward. However, capacities of flights are constrained and, thus, not all passenger demand can be met. Because each flight might be a leg of a connecting itinerary, this competition takes place between itineraries of different markets. The objective of the passenger allocation step is to satisfy passenger demand by providing an assignment of the demand to the itineraries (and its flights, respectively) without violating capacity constraints [27]. This task is commonly referred to as *"Spill & Recapture"*. The procedure in this study for calculating spill and recapture consists of the following steps.

1. Estimation and assignment of the total demand independently of the flight capacities.
2. Determination of excess demand per flight.
3. Calculation of "spilled" passengers for all itineraries affected by limited capacities. Passengers are spilled according to the attraction of the corresponding itineraries.
4. Recapture of spilled passenger by itineraries with free capacities left. The assignment follows the itineraries' attraction.
5. Determination of excess demand per flight and removal according to step 3.

Details can be found in [20].

Fitness Calculation

Given the number of passengers on the itineraries (from the passenger estimation model presented above), the calculation of the overall profit of a flight schedule F is as follows. Let c_f^{block} denote the block hour costs of the aircraft type assigned to flight $f \in F$, t_f^{block} the block time of flight f, y_m the passenger yield in market m, and P_k the number of passengers on itinerary $k \in K$. Then, the operational profit π_F of flight schedule F is calculated as:

$$\pi_F = \sum_{m} \sum_{\forall k \in K_m} P_k \cdot y_m - \sum_{\forall f \in F} c_f^{block} \cdot t_f^{block} \tag{4}$$

In the fitness function of a metaheuristic, additional properties of a solution can be considered like constraint violations that are no direct result of the objective function. As discussed before, the encoding of schedules and the corresponding operators (including the repair operators) result in feasible solutions except for maintenance considerations. Usually, aircrafts are required to undergo maintenance checks every three days. These maintenance restrictions are included as penalty term in the fitness function: for each aircraft not scheduled for appropriate maintenance at a proper airport, penalty costs reduce the fitness value by a certain value. For the experiments presented in Sect. 4, the fitness value π_F is reduced by 500,000 for every violated maintenance restriction.

3.3 Initialization

Since no problem-specific information about high-quality solutions is given, all solutions should have the same probability to be selected as initial solutions. This is accomplished by a random initialization of the decision variables. Thus, for initializing, randomly chosen airports are subsequently included in each segment until the corresponding flight program reaches the end of day. The ground times t assigned to each stop are chosen uniformly between 0 and t^{init}.

3.4 Search Strategy

In general, local and recombination-based search strategies are distinguished. Although the proper choice of the search strategy depends on the problem and its structure, both strategies are used here. Consequently, three metaheuristic techniques are implemented and evaluated:

1. threshold accepting (TA) as a representative example of local search,
2. a selecto-recombinative steady-state genetic algorithm (rGA) as example of pure recombination-based search, and
3. a standard steady-state genetic algorithm (GA) as an example of metaheuristic search with both, local and recombination-based search.[5]

Even when focusing on a single search strategy, an explicit control of the search operators is necessary, because for each type of search operator different variants were designed. In each search step, one of these variants has to be selected. This selection can be random or follow a given rule. In this study, an adaptive control of the operators is used which preferably applies those operators in each search step that were advantageous in the previous steps. This procedure not only reduces the number of decisions and parameters required to be manually set, but also increases the efficiency of the search process. The adaptive control used in this approach randomly selects one operator per iteration with a probability that depends on its contribution to the past search progress. If the application of the operator yielded in high-quality solutions (compared to the results of other operators), its selection probability is increased. For each operator $o \in O$ of the current search type (local or recombination-based), the progress of its last N applications is monitored. Its progress is evaluated according to the change in the fitness value $f_n^o = f(s_o^*) - f(s)$ between the original solution s and solution s_o^* resulting from the nth application of the operator o. The relative fitness contribution c_o of operator o is calculated as

$$c_o = \frac{max\left(\sum\limits_{n \in N} f_n^o, 0\right)}{\sum\limits_{q \in O} max\left(\sum\limits_{n \in N} f_n^q, 0\right)}. \tag{5}$$

[5] For an introduction and more information on TA see [10]; further information on GA can be found for example in [23, 14, 15, 30].

Based on this fitness contribution, the selection probability p_o of operator o can be calculated as

$$p_o = \frac{c_o}{\sum_{q \in O} c_q}. \tag{6}$$

To prevent diminishing operators, each operator has a minimum selection probability of 0.05. The initial setting for the selection probability of each operator is determined by applying the operator N times to a randomly generated solution. In the experiments presented in Sect. 4, $N = 5$ is used.

This adaptive procedure is also applied to the standard steady-state GA to choose among local and recombination-based search. Thus, the GA uses a two-step adaptation: first, the type of operator (local or recombination) is chosen, then, the operator itself is selected. The probability to select a crossover operator depends on the average progress contribution of all crossover operators. Let R denote the set of crossover operators and L the set of local search operators ($O = R \cup L$). Then, the probability p_R of using a crossover operator is calculated as

$$p_R = \frac{\frac{\sum_{r \in R} c_r}{|R|}}{\frac{\sum_{r \in R} c_r}{|R|} + \frac{\sum_{l \in L} c_l}{|L|}}. \tag{7}$$

The functionality of the three different search strategies is described in Algorithms 1, 2, and 3.

Algorithm 1. Threshold Accepting

```
 1: choose parameters:
 2:         initial threshold T ∈ [0, 1]
 3:         threshold reduction step size r < T
 4:         maximum number of iterations i_decrease between threshold reduction
 5:         maximum number of iterations i_max > i_decrease when T = 0
 6: create initial solution s with fitness f(s)
 7: calculate p_o for all operators o ∈ O
 8: iteration i = 0
 9: repeat
10:   i = i + 1
11:   select local search operator o ∈ O according to p_o
12:   create neighboring solution s*_o
13:   calculate new fitness value f(s*_o)
14:   Δf = f(s) − f(s*_o)
15:   if Δf < T f(s) then
16:     s = s*_o
17:     update p_o for all operators o ∈ O
18:   end if
19:   if T > 0 and i > i_decrease then
20:     T = T − r
21:     i = 0
22:   end if
23: until i = i_max
```

Algorithm 2. Selecto-Recombinative Steady-State Genetic Algorithm

```
1: choose parameters:
2:         population size n
3:         p_conv to determine convergence of the population
4: create initial population S_0 with n solutions s
5: calculate fitness f(s) for each s ∈ S_0
6: calculate p_o for all operators o ∈ O
7: iteration i = 0
8: repeat
9:    i = i + 1
10:   select recombination-based search operator o ∈ O according to p_o
11:   choose two solutions s_1 and s_2 randomly
12:   create solution s*_o from s_1 and s_2 using a crossover operator
13:   calculate new fitness f(s*_o)
14:   replace the worst solution in S by s*_o
15:   update p_o for all operators o ∈ O
16:   determine solution ŝ ∈ S with maximum fitness
17:   calculate average fitness f̄(S) of population
18: until f(ŝ) − f̄(S) < p_conv · f(ŝ)
```

Algorithm 3. Standard Steady-State Genetic Algorithm

```
1: choose parameters:
2:         population size n
3:         p_conv to determine convergence of the population
4: create initial population S_0 with n solutions s
5: calculate fitness f(s) for each s ∈ S_0
6: calculate p_o for all operators o ∈ O
7: calculate p_R and p_L
8: iteration i = 0
9: repeat
10:   i = i + 1
11:   if random(0, 1) < p_R then
12:      select recombination-based search operator o ∈ R according to p_o
13:      choose two solutions s_1 and s_2 randomly
14:      create solution s*_o from s_1 and s_2
15:   else
16:      select local search operator o ∈ L according to p_o
17:      create neighboring solution s*_o
18:   end if
19:   calculate new fitness f(s*_o)
20:   replace the worst solution in S by s*_o
21:   update p_o for all operators o ∈ O
22:   update p_R and p_L
23:   determine solution ŝ ∈ S with maximum fitness
24:   calculate average fitness f̄(S) of population
25: until f(ŝ) − f̄(S) < p_conv · f(ŝ)
```

4 Experimental Results

This section presents results from implementing and applying the simultaneous airline scheduling approach. First, calibration results focusing on the setting of the parameters for each metaheuristic are presented. Then, each calibrated metaheuristic is applied to the same test scenarios to compare the three different solution strategies. Finally, the search process and the obtained solutions of the search strategy that yields best results are studied.

All metaheuristics have been implemented in C++. The experiments presented here were conducted on different workstations with different processor and memory specifications. Thus, the number of fitness evaluations is used to compare the computational effort of different approaches.[6] Five different planning scenarios are used in the experiments. Each scenario models a different situation or planning problem an airline might be confronted with. Different situations are different with respect to

- the number of aircrafts available, fleet composition, and maintenance stations,
- the set of airports or markets that the airline is willing to accept in its schedule.

Table 2 gives an overview of the five different scenarios and lists the number of aircrafts, number of different fleets, and number of airports available.

Table 2. Scenarios

scenario	aircrafts	fleets	airports
A	30	4	62
B	30	3	29
C	8	4	55
D	20	2	50
E	10	1	90

Schedule evaluation is the computationally most expensive part of the optimization process. The required time depends on the number of flights and itineraries that have to be evaluated which in turn depends on the number of competing flights. To be able to conduct a sufficient number of experiments and to avoid statistical outliers, only 10% of the competing flights listed in given real-world flight schedules are considered. This reduction does not bias the fundamental results, because this reduction is applied to all experiments and – to keep a realistic estimation of passenger demand – the given market sizes are also reduced to 10% of their original value.

[6] As one example, calculating the fitness of a schedule with 15,000 flights requires about 60 seconds on a workstation with a 2.8 GHz processor and 1.0 GB RAM.

4.1 Calibration

Using an adaptive control of the search operators allows to reduce the number of parameters to be set for each metaheuristic. However, there are still some remaining parameters that have to be calibrated. In this section, experiments are presented that determine proper values of the remaining parameters.

For this calibration process, each metaheuristic is applied to the five different planning scenarios with different parameter settings. In each setting, one parameter is varied whereas the other parameters remain constant. The performance of a metaheuristic is measured by

1. the fitness value, and
2. the number of fitness evaluations until the best solution was found.

For each scenario and parameter setting, five independent optimization runs are performed. The different scenarios result in different fitness values (and number of fitness evaluations) because the scenarios consist of different numbers of aircrafts and airports. Thus, to compare parameter values between different scenarios, the performance results are normalized and aggregated. For a given parameter setting and scenario, the difference between the averaged results for one particular parameter setting and the averaged results for all parameter settings are used as measurement for the quality of the particular parameter setting. With $f_{p,s}$ denoting the average fitness value of the five runs with parameter setting $p \in P$ (P is the set of tested values of p) and scenario $s \in S$, the average fitness value $\bar{f}_s$ for all settings for scenario s becomes

$$\bar{f}_s = \frac{\sum\limits_{p \in P} f_{p,s}}{|P|}. \tag{8}$$

The influence $i_{p,s}$ of setting p in scenario s can be expressed as relation between the fitness values obtained with p compared to the average fitness values obtained with all parameter settings:

$$i_{p,s} = \frac{f_{p,s} - \bar{f}_s}{|\bar{f}_s|}. \tag{9}$$

Finally, aggregating over all scenarios yields the average influence $\bar{i}_p$ of setting p:

$$\bar{i}_p = \frac{i_{p,s}}{\sum\limits_{s \in S} i_{p,s}} \tag{10}$$

The following paragraphs present the results $\bar{i}_p$ for different settings p for each parameter.

Threshold Accepting

The TA algorithm was implemented as presented in Algorithm 1. There are four different parameters that have to be calibrated:

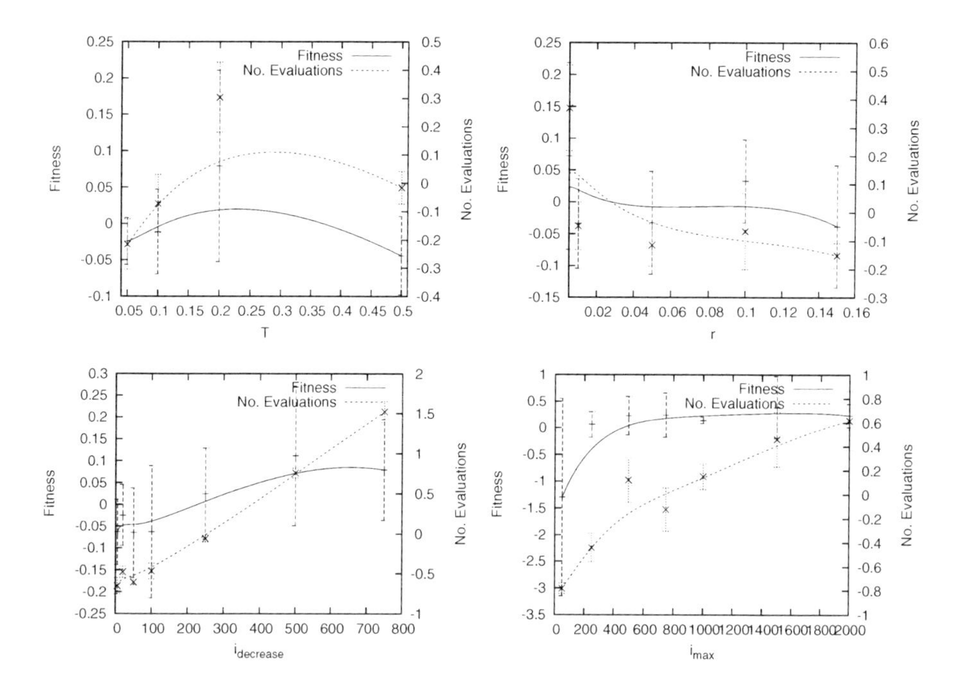

Fig. 7. Calibration results for TA parameters

- the initial threshold $T \in [0, 1]$,
- the threshold reduction step size $r < T$,
- the number of iterations $i_{decrease}$ between the threshold reductions, and
- the maximum number of iterations $i_{max} > i_{decrease}$ at the end of the algorithm when $T = 0$ (then the search process becomes a local hill climber).

Figure 7 presents the results for different values of the four different parameters. Bézier curves are shown as well as the mean and standard deviation for distinct parameter settings. The solution quality decreases for low and high values of T, it is highest for values around 0.2 – 0.25. Thus, these values seem to represent the best compromise between a random search (high T) and a hill-climbing technique that does not accept inferior solutions during search (low T). With lower r, the resulting solution quality decreases. If r is low, the threshold is reduced very slowly, allowing an explorative search. However, the computational effort also increases. In general, solution quality increases with higher $i_{decrease}$. The higher $i_{decrease}$, the more search steps are performed before the threshold is further reduced. This allows to explore larger areas of the solution space during optimization. However, the higher solution quality is obtained at the cost of increased computational effort. Increasing i_{max} yields higher solution quality, since the hill-climbing technique has more attempts to escape from local optima.

For each parameter, the value that resulted in the best fitness ist used for the subsequent experiments. The values are shown in Table 3. As a reference, this

Table 3. Initial and final setting of the TA parameters

parameter	final setting	initial setting
T	0.25	0.2
r	0.005	0.005
$i_{decrease}$	650	20
i_{max}	1500	500

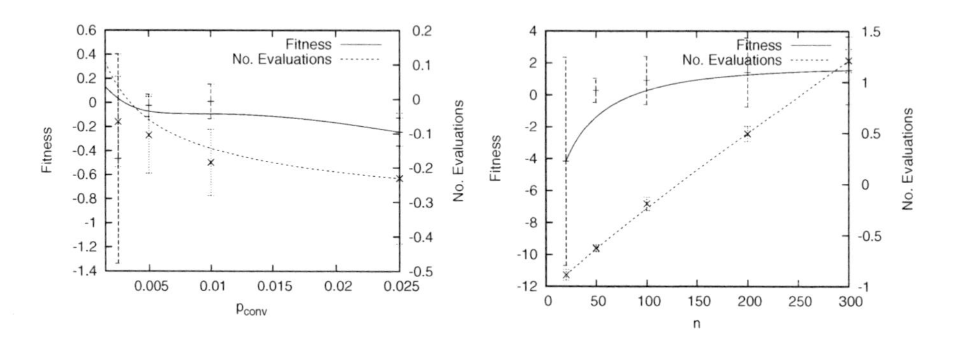

Fig. 8. Calibration results for rGA parameters

table also includes the initial setting of the parameters that remained constant while one parameter was varied.

Selecto-Recombinative Steady-State Genetic Algorithm

The selecto-recombinative steady-state GA described in Algorithm 2 has the following two parameters:

- population size n and
- p_{conv} (in percent) which measures the diversity of the fitness of the individuals in the population.

Figure 8 shows the calibration results for the two parameters. The final and initial parameter setting is presented in Table 4.

With smaller p_{conv}, a population must be more homogenous. To obtain more homogeneous populations, more search steps are necessary, each possibly creating a better solution. This results in overall better solution quality. On the other hand, more search steps require more computational effort. Increasing n yields higher solution quality, since more solutions are processed during search. The more solutions, the higher the chance of a search step to find a better solution. As Fig. 8 shows, for increasing n the fitness value asymptotically approximates a maximum value, whereas the required number of schedule evaluations constantly increases.

Table 4. Initial and final parameter setting for the rGA

parameter	final setting	initial setting
n	200	50
p_{conv}	0.00125	0.01

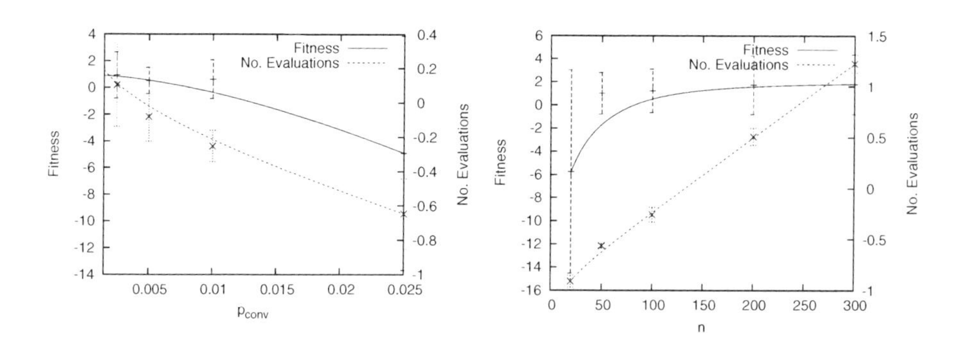

Fig. 9. Calibration results for GA parameters

Table 5. Initial and final parameter setting for GA

parameter	final setting	initial setting
n	200	50
p_{conv}	0.00125	0.01

Standard Steady-State Genetic Algorithm

The standard stead-state GA uses the same parameters as the selecto-recombinative GA. Calibration results are presented in Fig. 9, the initial and final parameter setting in Table 5. The results for the standard GA are almost the same as for the selecto-recombinative GA.

4.2 Comparison of Metaheuristics

The final parameter settings obtained by the calibration allow the metaheuristics to return high-quality results. To decide which type of metaheuristic performs the best, each strategy using the final parameters is applied to the different planning scenarios. The resulting (average) fitness values and number of required fitness evaluations are presented in Fig. 10.

For all scenarios, using the GA yields the highest solution quality. Except for scenarios A and E, the selecto-recombinative GA produces better results than TA. Because the two types of genetic algorithms use a population of solutions they require more fitness evaluations in comparison to TA which processes only one solution. The results indicate that with regard to solution quality combining local and recombination-based search outperforms search strategies using

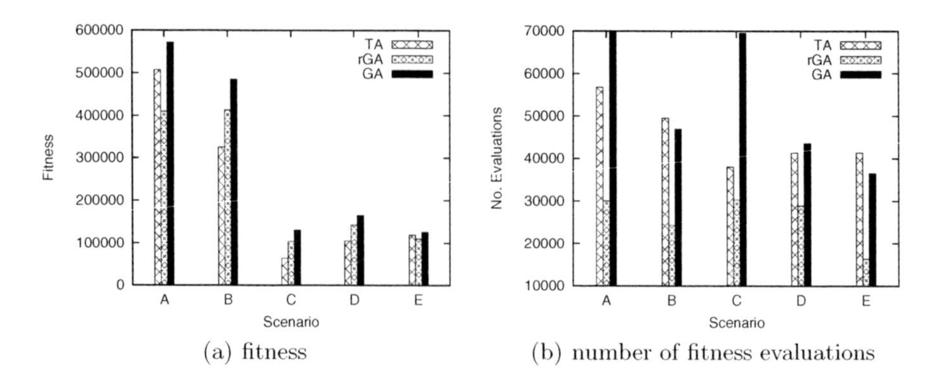

(a) fitness (b) number of fitness evaluations

Fig. 10. Comparing the performance of the different metaheuristics for the five different scenarios

only on one type of search operator, but also requiring a higher number of fintess evaluations. This finding is not surprising, since most problems of practical importance inherit properties applicable to both search concepts, local and recombination-based search [28, 31].

To validate the results, an unpaired t-test is conducted.[7] The null hypothesis H_0 is that the observed differences in the fitness values are random. H_α says that the differences are a result of the model specification. The critical t-value for $p = 0.975$ is 2.306. The results in Table 6 for the three models and five different scenarios show that the t-values always exceed the critical t-value of the level of significance. Thus, H_0 can be rejected on the 97.5%-level. The GA represents the search strategy that works best using the presented airline scheduling approach.

4.3 Solution Process

In the remainder of the section, the solution process of the GA is analyzed. Detailed results for scenario A for the five different runs performed are presented. The results for the other scenarios are analogous.

Search Progress

Figure 11 plots the fitness of the best solution over the number of evaluations for all five runs. It shows the typical GA behavior. The improvement rate is maximal at the beginning and continuously decreases during optimization. There are two reasons for this: at the beginning of the search, the search operators have much room for improvements, since the early solutions inherit random elements due to the random initialization; second, the population converges during the GA run, reducing the potential capability of the recombination-based operators with its

[7] The required test of the results for normal distribution was conducted using a Kolmogorov-Smirnov test.

Table 6. t-values for the comparison of different search strategies

models	scenario A	B	C	D	E
TA vs. rGA	6.405	20.136	13.222	13.637	3.277
TA vs. GA	7.057	79.604	16.775	16.127	2.400
rGA vs. GA	10.554	17.641	5.645	6.251	4.811

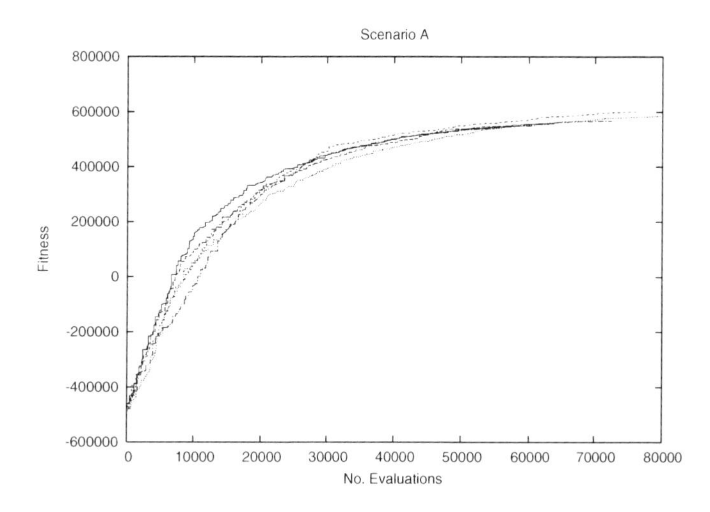

Fig. 11. Fitness of optimal solution over number of evaluation for scenario A

rather large modifications to solutions, and leaving only the possibility to make local improvements.

Figure 12 plots the number of flights (Fig. 12(a)) in the best schedule and the total number of passengers (Fig. 12(b)) expected to travel on these flights. The plots show that the fitness increase observed in Fig. 11 is due to an increase of the seat load factor during optimization, which itself results from both an increase of the number of passengers and a reduction of the number of total flights. Apparently, in each GA run unprofitable flights are successively removed from the schedule. Since the number of flights is higher in the beginning of each run and assuming that the average block time of all flights remains constant, the final schedules must include some idle ground times. Thus, an additional increase in profit would become possible by increasing the number of airports available for planning. Then additional (profitable) flights can be scheduled.

Parameter Adaptation

Since the operator probabilities are self-adjusted, the change of the relevance of the operators changes during a GA run is studied. Figure 13 shows the probabilities of applying a crossover operator over the number of evaluations. In

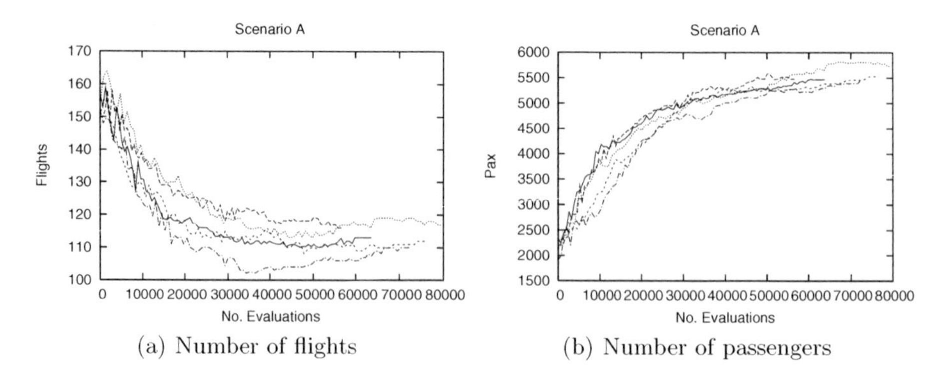

(a) Number of flights (b) Number of passengers

Fig. 12. Number of flights and passengers over the number of evaluations for scenario A

general, the five different runs show a similar behavior. The crossover probabilities are at a minimum at the start of the optimization but crossover becomes the main search operator after a few evaluations. Then, the probability continuously decreases until the end of the optimization. The continuous shift from crossover to local search is due to the different character of the search operators. The crossover operator performs a global search which is useful for the exploration of the search space at the early stages of optimization. With increasing number of evaluations, the solutions in the population become more similar and search focuses more on finding high-quality solutions in the neighborhood of solutions already in the population (exploitation).

Finally, Fig. 14 plots the application probabilities of the different variants of the crossover operators (Fig. 14(a)) and the local search operators (Fig. 14(b)).

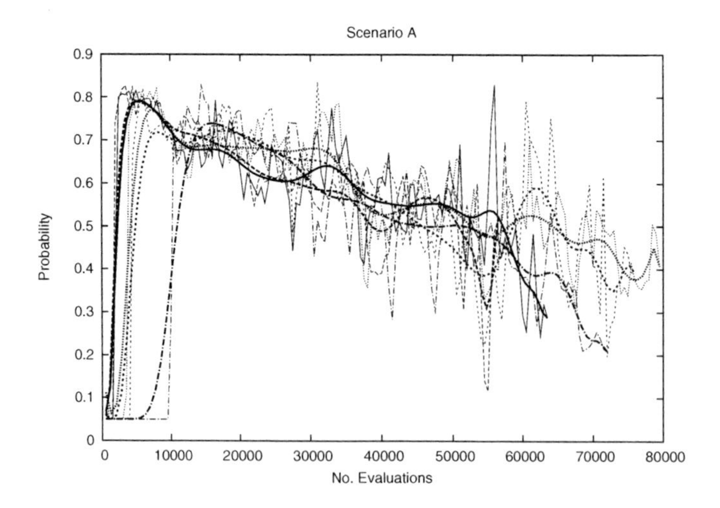

Fig. 13. Application probability of recombination (scenario A)

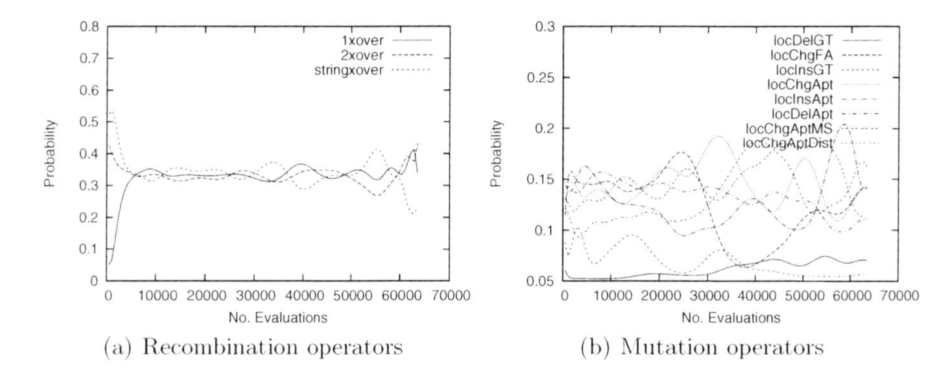

(a) Recombination operators
(b) Mutation operators

Fig. 14. Application probability of crossover and mutation operators for scenario A

For crossover operators, after an initialization phase all three different crossover operators are applied with about the same probability. For mutation operators, there is no clear dominant operator and all operators are used during the optimization in a different and volatile extent. However, on average, locDelGT, locInsGT, and locInsApt are more often used than the other operators.

5 Summary, Conclusions, and Future Work

The airline scheduling problem is one of the most complex optimization problem an airline has to solve. Until now, no integrated solution approaches exist but the overall problem is solved by decomposing it into smaller and less complex subproblems. In this chapter, a metaheuristic approach is presented to integrate different partial problems into one single optimization approach (except crew scheduling). Based on a problem-specific representation that is able to encode complete airline schedules, corresponding local and crossover operators were developed. The fitness of a flight schedule is the estimated operational profit of all scheduled flights. Three metaheuristic techniques were implemented and calibrated as representative examples for different search strategies: a threshold accepting algorithm (local search), a selecto-recombinative genetic algorithm (recombination-based search), and a genetic algorithm (local and recombination-based search). The application of the search operators is adaptively controlled according to their fitness contributions in past iterations. The genetic algorithm with local and recombination-based search yields the best results. At the beginning of its search process, mainly recombination is applied to explore the search space. With ongoing progress, exploitation gets more important and local search operators are applied to a greater extent.

These findings further advise to use both local and recombination-based search when solving real-world problems. Usually the properties of these problems that favour one search concept are hard to identify in advance or they are applicable to both search concepts, thus, techniques using both search strategies like genetic algorithms could be the proper choice to tackle real-world problems. In general, metaheuristics have to be designed problem-specific to be successful, however, usually there still is much freedom in designing the individual components like for example the representation or search operators. In this chapter, different search operators were developed each focusing on different elements of airline schedules. To choose among the different operators in each iteration of the metaheuristic, an adaptive procedure is presented. This procedure is not limited to the airline scheduling problem but can also be used in other applications of metaheuristics.

The presented approach makes it possible to integrate different subproblems of the airline scheduling problem into one single model. By processing complete airline schedules at once, the several different subproblems and their interdependencies are implicitly included and the elements of the schedules are optimized with respect to the overall objective. In addition, airline operations can be modelled on a higher level of detail. Simplifying assumptions of existing approaches (like for example only one fleet type, negligence of operating restrictions, no competition, single hub network, and others) can be avoided and additional elements of airline scheduling can be easily included in future work. Significant cost reductions could be generated if the crew scheduling would be integrated into the presented planning approach. If a schedule is evaluated according to the extent of possible delays caused by disruptions, the robustness of airline schedules can also be increased. Other operational or managerial objectives in scheduling can also be easily included by modifying the fitness function. If airport slots need to be considered, penalty costs can be adjusted so that they represent the efforts to obtain a slot.

Until now, this study represents a theoretic framework; the postulated advantageousness compared to a decomposition-approach and its applicability in real-world airline scheduling still have to be assessed. Thus, benchmark tests with airline scheduling models using the sequential planning paradigm have to be conducted. Even more desirable would be to replace the planning scenarios and all input used in this study by real existing data from an airline. Then, if possible and applicable, using the same scenario and prerequisites for optimization real airline schedules were based on, the presented approach can be further evaluated and compared to the corresponding real-world schedules.

References

1. Antes, J.: Structuring the process of airline scheduling. In: Kischka, P. (ed.) Operations Research Proceedings 1997, Selected Papers of the Symposium on Operations Research (SOR 1997), pp. 515–520. Springer, Berlin (1998)
2. Antes, J., Campen, L., Derigs, U., Titze, C., Wolle, G.-D.: SYNOPSE: a model-based decision support system for the evaluation of flight schedules for cargo airlines. Decision Support Systems 22(4), 307–323 (1998)
3. Barnhart, C., Belobaba, P.P., Odoni, A.R.: Applications of operations research in the air transport industry. Transportation Science 37(4), 368–391 (2003)
4. Barnhart, C., Boland, N.L., Clarke, L.W., Johnson, E.L., Nemhauser, G.L., Shenoi, R.G.: Flight string models for aircraft fleeting and routing. Transportation Science 32(3), 208–220 (1998)

5. Barnhart, C., Talluri, K.T.: Airline operations research. In: ReVelle, C., McGarity, A.E. (eds.) Design and Operation of Civil and Environmental Engineering Systems, pp. 435–469. Wiley, New York (1997)
6. Clarke, L.W., Johnson, E., Nemhauser, G.L., Zhu, Z.: The aircraft rotation problem. Annals of Operations Research 69, 33–46 (1997)
7. Coldren, G.M., Koppelman, F.S.: Modeling the competition among air-travel itinerary shares: GEV model development. Transportation Research Part A: Policy and Practice 39, 345–365 (2005)
8. Coldren, G.M., Koppelman, F.S., Kasturirangan, K., Mukherjee, A.: Modeling aggregate air-travel itinerary shares: Logit model development at a major US airline. Journal of Air Transport Management 9, 361–369 (2003)
9. Doganis, R.: Flying Off Course – The Economics of International Airlines, 3rd edn. Routledge, London (2004)
10. Dueck, G., Scheuer, T.: Threshold accepting: A general purpose optimization algorithm appearing superior to simulated annealing. Journal of Computational Physics 90(1), 161–175 (1990)
11. Erdmann, A., Nolte, A., Noltemeier, A., Schrader, R.: Modeling and solving an airline schedule generation problem. Annals of Operations Research 107(1-4), 117–142 (2001)
12. Etschmaier, M.M., Mathaisel, D.F.X.: Airline scheduling: An overview. Transportation Science 19(2), 127–138 (1985)
13. Feo, T.A., Bard, J.F.: Flight scheduling and maintenance base planning. Management Science 35(12), 1415–1432 (1989)
14. Goldberg, D.E.: Genetic Algorithms in Search, Optimization and Machine Learning. Addison-Wesley, Reading (1989)
15. Goldberg, D.E.: The Design of Innovation. Genetic Algorithms and Evolutionary Computation. Kluwer, Dordrecht (2002)
16. Gopalan, R., Talluri, K.T.: The aircraft maintenance routing problem. Operations Research 46(2), 260–271 (1998)
17. Gopalan, R., Talluri, K.T.: Mathematical models in airline schedule planning: A survey. Annals of Operations Research 76, 155–185 (1998)
18. Grandeau, S.C., Clarke, M.D., Mathaisel, D.F.X.: The processes of airline system operations control. In: Yu, G. (ed.) Operations Research in the Airline Industry, pp. 312–369. Kluwer Academic Publishers, Boston (1998)
19. Graves, G.W., Mcbride, R.D., Gershkoff, I., Anderson, D., Mahidhara, D.: Flight crew scheduling. Management Science 39(6), 736–745 (1993)
20. Grosche, T.: Integrated Airline Scheduling. PhD thesis, University of Mannheim (2007)
21. Grosche, T., Rothlauf, F.: Air travel itinerary market share estimation. Working Papers in Information System I, University of Mannheim (2007)
22. Gu, Z., Johnson, E.L., Nemhauser, G.L., Wang, Y.: Some properties of the fleet assignment problem. Operations Research Letters 15, 59–71 (1994)
23. Holland, J.H.: Adaption in natural and artificial systems. University of Michigan Press, Ann Arbor (1975)
24. ICAO. Annual review of civil aviation 2005. ICAO Journal 61(5) (2006)
25. Jarrah, A.I., Goodstein, J., Narasimhan, R.: An efficient airline re-fleeting model for the incremental modification of planned fleet assignments. Transportation Science 34(4), 349–363 (2000)
26. Lohatepanont, M., Barnhart, C.: Airline schedule planning: Integrated models and algorithms for schedule design and fleet assignment. Transportation Science 38(1), 19–32 (2004)

27. Mathaisel, D.F.X.: Decision support for airline schedule planning. Journal of Combinatorial Optimization 1, 251–275 (1997)
28. Michalewicz, Z., Fogel, D.B.: How to Solve It: Modern Heuristics. Springer, Berlin (2000)
29. Papadakos, N.: Integrated airline scheduling. Computers and Operations Research (to appear, 2008)
30. Reeves, C.R., Rowe, J.E.: Genetic Algorithms: Principles and Perspectives. Kluwer Academic Publishers, Boston (2003)
31. Rothlauf, F.: Design and Application of Metaheuristics. Habilitationsschrift, University of Mannheim (2006)
32. Rothlauf, F.: Representations for Genetic and Evolutionary Algorithms, 2nd edn. Springer, Heidelberg (2006)
33. Rushmeier, R.A., Kontogiorgis, S.A.: Advances in the optimization of airline fleet assignment. Transportation Science 31(2), 159–169 (1997)
34. Suhl, L.: Computer-Aided Scheduling – An Airline Perspective. Gabler, Wiesbaden (1995)
35. Talluri, K.T.: The four-day aircraft maintenance routing problem. Transportation Science 32(1), 43–53 (1998)
36. Teodorovic, D.: Airline Operations Research. Transportation Studies. Gordon and Breach, New York (1988)

Part II

Vehicle Routing

GRASP with Path Relinking for the Capacitated Arc Routing Problem with Time Windows

Nacima Labadi, Christian Prins, and Mohamed Reghioui

Institute Charles Delaunay, University of Technology of Troyes, BP 2060, 10010 Troyes Cedex, France
{nacima.labadi,christian.prins,mohamed.reghioui_hamzaoui}@utt.fr

Summary. Greedy randomized adaptive search procedures (GRASP) are proposed for the undirected capacitated arc routing problem with time windows (CARPTW). Contrary to the vehicle routing problem with time windows (VRPTW), this problem has received little attention. The procedures combine simple components: a greedy randomized heuristic to compute either a giant tour (three versions) or a feasible CARPTW solution (one version), a tour-splitting algorithm for the giant tours (two versions), a local search, and an optional path relinking process. Computational experiments on the CARPTW indicate that proposed metaheuristics compete with the best published algorithms: on a set of 24 instances, the best combination of components finds 17 optima (including 4 new ones) and improves 5 best-known solutions. Moreover, when applied to the CARP without time windows, these GRASPs are as effective as state-of-the-art algorithms, while being significantly faster.

Keywords: Capacitated arc routing problem, Time windows, GRASP, Path relinking.

1 Introduction

The NP-hard capacitated arc routing problem (CARP) is raised by applications like urban waste collection and winter gritting. A natural extension is the CARP with time windows or CARPTW, motivated for instance by the interruption of operations during rush hours or maintenance interventions on gas, power or telecom networks. The CARPTW can be defined on an undirected network $G = (V, E)$ with $|V| = n$ and $|E| = m$. The node set V includes one depot node s with a fleet of identical vehicles of capacity Q. Each edge e has a non-negative demand q_e, a length c_e and a deadheading time t_e (traversal time without service). The τ edges with non-zero demands (called *required edges* or *tasks*) must be serviced by a vehicle. Each task e has also a processing time p_e (when e is traversed by a vehicle to be serviced) and a time window $[a_e, b_e]$ in which service must begin. Arriving earlier than a_e induces a waiting time, while arriving later than b_e is not allowed. In this chapter all data are integers.

The CARPTW consists of computing a set of vehicle trips minimizing the total distance travelled. Each trip must start and end at the depot and service a subset of required edges while respecting their time windows. The total demand

A. Fink and F. Rothlauf (Eds.): Advances in Computational Intelligence, SCI 144, pp. 111–135, 2008.
springerlink.com

satisfied by a vehicle must not exceed its capacity and each required edge must be serviced by one single vehicle (split service is not allowed). The total distance is minimized in this chapter because this is the usual objective for the vehicle routing problem with time windows (VRPTW). Most instances of literature assume that $t_e = c_e = p_e$ for each required edge e.

To the best of our knowledge, the only publicly available works that deal with the CARPTW are three Ph.D. dissertations. Mullaseril [18] considered the directed CARPTW and proposed heuristics and a transformation into a VRPTW. Later, Gueguen [10] described an integer linear programming model for the undirected CARPTW and another transformation into a VRPTW, but without giving numerical results. More recently, Wøhlk [23] presented two ways of modeling the undirected CARPTW (one arc routing and one node routing formulation), several heuristics and a dynamic programming algorithm combined with simulated annealing (DYPSA). She also designed a column generation method to get tight lower bounds.

Tagmouti et al. [22] studied a closely related problem, the CARP with time-dependent service costs, with a column generation method based on a node routing model. However, such a problem involves time-dependent piecewise-linear service cost functions rather than hard time windows.

Powerful metaheuristics are available for the VRP [4], the VRPTW [17] and the CARP [2, 3, 11, 12, 15]. However, apart from the simulated annealing embedded in Wøhlk's DP method, no metaheuristic has been proposed for the CARPTW. This chapter bridges the gap by presenting simple but very effective greedy randomized adaptive search procedures (GRASP), strengthened by a path relinking process (PR). To the best of our knowledge, only two very recent journal papers describe applications of PR to vehicle routing problems [13, 19], but both deal with node routing problems (respectively the VRP and a school bus routing problem).

The chapter is structured as follows. Section 2 presents GRASP components: greedy randomized heuristics to build either CARPTW solutions or giant tours, splitting algorithms to convert giant tours into feasible CARPTW solutions and a local search procedure to improve resulting solutions. The path relinking is described in Sect. 3. Section 4 is devoted to computational experiments on the CARPTW and CARP.

2 GRASP Components and General Structure

Greedy randomized adaptive search procedures or GRASP were introduced in 1989 by Feo and Bard [6], see also [7] for a general presentation of the method. Each GRASP iteration computes one solution in two phases. The first phase builds one trial solution using a greedy randomized heuristic. Each step of this heuristic extends a partial solution by adding one element randomly selected among the best candidates (restricted candidate list or RCL). The second phase improves the trial solution with a local search procedure. The method stops after a fixed number of iterations and returns the best solution found.

The following subsections describe ad hoc components for the CARPTW. Sect. 2.1 describes one greedy randomized heuristic which directly builds a CARPTW solution, and three other methods which relax the vehicle capacity and time windows constraints to compute one single tour servicing all customers (giant tour). Two splitting algorithms to convert giant tours into feasible CARPTW solutions are presented in Sect. 2.2. The local search is introduced in Sect. 2.3 and the resulting GRASP structure in Sect. 2.4.

Some additional notation is required for this section. A dummy loop (index 0) is added on the depot, each required edge is replaced by two opposite arcs (possible traversal directions) and the resulting arcs are indexed from 1 to $2\tau + 1$. The opposite arc associated with one arc u is denoted u'. A distance matrix D, $(2\tau+1) \times (2\tau+1)$ is pre-computed. The distance d_{uv} between two arcs $u = (i, j)$ and $v = (k, l)$ is defined as the length of a shortest path from node j to node k.

During the construction of a trip, u denotes the last edge treated and Z the set of edges not yet serviced whose demands fit residual vehicle capacity. $d_{min} = \min\{d_{uz} | z \in Z\}$ is the minimum distance between u and the edges of Z and $L = \{z \in Z | d_{uz} = d_{min}\}$ is the set of edges of Z at distance d_{min}.

2.1 Greedy Randomized Heuristics

It is not an easy task to design a greedy randomized heuristic able to provide good but diverse solutions. On one hand, the same trial solution can be generated several times if the RCL is too small: the heuristic tends to behave like a deterministic greedy heuristic. On the other hand, a large RCL leads to a random heuristic with low-quality solutions on average. Another difficulty is that the local search can lead to the same local optimum when applied to several trial solutions located in the same attraction basin, and the probability to obtain duplicate solutions after improvement increases with the number of GRASP iterations.

For a given number of GRASP iterations, a better diversity can be expected if several constructive heuristics are used and this was confirmed during preliminary testing. For the CARPTW, our GRASPs call two heuristics among the ones described in the two following subsections. In practice, each heuristic is used during one half of GRASP iterations. Some results which prove the effectiveness of this strategy are given in Sect. 4. All the proposed heuristics are adaptive, i.e. they update at each iteration the benefits associated with the choice of each task, instead of working with a list of decisions prepared in advance.

Randomized Path-Scanning Heuristic

Path-Scanning (PS) is a classical greedy heuristic for the CARP [9], which builds one trip at a time using the nearest neighbour principle and a set of five rules to break ties. We recall it first for the CARP.

Starting from the depot, each iteration extends the incumbent trip by adding the closest edge in Z. The vehicle returns to the depot when Z is empty. The algorithm stops if all required edges are serviced, otherwise a new trip is started.

However, contrary to node routing problems, d_{min} is often null and the set L contains several edges. For instance, consider a crossroad with four streets. If a vehicle treats one street, it may choose one of the three adjacent streets to continue: in that case, $d_{min} = 0$ and $|L| = 3$.

When $|L| > 1$, PS evaluates each edge e in L and uses five rules to select the next edge to be serviced: 1) minimize the distance d_{e0} to return to the depot, 2) maximize this distance, 3) minimize the ratio q_e/c_e (a kind of productivity), 4) maximize this ratio, 5) use rule 2 if vehicle is less than half-full, otherwise use rule 1. One solution is computed for each rule and the best one is returned at the end.

For the same running time (5 solutions), Pearn [20] noticed that better average results are obtained if one of the 5 rules is randomly selected in each iteration. We adapted Pearn's version for the CARPTW: the set Z now gathers the edges not yet serviced, with compatible demands, *and* which can be reached from the incumbent edge without violating time windows.

However, we added 12 other rules proposed by Wøhlk [23] and specific to time windows. These rules consists in selecting the next edge e in order to minimize the following criteria, in which i is the incumbent edge and μ_i the current time when the service of i terminates: 1) $b_e \cdot d_{ie}$, 2) b_e, 3) $(b_e - \mu_i) \cdot d_{ie}$, 4) $a_e \cdot d_{ie}$, 5) $(b_e - a_e) \cdot d_{ie}$, 6) $b_e \cdot d_{ie}/q_e$, 7) $(b_e - a_e) \cdot d_{ie}$, 8) $a_e - \mu_i$, 9) $b_e + d_{ie}$, 10) $a_e - \mu_i - d_{ie}$, 11) $d_{ie} \cdot (a_e - \mu_i - d_{ie})$, 12) a_e. For instance, rule 2 is equivalent to the earliest due date priority rule used in scheduling heuristics.

Like in Pearn [20], the heuristic is randomized by randomly selecting one of the rules (now 17) in each iteration. L is used as restricted candidate list. A second randomization consists of choosing randomly one edge among the k best candidates in L for the selected rule.

Preliminary tests on the CARPTW confirmed that using the 17 rules gives better results than using only the 5 rules used in PS for the CARP (but adapted here to time windows) or the 12 rules specifically designed by Wøhlk for time windows. The resulting greedy randomized heuristic is called RPS in the sequel. When applied to the basic CARP (without time windows), all time windows are initialized to $[0, \infty]$ but Wøhlk's rules are deactivated for the sake of efficiency.

Randomized Heuristics Producing Giant Tours

Three randomized heuristics called RR, RT and RTF were also designed. They ignore vehicle capacity and time windows to build one giant tour covering all required edges. To obtain a feasible CARPTW solution, they must be followed by one of the splitting algorithms described in the next subsection. RR (like *Random Rule*) is equivalent to RPS, but with a fictitious vehicle capacity $Q = \sum_{e=1}^{m} q_e$ and without time windows. One single tour is constructed. In each iteration, one of the 5 rules not designed for time windows is selected at random and the next edge of the trip is selected randomly in L using this rule. Note that Q is set to the sum of demands to guarantee a correct behavior of rule 5 of PS, i.e. the vehicle must start getting closer to the depot when it is half-full.

RT (like *Random Task*) draws the next edge in the candidate list $L = \{z \in Z | d_{uz} = d_{min}\}$. As already mentioned for RPS, this simple strategy works well for arc routing problems because d_{min} is often null and L contains several edges in general.

In RTF (like *RT Flower*), candidate edges are partitioned in two lists L_1 and L_2. L_1 gathers the edges which drive the vehicle away from the depot, i.e., $L_1 = \{z \in Z | d_{uz} = d_{min} \wedge d_{z0} > d_{u0}\}$. The other tasks are stored in L_2. Let *load* denote the current load of the tour. If one list is empty, RTF selects one task at random in the other list. If no list is empty, RTF randomly draws the next task in L_1 if $(load \texttt{ mod } Q) \leq Q/2$, and in L_2 otherwise. In other words, the giant tour tends to go away from the depot if its load is in one interval $[iQ, iQ + Q/2]$ for some integer $i \geq 0$, otherwise it nears the depot. The name RTF comes from the flower-shaped resulting tour. The goal of this strategy inspired by rule 5 of Path-Scanning is to help the splitting algorithms: the most likely splitting points will be closer to the depot.

2.2 Tour Splitting Algorithms

Basic Splitting Algorithm

Beasley [1] introduced route-first cluster-second heuristics for the VRP, but without providing numerical results. The principle is to relax vehicle capacity and to build one giant tour, using any TSP algorithm. Then an optimal partition (subject to the sequence) of the giant tour is computed using a tour splitting algorithm. The idea is also valid for arc routing problems. For instance, Lacomme et al. [15] designed an efficient memetic algorithm (MA) for the CARP, in which the chromosomes are giant tours: they do not contain special symbols to delimit successive trips. Each chromosome is evaluated using a tour splitting algorithm called *Split*. In this MA, the algorithm to solve the uncapacitated problem and provide giant tours is replaced by a crossover operator. *Split* is adapted here for the CARPTW, the giant tours being provided by the greedy randomized heuristics RR, RT or RTF presented before.

Figure 1 shows a giant tour $T = (T_1, T_2, T_3, T_4, T_5)$ with $\tau = 5$ required edges (thick segments), with demands in brackets and time windows in square brackets. Thin lines represent intermediate shortest paths between successive edges or between each edge and the depot. It is assumed that $Q = 9$ and that $t_e = c_e = p_e$ for each edge e. *Split* builds an auxiliary graph H with $\tau + 1$ nodes indexed from 0 onwards. This graph contains one arc $(i-1, j)$ if trip $(0, T_i, T_{i+1}, \ldots, T_j, 0)$ is feasible. For instance, the leftmost arc corresponds to the trip reduced to edge T_1: its cost 35 includes the distance from the depot to edge T_1, the length of T_1 and the distance to return to the depot. Once the auxiliary graph is completed, *Split* computes a min-cost path from the first node 0 to the last node $\tau = 5$. The arcs along this path indicate the optimal decomposition of the giant tour (lower part of the figure).

This example illustrates the flexibility of the *Split* algorithm: all constraints at the trip level are only used to filter the trips represented in H, but the shortest

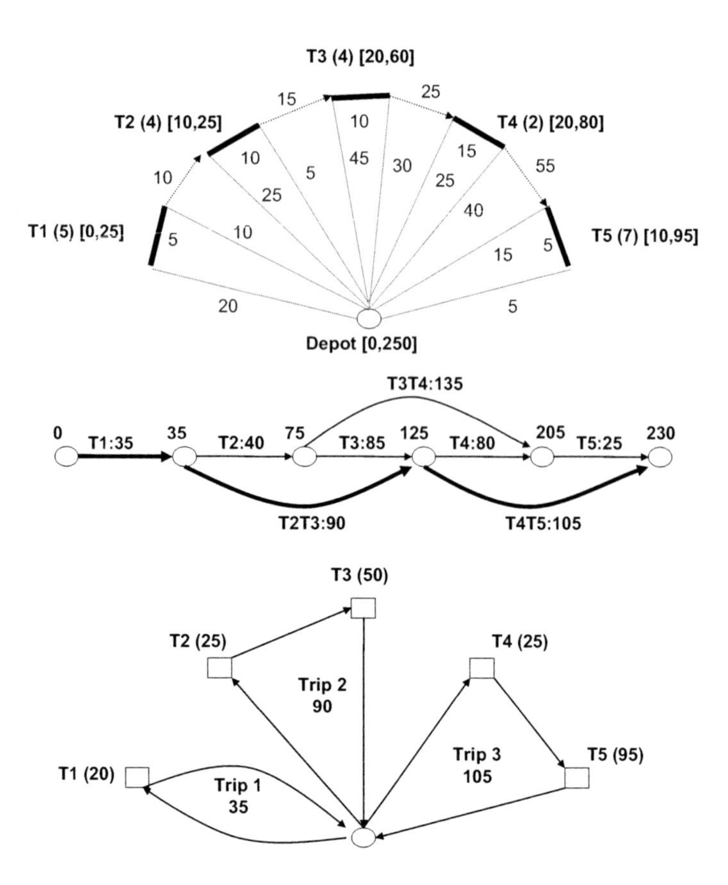

Fig. 1. A five-edge example with $Q = 9$ for *Split*

path computation does not change. For instance, there is no time window in the CARP MA [15] and the trip $(0, T_1, T_2, 0)$ is modelled by one arc in H. For the CARPTW, this trip is not represented because time windows are violated.

A range constraint $L = 130$ (i.e. maximum trip length) could be added easily: in that case, the trip $(0, T_3, T_4, 0)$ with length 135 would not be considered.

By construction, H is acyclic and contains $h \leq \tau(\tau+1)/2$ arcs if the CARPTW involves τ required edges. Using the version of Bellman's algorithm for directed acyclic graphs [5], whose complexity is linear in h, the shortest path can be computed in $O(h)$. In the worst case, $h = O(\tau^2)$. Let q_{min} denote the minimum demand. A trip contains at most $r = \lfloor Q/q_{min} \rfloor$ tasks, there are at most r trips starting with a given edge T_i and so r outgoing arcs per node of H. H contains $O(r\tau)$ arcs and then the complexity of *Split* is $O(r\tau)$. Hence, *Split* is faster when q_{min} represents a substantial fraction of vehicle capacity.

It is possible to implement *Split* without generating H explicitly, like in Algorithm 1. For each task T_j of the giant tour T, the algorithm computes two labels: W_j, the cost of a shortest path from node 0 to node j in H, and

Algorithm 1. A compact implementation of *Split* for the CARPTW

1: $W_0 := 0$
2: $P_0 := 0$
3: **for** $i := 1$ to τ **do**
4: $\quad W_i := \infty$
5: **end for**
6: **for** $i := 1$ to τ **do**
7: $\quad j := i$
8: $\quad load, length := 0$
9: $\quad ArrivalTime, DepartureTime := 0$
10: $\quad u := 0$
11: $\quad$ **repeat**
12: $\quad\quad v := T_j$
13: $\quad\quad load := load + q_v$
14: $\quad\quad length := length - d(u,0) + d(u,v) + c_v + d(v,0)$
15: $\quad\quad ArrivalTime := DepartureTime + \delta_{uv}$
16: $\quad\quad DepartureTime := ArrivalTime + \max(0, a_v - ArrivalTime) + p_v$
17: $\quad\quad$ **if** $(load \leq Q)$ and $(W_{i-1} + length < W_j)$ and $(ArrivalTime \leq b_v)$ **then**
18: $\quad\quad\quad W_j := W_{i-1} + length$
19: $\quad\quad\quad P_j := i - 1$
20: $\quad\quad$ **end if**
21: $\quad\quad j := j + 1$
22: $\quad\quad u := v$
23: $\quad$ **until** $(j > m)$ or $(load > Q)$ or $(ArrivalTime > b_u)$
24: **end for**

P_j, the predecessor of T_j on this path. At the beginning, the labels in W are null for node 0 and infinite for all other nodes. The main *for* loop scans T using index i. For each i, the *repeat* loop inspects each possible trip $(0, T_i, \ldots, T_j, 0)$ starting with task T_i and evaluates its *length*, its *load* and its feasibility for time windows. To avoid nested indices, v denotes the incumbent arc (T_j) and u the previous arc of the trip (T_{j-1}, or the depot loop 0 if $i = j$). δ_{uv} is used to check time windows, it denotes the duration of the shortest path from u to v (the one with minimum length d_{uv}). The loop stops when *load* exceeds vehicle capacity, the vehicle arrives too late at T_j or when j reaches the end of T. A feasible trip $(0, T_i, \ldots, T_j, 0)$ corresponds to one arc $(i-1, j)$ in H. Instead of storing this arc, the label of node j is directly updated when improved. For each j, note that *load*, *length* and *ArrivalTime* are updated in $O(1)$ from their values for $j-1$ (lines 13-14-15).

This avoids an additional loop to compute the length of the incumbent trip or to check its feasibility for time windows. This trick is essential to achieve the $O(h)$ complexity, coming from the two nested loops which are equivalent to inspecting all arcs of H. Moreover, the algorithm runs in $O(\tau)$ space only, instead of $O(h)$ if H were generated explicitly. At the end, the cost of the optimal CARPTW solution, subject to the sequence imposed by the giant tour T, is given by W_τ.

The trips of the associated CARPTW solution can be easily extracted, using the predecessors stored in vector P.

Improved Splitting Algorithm

For each sub-sequence $(T_i, T_{i+1}, \ldots, T_j)$ of a giant tour T, a rotation (circular left shift) can give a better trip. It is then possible to evaluate all rotations and to model only the best one in the auxiliary graph. We call this version *Split-Shift*.

One example with four required edges is given in Fig. 2. It is assumed that $Q = 9$, $p_e = c_e = t_e$ for each edge e, and that the shortest paths from T_3 to T_2, T_4 to T_3 and T_4 to T_2 have respective lengths 20, 30 and 10. To better visualize rotations, these paths are drawn on the giant tour. The trip without rotation (T_2, T_3)

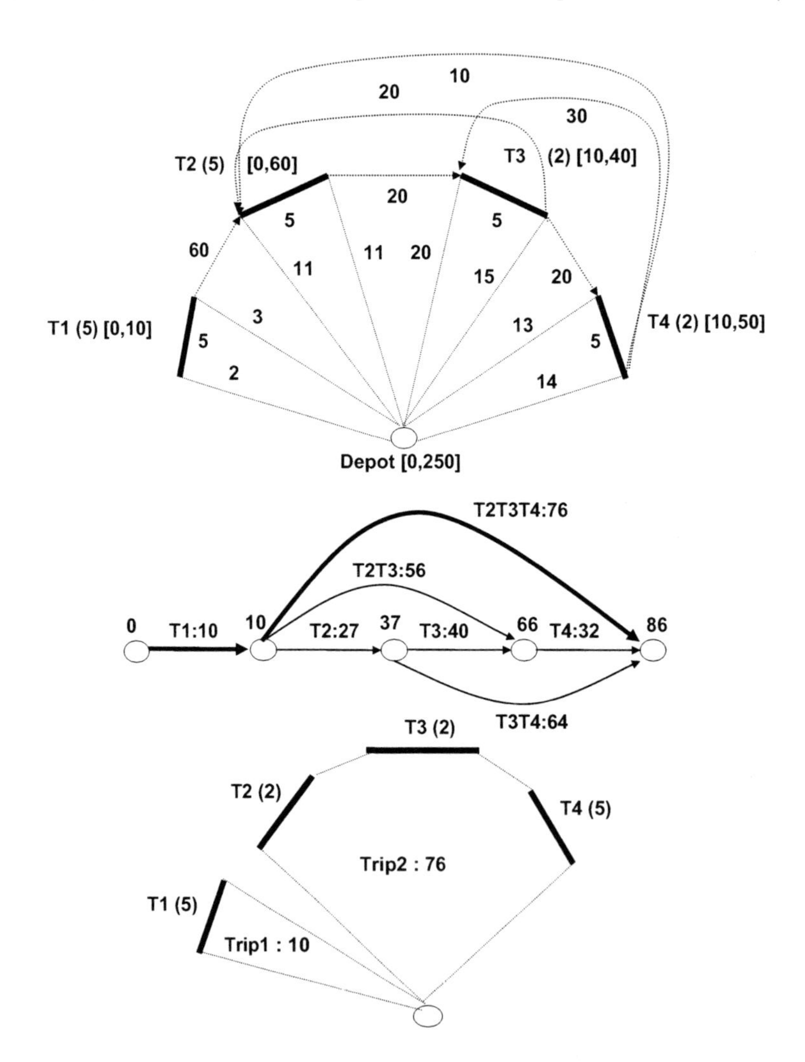

Fig. 2. A four-edge example for *Split-Shift*

(length 56) is better than its rotation (T_3, T_2) (length 61): hence, only the first trip is modelled in the auxiliary graph H. The rotation (T_3, T_4, T_2) (length 76) is included in H because the two other possibilities (T_2, T_3, T_4) and (T_4, T_2, T_3) violate time windows. Finally, (T_3, T_4) (length 76) is kept while (T_4, T_3) is infeasible.

Like *Split*, it is possible to implement *Split-Shift* without generating the auxiliary graph explicitly. The algorithm is like the one for *Split*, with the same two nested loops which enumerate all feasible sub-sequences $(T_i, T_{i+1}, \ldots, T_j)$. However, one internal loop is added to evaluate the cost of the trip if it starts with one of the tasks and browses the other tasks cyclically. Only the trip corresponding to the best rotation is modelled in the auxiliary graph. Since *Split* runs in $O(r\tau)$ if the trips contains at most r required edges, *Split-Shift* with the additional loop in $O(r)$ runs in $O(r^2\tau)$. In practice, the algorithm is fast enough because r is in general much smaller than τ.

2.3 Local Search

Like in the VRP, possible moves in a local search for arc routing problems include relocating, swapping and inverting chains of edges. However, their implementation is more difficult in case of time windows. Feasibility tests for capacity can be done in $O(1)$ but the insertion of one new edge in a trip with ξ edges may shift all arrival times after the insertion point: apparently, the modified trip must be scanned in $O(\xi)$ to check time windows. In fact, Kindervater and Savelsbergh [14] have shown how to detect time window violations in $O(1)$ for the VRPTW: the slack or maximum allowed delay for an insertion must be kept for each customer. We have adapted this technique to the CARPTW, with the following moves:

- OR-OPT moves: relocate a chain of 1 or 2 required edges.
- SWAP moves: swap two different edges.
- 2-OPT moves adapted to the arc routing context.

These moves may involve one or two trips, and the two possible traversal directions of a required edge are tested in the reinsertions. Remember (beginning of Sect. 2) that each edge is coded by two opposite arcs. For one arc (i, j) indexed by u, let u' denote the index of the opposite arc (j, i). For instance, an OR-OPT move may remove u from its current trip to reinsert it as u or u', in the same trip or in another trip.

What we call a 2-OPT move is defined by two edges u and v. Assume that a trip X is encoded as a list of α edges including one copy of the depot loop at the beginning and at the end, i.e., $X = (x_1 = 0, x_2, \ldots, x_{\alpha-1}, x_\alpha = 0)$. If u and v belong to the same trip X with $x_i = u$, $x_j = v$ and $i \leq j$, the 2-OPT move consists of inverting sub-sequence $(x_i, x_{i+1}, \ldots, x_j)$: it is replaced by sub-sequence $(x'_j, x'_{j-1}, \ldots, x'_i)$.

If u and v belong to two distinct trips X (α edges) and Y (β edges) with $x_i = u$ and $y_j = v$, the 2-OPT moves consists of cutting the shortest paths from

Algorithm 2. GRASP for the CARPTW

```
1: f* := ∞
2: iter := 0
3: repeat
4:    iter := iter + 1
5:    if iter ≤ maxiter/2 then
6:       H1(S)
7:    else
8:       H2(S)
9:    end if
10:   LocalSearch(S)
11:   if f(S) < f* then
12:      f* := f(S)
13:      S* := S
14:   end if
15: until (iter = maxiter) or (f* = LB)
```

x_i to x_{i+1} and from y_j to y_{j+1}. Then, there are two different ways of reconnecting the trips. The first one preserves edge directions: the two trips are replaced by $(x_1, \ldots, x_i, y_{j+1}, \ldots, y_\beta)$ and by $(y_1, \ldots, y_j, x_{i+1}, \ldots, x_\alpha)$. Some edges are inverted in the second way, which replaces X and Y by $(x_1, \ldots, x_i, y'_j, \ldots, y'_1)$ and $(x'_\alpha, \ldots, x'_{i+1}, y_{j+1}, \ldots, y_\beta)$. The second version is less productive, because of a higher probability of time window violation.

Each iteration of the local search evaluates all feasible moves and executes the best one found. The local search ends when no more improvement can be found.

2.4 General Structure of the GRASP

Algorithm 2 summarizes the general structure of our GRASPs, without the PR procedure. Two randomized greedy heuristics H_1 and H_2 must be selected among the ones described in Sect. 2.1. In each iteration, the solution S resulting from one of the two heuristics is improved by local search. $f(S)$ denotes the cost (total length of the trips) of S. The algorithm stops after a fixed number of iterations *maxiter* or when a known lower bound LB is reached. Its results are the best solution found S^* and its cost f^*.

3 Path Relinking

Path relinking (PR) is an intensification strategy which can be added to any metaheuristic. The method was used for a first time by Glover and Laguna [8] to improve solutions obtained by tabu search. Its principle is to explore a trajectory linking two solutions in solution space, to try finding a better one. The attributes of the target solution are progressively introduced in the source solution, which generates a sequence of intermediate solutions. In each iteration,

```
S:  2   5  10 | 7   6 | 8   1 | 9 | 4   3
T:  9 | 6'  7'| 2   5  10 | 8   1 | 4   3
```

Fig. 3. Example of distance between two solutions

the attributes added are selected to minimize the cost of the next intermediate solution.

In spite of its simplicity and speed, GRASP is often less effective than more aggressive metaheuristics like tabu search. A possible explanation is the independency of GRASP iterations, which perform a kind of random sampling of solution space. The addition of a PR procedure working on a small pool of elite solutions collected during the GRASP is a good way to remedy this relative weakness. The following subsections describe a distance measure for the CARPTW, show how to use it to explore a trajectory between two solutions, and present two possible ways of integrating the PR and the GRASP.

3.1 Distance Measure

Our distance for the CARPTW extends a distance proposed by Martí et al. [16], for permutation problems in which the relative position of each element is more important than its absolute position (R-permutation problems). Given two permutations $S = (p_1, p_2, \ldots, p_m)$ and $T = (q_1, q_2, \ldots, q_m)$, this distance $D(S,T)$ is the number of times p_{i+1} does not immediately follow p_i in T, for $i = 1, 2, \ldots, m-1$. In other words, D is the number of pairs of consecutive elements in P that are "broken" in Q, it varies between 0 and $m-1$.

A similar strategy is used to compute a distance between CARPTW solutions, represented as giant tours to have a constant length of τ required edges. The *Split* procedure of Sect. 2.2 is used to extract the true CARPTW solutions. We have also extended the original distance to make it reversal-independent: a pair of consecutive arcs (u,v) of S is counted as broken if neither (u,v) nor (v',u') are found in T. Like Martí's distance, the new version can be computed in $O(\tau)$. Figure 3 provides an example with $D(S,T) = 4$. Vertical bars in S (resp. T) indicate the pairs broken in the other solution.

3.2 Exploration of a Trajectory

To generate a path from one solution S to one solution T, we first compute $D(S,T)$. Then, starting from S, the incumbent solution undergoes a move that reduces its distance to T. The moves include the inversion, the relocation and the relocation with inversion of a chain of consecutive edges without broken pairs, called *block*. The concept of block is used to avoid repairing one pair while breaking another. In Fig. 3 for instance, moving arc 1 after arc 9 in S repairs the pair $(1,4)$ but breaks $(8,1)$. This can be avoided if the whole block $(8,1)$ is moved. To avoid too long trajectories and the generation of low-quality solutions, all possible moves are evaluated and the least-cost one is executed to obtain the next solution on the path. This new solution is immediately decoded with *Split*.

It is well-known that the PR mechanism alone provides few improved solutions: it must be helped by a local search. However, consecutive solutions on the path often lead to the same local optimum because their structures are too close. For the CARPTW, the local search already described for the GRASP is applied only every λ iterations, λ being a given parameter. This accelerates the path relinking with a negligible loss in solution quality.

3.3 Integrating GRASP and PR

Resende and Ribeiro [21] describe two main ways of combining GRASP and PR. The first one (*external PR*) is to use PR for post-optimization, on a small pool of elite solutions collected during the GRASP. The second one (*internal PR*) is

Algorithm 3. GRASP with internal PR for the CARPTW

1: $iter := 0$
2: **repeat**
3: $\quad iter := iter + 1$
4: $\quad$ **if** $iter \leq maxiter/2$ **then**
5: $\quad\quad H_1(S)$
6: $\quad$ **else**
7: $\quad\quad H_2(S)$
8: $\quad$ **end if**
9: $\quad LocalSearch(S)$
10: $\quad$ **if** $iter = 1$ **then**
11: $\quad\quad \mathcal{P} := \{S\};\ S^* := S;\ f_{max} := f(S)$
12: $\quad$ **else**
13: $\quad\quad$ select any T in $\arg\max\ \{D(S,X) : X \in \mathcal{P}\}$
14: $\quad\quad k := 0$
15: $\quad\quad Q := S$
16: $\quad\quad$ **while** $Q \neq T$ **do**
17: $\quad\quad\quad MoveOnPath(Q)$
18: $\quad\quad\quad k := k + 1$
19: $\quad\quad\quad$ **if** $(k \bmod \lambda) = 0$ **then**
20: $\quad\quad\quad\quad LocalSearch(Q)$
21: $\quad\quad\quad$ **end if**
22: $\quad\quad\quad$ **if** $|\mathcal{P}| < \rho$ **then**
23: $\quad\quad\quad\quad \mathcal{P} := \mathcal{P} \cup \{Q\}$
24: $\quad\quad\quad$ **else if** $f(Q) < f_{max}$ **then**
25: $\quad\quad\quad\quad$ select any Y in $\arg\min\ \{D(Q,X) : X \in \mathcal{P} \wedge f(X) > f(Q)\}$
26: $\quad\quad\quad\quad \mathcal{P} := (\mathcal{P} \setminus \{Y\}) \cup \{Q\}$
27: $\quad\quad\quad\quad$ update best solution S^* and worst cost f_{max}
28: $\quad\quad\quad$ **end if**
29: $\quad\quad$ **end while**
30: $\quad$ **end if**
31: **until** $(iter = maxiter)$ or $(f(S^*) = LB)$

Algorithm 4. GRASP with external PR for the CARPTW

1: $iter := 0$
2: **repeat**
3: $iter := iter + 1$
4: **if** $iter \leq maxiter/2$ **then**
5: $H_1(S)$
6: **else**
7: $H_2(S)$
8: **end if**
9: $LocalSearch(S)$
10: **if** $iter = 1$ **then**
11: $\mathcal{P} := \{S\}$; $S^* := S$; $f_{max} := f(S)$
12: **else if** $|\mathcal{P}| < \rho$ **then**
13: $\mathcal{P} := \mathcal{P} \cup \{Q\}$
14: **else if** $f(Q) < f_{max}$ **then**
15: select any Y in $\arg\min\ \{D(Q, X) : X \in \mathcal{P}\ \wedge\ f(X) > f(Q)\}$
16: $\mathcal{P} := (\mathcal{P} \setminus \{Y\}) \cup \{Q\}$
17: update best solution S^* and worst cost f_{max}
18: **end if**
19: **until** $(iter = maxiter)$ or $(f(S^*) = LB)$
20: **for** each pair (S, T) of distinct solutions of $\mathcal{P}$ **do**
21: $k := 0$
22: $Q := S$
23: **while** $(Q \neq T)$ and $(f(S^*) \neq LB)$ **do**
24: $MoveOnPath(Q)$
25: $k := k + 1$
26: **if** $(k \bmod \lambda) = 0$ **then**
27: $LocalSearch(Q)$
28: **end if**
29: update best solution S^*
30: **end while**
31: **end for**

to apply PR to each local optimum during the GRASP, by exploring the path which links it to an elite solution randomly chosen in the pool.

The GRASP with internal PR is sketched in Algorithm 3. It works on a pool $\mathcal{P}$ limited to ρ solutions. In the first GRASP iteration, the pool is initialized with the first solution. In each subsequent iteration, the PR generates a path between the incumbent GRASP solution S and the solution T most different from S in $\mathcal{P}$.

The incumbent solution Q on the trajectory is modified by *MoveOnPath*. This procedure evaluates all ways of moving or inverting a block in Q and performs the least-cost move among the ones which reduce the distance between Q and T. Remember that distance computations consider solutions encoded as strings without delimiters. The string corresponding to one detailed solution is obtained by concatenating the sequences of arcs traversed by the trips. Conversely, a string can be decoded using the *Split* procedure. These conversions are not detailed in Algorithm 3, to avoid overloading its general structure.

The counter k counts the solutions generated along the path. As explained in the previous subsection, the local search is applied only every λ solutions. If $\mathcal{P}$ has not yet reached its nominal size ρ, the intermediate solution Q is added to the pool. Otherwise, Q enters the pool only if it is better than the worst solution in $\mathcal{P}$. f_{max} denotes the cost of this worst solution. To increase pool diversity, the solution Y most similar to Q (in terms of distance) is replaced. This solution is selected among the solutions of $\mathcal{P}$ outperformed by Q. Note that the best pool solution S^* is preserved or improved, but never degraded.

The GRASP with external PR is detailed in Algorithm 4. PR is applied to the pool $\mathcal{P}$ of elite solutions obtained at the end of the GRASP iterations. This pool contains the best and most diverse GRASP solutions.

All pairs (S,T) of solutions in pool $\mathcal{P}$ are combined using PR. To preserve S for the next iterations, a copy of S is assigned to Q, then Q is modified by *MoveOnPath* to explore the path between S and T. As for the internal PR, local search is applied periodically on solutions generated along the path. If a new best solution is produced, S^* is updated. The algorithm stops if the lower bound is reached in the GRASP iterations or in the PR part, and returns the best solution S^*.

4 Computational Results

4.1 Implementation and Instances

Our algorithms were implemented in Delphi and executed on a 3 GHz PC. They were evaluated first on 3 sets of CARPTW problems A, B, C proposed by Wøhlk [23] and containing 8 instances each. The number of nodes ranges from 10 to 60, the number of edges m from 15 to 90 and the number of required edges τ from 11 to 81. These sets mainly differ in the width of their time windows, which is tight for set A (between 5 and 20 time units), medium for B (15-30 time units) and wider for C (30-45). The GRASPs were compared to the best heuristic proposed by Wøhlk, the Preferable Neighbor Heuristic (PNH), and to her lower bound obtained by column generation.

The efficiency of the GRASPs was also appraised for the CARP, although the algorithms were not designed for problems without time windows. We wish to underline here that no published paper dealing with the VRPTW provides results for the VRP. It is even well known that the most difficult VRPTW instances for exact algorithms are the ones that are closest to a VRP, i.e. with the widest windows.

The three sets of standard CARP instances used can be downloaded at `http://www.uv.es/~belengue/carp.html`. The 23 *gdb* files have 7 to 27 nodes and 11 to 55 edges, all required. The 34 *val* files have 24 to 50 nodes and 34 to 97 edges, all required. The last set (*egl* files) comprises 24 larger instances with 77 to 140 nodes, 98 to 190 edges, and 51 to 190 required edges.

The GRASPs were compared to three state-of-the-art metaheuristics for the CARP: the memetic algorithm from Lacomme et al. [15], the tabu search from Hertz et al. [11] and the very recent tabu search described by Brandão and Eglese [3]. These algorithms are respectively called MA, TSH and TSB in the sequel. We have also evaluated the deviations to the most recent lower bounds, compiled by Brandão and Eglese.

4.2 Evaluation Method

In order to have a uniform comparison, all algorithms were evaluated with $maxiter = 500$ iterations. The first step consisted of evaluating the greedy randomized heuristics RPS (with $k = 5$), RR, RT and RTF without the local search. The giant tours produced by RR, RT and RTF were partitioned using either the basic splitting algorithm (*Split*) or its version with rotations (*Split-Shift*). RPS directly generates a CARPTW solution, but we observed a significant improvement if the resulting trips are concatenated to get one giant tour which is finally split to get a definitive CARPTW solution. This is why the solutions of RPS are re-evaluated by *Split* or *Split-Shift*, like the other greedy heuristics.

In a second step, we added local improvement to get a basic GRASP without PR. The goal of these tests was to evaluate the gain brought by the local search and to select the greedy randomized heuristics which benefit best from local improvement. The aim of the third step was to reinforce the best GRASP by calling two heuristics instead of one, like in Algorithm 2. Finally, the best GRASP with two heuristics was strengthened even further by adding either an internal or external PR. The two kinds of PR use a pool size $\rho = 8$ and local search is applied every $\lambda = 2$ solutions on PR trajectories for an internal PR and every $\lambda = 1$ for an external one.

4.3 Results for CARPTW Instances

For the 24 CARPTW instances, Table 1 summarizes the results obtained by each greedy randomized heuristic, followed by *Split* or *Split-Shift* and improved or not by local search. The selected performance indicators are the average deviation to Wøhlk's lower bound in percent, the worst deviation to this bound, the average running time in seconds and the number of proven optima (when the bound is reached). The greedy heuristic RTF with its flower-shaped giant tours gives the worst results: due to time windows, the splitting points are not often close to the depot (see Sect. 2.1). The three other heuristics provide very similar results. Better CARPTW solutions are obtained if the giant tours are evaluated using *Split-Shift* instead of *Split*. Of course, adding the local search brings a strong improvement. The best simple GRASP, based on RT, has an average solution gap below 4% and is able to solve one quarter of instances to optimality.

Table 2 illustrates the kind of synergy obtained by using two greedy randomized heuristics in the best simple GRASP: by combining the two best heuristics RR and RT, two more optima are obtained and the average deviation to the lower bound drops to 3.66%.

Table 1. Impact of basic components for the CARPTW

		RPS	RR	RT	RTF
	Avg. dev. to LB (%)	25.48	25.94	25.72	28.41
Heuristic + Split	Worst dev. to LB (%)	55.67	56.06	55.07	57.65
	Time (s)	0.07	0.06	0.05	0.04
	Nb of optima	0	1	0	0
	Avg. dev. to LB (%)	22.52	22.73	22.47	24.83
Heuristic + Split-Shift	Worst dev. to LB (%)	47.51	47.12	49.11	48.31
	Time (s)	0.08	0.08	0.07	0.05
	Nb of optima	0	1	0	0
	Avg. dev. to LB (%)	4.24	3.96	3.84	4.83
Heuristic + Split-Shift	Worst dev. to LB (%)	16.30	14.71	17.17	16.90
+ local search (GRASP)	Time (s)	7.10	6.43	6.31	4.54
	Nb of optima	3	6	6	4

Table 2. Combination of heuristics in the best simple GRASP (CARPTW)

	RR	RT	RR+RT
Avg. dev. to LB (%)	3.96	3.84	3.66
Worst dev. to LB (%)	14.71	17.17	17.10
Nb of optima	6	6	8

Table 3. Impact of PR on the best simple GRASP for the CARPTW

PR	None	Internal	External
Avg. dev. to LB (%)	3.66	0.71	1.54
Worst dev. to LB (%)	17.10	8.35	11.73
Time (s)	7.82	79.02	30.06
Nb of optima	8	17	12

Table 3 gives a comparison between the results obtained by the best simple GRASP and a GRASP with internal or external PR. The basic GRASP is very fast but finds few optima. The PR brings a strong improvement at the expense of a running time multiplied by 4 for an external PR and by 10 for an internal one. Using PR inside the GRASP appears to be more effective but more time-consuming than using it as post-optimization. In the first case, PR is applied to each local optimum and the local search procedure is more frequently called. However, priority was given to the most effective but slowest option, *since Wøhlk provides no running times for PNH.*

A detailed comparison of our method with PNH is provided by Table 4. PNH is a kind of two-phase heuristic column generation method. The first phase generates a large set of promising feasible tours. Using a commercial IP solver (CPLEX), the second phase solves a set covering problem defined by these routes.

Table 4. Results for each CARPTW instance

Instance	n	m	τ	LB	PNH	Grasp	CPU	GIPR	CPU	Best GIPR
A10A	10	15	11	107	*107	*107	0.00	*107	0.03	*107
A13A	13	23	22	202	*202	*202	0.17	*202	0.58	*202
A13B	13	23	22	171	173	175	1.31	***171**	0.89	***171**
A13C	13	23	22	163	*163	*163	0.22	*163	4.02	*163
A20B	20	31	29	260	264	264	0.98	***260**	6.81	***260**
A40C	40	69	63	660	*660	708	29.55	*660	152.58	*660
A40D	40	69	63	807	*807	852	27.23	815	265.78	*807
A60A	60	90	81	1822	*1822	1884	52.59	1837	247.24	1830
B10A	10	15	11	87	*87	*87	0.02	*87	0.05	*87
B13A	13	23	22	167	*167	*167	0.97	*167	0.39	*167
B13B	13	23	22	152	158	158	0.06	***152**	0.44	***152**
B13C	13	23	22	141	*141	143	1.13	*141	0.84	*141
B20B	20	31	29	214	*214	220	1.91	*214	2.64	*214
B40C	40	69	63	588	602	632	26.03	602	192.55	602
B40D	40	69	63	730	*730	777	26.05	*730	229.22	*730
B60A	60	90	81	1554	*1554	1644	28.73	1567	272.08	1565
C10A	10	15	11	73	*73	*73	0.02	*73	0.00	*73
C13A	13	23	22	142	148	*142	0.95	***142**	0.53	***142**
C13B	13	23	22	132	*132	*132	0.28	*132	0.51	*132
C13C	13	23	22	121	*121	123	0.27	*121	2.66	*121
C20B	20	31	29	186	*186	192	0.51	*186	1.22	*186
C40C	40	69	63	503	563	589	17.97	**545**	161.94	**545**
C40D	40	69	63	611	626	672	19.70	631	148.14	631
C60A	60	90	81	1283	*1283	1384	22.95	1289	205.77	1289
Average					1.20%	3.66%	10.82s	0.71%	79.02s	0.64%
Worst					11.92%	17.10%	52.59s	8.35%	272.08s	8.35%
Optima					17	8		17		18

The first four columns of Table 4 respectively indicate the instance name, then the number of nodes, the number of edges and the number of tasks. The 5th and 6th columns show Wøhlk's lower bound (LB) and the total distance obtained by PNH. Remember that Wøhlk provides no running times. The four next columns respectively represent the solution cost and the running time in seconds found by the best simple GRASP of Table 3 (RR and RT heuristics, *Split-Shift*) and by the same GRASP with internal PR (GIPR). The last column gives the best cost achieved by the GRASP with PR using several settings of parameters. The last rows give for each column the average value (a deviation to LB in % when the column concerns solution costs), the worst value and the number of optima.

The GRASP with PR finds the same number of optimal solutions as PNH (17 out of 24 instances) but displays a smaller deviation to the lower bound: 0.71% versus 1.2%. Four new optima are found (instances A13B, A20B, B13B, C13A) and the best-known solution for C40C is improved. By modifying the pool size and the periodicity of local search in the PR, another optimum was achieved (A40D), reducing the average deviation to the lower bound to 0.64%. The running times are irregular because all algorithms stop after a fixed number of iterations or when they reach the lower bound.

4.4 Results for CARP Instances

In Table 5, the performances of the heuristics for the CARP are compared in the same way as for the CARPTW. The comparison to select the best components is based on the 34 *val* files. RTF seems to be the most appropriate for the CARP, by achieving 16 optima and giving the best deviation to the lower bound: 5.72%.

Confirming the results achieved for the CARPTW, the GRASP performs better when two greedy randomized heuristics are used. Table 6 gives some statistics about the best combination obtained for the CARP, RPS with RTF. Compared to a GRASP with RPS, the combination saves 0.59% and finds three additional

Table 5. Impact of basic components for the CARP (*val* files)

		RPS	RR	RT	RTF
	Avg. dev. to LB (%)	7.65	7.27	7.21	5.72
Heuristic + Split	Worst dev. to LB (%)	15.77	16.75	15.41	15.09
	Time (s)	0.32	0.29	0.30	0.31
	Nb of optima	0	0	1	1
	Avg. dev. to LB (%)	6.21	5.94	5.80	4.55
Heuristic + Split-Shift	Worst dev. to LB (%)	12.77	15.71	13.61	12.38
	Time (s)	0.38	0.33	0.36	0.32
	Nb of optima	0	1	2	2
	Avg. dev. to LB (%)	1.77	1.95	1.80	1.42
Heuristic + Split-Shift	Worst dev. to LB (%)	7.53	7.88	8.45	8.56
+ local search (GRASP)	Time (s)	1.32	2.07	1.75	1.60
	Nb of optima	14	14	14	16

Table 6. Combination of heuristics in the best simple GRASP (CARP)

	RPS	RTF	RPS+RTF
Avg. dev. to LB (%)	1.77	1.42	1.18
Worst dev. to LB (%)	7.53	8.56	5.71
Nb of optima	14	16	17

Table 7. Impact of PR on the best simple GRASP for the CARP

PR	None	Internal	External
Avg. dev. to LB (%)	1.18	0.40	0.57
Worst dev. to LB (%)	5.71	4.19	4.54
Time (s)	2.05	62.11	13.68
Nb of optima	17	27	26

optima. Compared to a GRASP with the best heuristic (RTF), one more optimum is found and the average deviation to the lower bound decreases by 0.24%.

Table 7 confirms our previous observations for the two ways of combining the GRASP and the PR. Again, the addition of PR improves the GRASP significantly, especially for the number of optimal solutions. Compared to the GRASP with an external PR, the internal implementation finds one more optima and gives a slightly better average deviation to the lower bound, but achieves its solutions more slowly. Contrary to the CARPTW case, the external implementation was selected for the CARP to offer a good tradeoff between speed and solution quality.

First, we compare our results to those obtained by TSH, TSB and MA on the 23 *gdb* files. In Table 8, the number of edges m is omitted because all edges are required ($m = \tau$). The column (LB) gives the best-known lower bounds listed in [3]. The solution costs and the running times obtained with standard parameters by TSH, TSB and MA are given by the next columns. The running time is scaled for the 3 GHz Pentium-IV PC used for the GRASP. For TSB, we cannot be more precise for the running times because Brandão and Eglese provide one decimal only. The GRASP column concerns a GRASP with $maxiter = 500$, $\rho = 8$ and using the two best heuristics RPS and RTF backed by *Split-Shift*. An external PR with $\lambda = 1$ is added to this GRASP to obtain costs given by the GEPR column. The last column provides the best cost found by this GRASP version, using several settings of parameters.

The simple GRASP is not as good as TSH, TSB and MA in terms of average deviation to the bound but it already outperforms TSH for the worst deviation and it is much quicker than the other metaheuristics: 28 times faster than TS, 16 times faster than MA and 13 times faster than TSB. When PR is added, the GRASP takes place between TSH and TSB, while remaining the fastest method (4.4 times faster than the fastest reference method, TSB).

The same table format is used to present the results for *val* files in Tables 9 and 10, using the same parameters as for *gdb* files. Conclusions drawn for *gdb* files remain valid but, this time, even the simple GRASP becomes competitive: it outperforms TSH for the average and worst deviations to lower bound, and finds the same number of optima. Concerning the running time, the basic GRASP can find a solution in 2 seconds on average versus 29 seconds for TSH, 17 for MA and 15 for the fastest reference metaheuristic (TSB).

The GRASP with PR finds 9 more optima than TSH and 2 less than MA and TSB. Compared to the *gdb* instances the algorithm gets closer to MA and TSB performance, while remaining faster than both methods. Using several settings of parameters, it is able to retrieve all proven optima.

The *egl* instances are the largest and the hardest ones. The lower bounds are seldom reached. The MA finds five optima, but TSH none. The parameters are identical except in the PR where the size ρ of the pool of elite solutions is set to 4. For those files no running times are known for TSH.

Once again, the quality of the solutions obtained by the GRASP with PR are better than those obtained by TSH, but not as good as those obtained by MA

Table 8. Computational results for CARP instances – gdb files

File	n	τ	LB	TSH	CPU	TSB	CPU	MA	CPU	Grasp	CPU	GEPR	CPU	Best GEPR
gdb1	12	22	316	*316	1.45	*316	< 0.1	*316	0.00	*316	0.00	*316	0.02	*316
gdb2	12	26	339	*339	2.38	*339	< 0.1	*339	0.20	*339	0.03	*339	0.02	*339
gdb3	12	22	275	*275	0.03	*275	< 0.1	*275	0.03	*275	0.00	*275	0.00	*275
gdb4	11	19	287	*287	0.04	*287	< 0.1	*287	0.00	*287	0.02	*287	0.02	*287
gdb5	13	26	377	*377	2.57	*377	< 0.1	*377	0.05	*377	0.03	*377	0.03	*377
gdb6	12	22	298	*298	0.39	*298	< 0.1	*298	0.08	*298	0.02	*298	0.02	*298
gdb7	12	22	325	*325	0.00	*325	< 0.1	*325	0.02	*325	0.02	*325	0.00	*325
gdb8	27	46	348	352	28.04	*348	1.2	350	18.31	357	0.11	351	1.66	*348
gdb9	27	51	303	317	24.78	*303	20.3	*303	3.26	313	1.77	309	5.55	*303
gdb10	12	25	275	*275	0.71	*275	< 0.1	*275	0.03	*275	0.00	*375	0.02	*375
gdb11	22	45	395	*395	1.05	*395	< 0.1	*395	0.58	*395	0.36	*395	0.34	*395
gdb12	13	23	458	*458	9.48	*458	0.6	*458	4.50	*458	0.14	*458	0.12	*458
gdb13	10	28	536	544	1.11	540	3.7	*536	3.41	544	0.03	544	0.05	544
gdb14	7	21	100	*100	0.22	*100	< 0.1	*100	0.02	*100	0.00	*100	0.00	*100
gdb15	7	21	58	*58	0.00	*58	< 0.1	*58	0.00	*58	0.00	*58	0.00	*58
gdb16	8	28	127	*127	0.78	*127	< 0.1	*127	0.03	*127	0.11	*127	0.11	*127
gdb17	8	28	91	*91	0.00	*91	< 0.1	*91	0.02	*91	0.00	*91	0.00	*91
gdb18	9	36	164	*164	0.13	*164	< 0.1	*164	0.05	*164	0.00	*164	0.00	*164
gdb19	8	11	55	*55	0.09	*55	< 0.1	*55	0.00	*55	0.00	*55	0.00	*55
gdb20	11	22	121	*121	4.37	*121	< 0.2	*121	0.15	*121	0.16	*121	0.16	*121
gdb21	11	33	156	*156	0.52	*156	< 0.1	*156	0.08	*156	0.33	*156	0.33	*156
gdb22	11	44	200	*200	1.55	*200	< 0.1	*200	1.54	201	0.00	*200	1.39	*200
gdb23	11	55	233	235	15.80	235	17.4	*233	23.53	235	0.25	235	0.25	235
Average				0.35%	4.15s	0.07%	1.95s	0.02%	2.43s	0.38%	0.15s	0.23%	0.44s	0.10%
Worst				4.62%	28.04s	0.86%	20.30s	0.57%	23.53s	3.30%	1.77s	1.98%	5.55s	1.49%
Optima				19		21		22		18		19		21

Table 9. Computational results for CARP instances – val files

File	n	τ	LB	TSH	CPU	TSB	CPU	MA	CPU	Grasp	CPU	GEPR	CPU	Best GEPR
val1a	24	39	173	*173	0.01	*173	< 0.1	*173	0.00	*173	0.05	*173	0.05	*173
val1b	24	39	173	*173	4.26	*173	0.7	*173	3.69	179	0.02	*173	0.02	*173
val1c	24	39	245	*245	42.85	*245	9.4	*245	13.18	*245	0.67	*245	0.61	*245
val2a	24	34	227	*227	0.08	*227	< 0.1	*227	0.02	*227	0.00	*227	0.02	*227
val2b	24	34	259	260	5.99	*259	0.2	*259	0.10	*259	0.19	*259	0.17	*259
val2c	24	34	457	494	14.56	*457	6.1	*457	10.00	465	0.14	463	2.19	*457
val3a	24	35	81	*81	0.35	*81	< 0.1	*81	0.02	*81	0.00	*81	0.00	*81
val3b	24	35	87	*87	1.28	*87	< 0.1	*87	0.00	*87	0.53	*87	0.48	*87
val3c	24	35	138	*138	19.15	*138	1.0	*138	12.98	*138	0.62	*138	0.55	*138
val4a	41	69	400	*400	13.02	*400	0.3	*400	0.33	*400	1.92	*400	1.73	*400
val4b	41	69	412	416	34.78	*412	4.3	*412	0.56	*412	2.69	*412	2.42	*412
val4c	41	69	428	453	32.21	*428	29.6	*428	8.79	440	3.05	*428	31.61	*428
val4d	41	69	526	556	107.38	530	85.8	541	47.47	546	2.47	538	30.47	536
val5a	34	65	423	*423	1.75	*423	0.2	*423	0.86	*423	0.61	*423	0.56	*423
val5b	34	65	446	448	19.03	*446	< 0.1	*446	0.48	*446	0.39	*446	0.36	*446
val5c	34	65	473	476	24.49	474	8.3	474	46.44	474	2.11	474	1.92	474
val5d	34	65	573	607	103.03	583	57.2	581	41.72	605	1.44	599	13.97	585
val6a	31	50	223	*223	1.79	*223	1.3	*223	0.08	*223	0.02	*223	0.02	*223
val6b	31	50	233	241	12.39	*233	9.9	*233	30.96	*233	0.78	*233	0.70	*233
val6c	31	50	317	329	39.16	*317	17.9	*317	24.01	323	0.47	*317	10.30	*317
val7a	40	66	279	*279	3.03	*279	0.8	*279	0.91	*279	0.48	*279	0.44	*279
val7b	40	66	283	*283	0.01	*283	0.4	*283	0.20	*283	0.05	*283	0.05	*283
val7c	40	66	334	343	55.83	*334	28.9	*334	46.51	340	2.25	335	14.58	*334
val8a	30	63	386	*386	1.77	*386	0.2	*386	0.30	*386	0.59	*386	0.55	*386
val8b	30	63	395	401	37.45	*395	1.4	*395	4.57	*395	3.22	*395	2.94	*395
val8c	30	63	518	533	67.77	529	43.4	527	32.85	547	1.84	540	21.59	531

Table 10. Computational results for CARP instances – val files (continued)

File	n	τ	LB	TSH	CPU	TSB	CPU	MA	CPU	Grasp	CPU	GEPR	CPU	Best GEPR
val9a	50	92	323	*323	13.11	*323	< 0.1	*323	8.41	325	6.27	*323	10.17	*323
val9b	50	92	326	329	27.53	*326	0.4	*326	13.51	330	5.48	*326	96.22	*326
val9c	50	92	332	*332	25.95	*332	0.3	*332	32.73	334	5.53	*332	62.20	*332
val9d	50	92	385	409	162.42	391	47.1	391	97.06	407	7.41	401	12.39	393
val10a	50	97	428	*428	2.54	*428	2.5	*428	11.71	430	5.11	*428	11.00	*428
val10b	50	97	436	*436	8.47	*436	1.4	*436	2.15	438	1.33	*436	24.48	*436
val10c	50	97	446	451	42.97	*446	5.8	*446	7.95	447	6.23	*446	26.37	*446
val10d	50	97	525	544	71.86	530	170.1	530	98.86	546	5.72	537	104.19	530
Average				1.59%	29.36s	0.21%	15.76	0.26%	17.63s	1.18%	2.05s	0.57%	13.68s	0.29%
Worst				8.10%	162.42s	2.12%	170.10s	2.85%	98.86s	5.71%	7.41s	4.54%	104.19s	2.51%
Optima				17		28		28		17		26		28

Table 11. Computational results for CARP instances – egl files

File	n	m	τ	LB	TSH	TSB	CPU	MA	CPU	Grasp	CPU	GEPR	CPU	Best GEPR
egl-e1-A	77	98	51	3548	3625	*3548	17.2	*3548	34.14	3587	1.03	*3548	2.63	*3548
egl-e1-B	77	98	51	4498	4532	4533	21.8	*4498	31.94	4543	1.89	4543	1.84	4525
egl-e1-C	77	98	51	5566	5663	5595	18.8	5595	32.72	5761	0.44	5687	2.76	5640
egl-e2-A	77	98	72	5018	5233	*5018	49.5	*5018	70.15	5032	2.69	5027	9.84	*5018
egl-e2-B	77	98	72	6305	6422	6343	52.0	6340	70.53	6492	3.25	6446	10.48	6351
egl-e2-C	77	98	72	8234	8603	8347	61.4	8415	59.60	8626	0.27	8574	9.81	8393
egl-e3-A	77	98	87	5898	5907	5902	60.3	*5898	111.26	5938	0.26	5933	25.17	5902
egl-e3-B	77	98	87	7704	7921	7816	88.5	7822	117.39	7953	4.11	7853	27.92	7823
egl-e3-C	77	98	87	10163	10805	10309	104.8	10433	94.87	10577	4.23	10577	4.25	10386
egl-e4-A	77	98	98	6408	6489	6473	105.7	6461	134.18	6663	8.67	6506	36.27	6476
egl-e4-B	77	98	98	8884	9216	9063	130.7	9021	143.83	9379	11.89	9276	41.94	9101
egl-e4-C	77	98	98	11427	11824	11627	147.1	11779	116.03	11946	9.72	11946	9.80	11798
egl-s1-A	140	190	75	5018	5149	5072	51.9	*5018	95.91	5127	1.97	5127	1.98	5019
egl-s1-B	140	190	75	6384	6641	6388	63.0	6435	95.98	6621	0.80	6566	6.08	6490
egl-s1-C	140	190	75	8493	8687	8535	61.8	8518	76.11	8674	5.11	8585	10.97	8561
egl-s2-A	140	190	147	9824	10373	10038	308.2	9995	401.98	10291	31.13	10291	31.36	10155
egl-s2-B	140	190	147	12968	13495	13178	349.7	13174	349.63	13615	39.05	13602	156.39	13441
egl-s2-C	140	190	147	16353	17121	16505	402.3	16795	343.39	17227	37.17	17064	173.83	16809
egl-s3-A	140	190	159	10143	10541	10451	432.3	10296	492.15	10625	43.67	10532	227.11	10469
egl-s3-B	140	190	159	13616	14291	13981	445.1	14053	489.17	14363	39.73	14300	219.03	13980
egl-s3-C	140	190	159	17100	17789	17346	465.2	17297	401.95	18018	34.97	17854	277.37	17595
egl-s4-A	140	190	190	12143	13036	12462	543.5	12442	706.89	12941	52.47	12665	286.09	12708
egl-s4-B	140	190	190	16093	16924	16490	744.6	16531	657.55	17088	25.77	16772	276.77	16638
egl-s4-C	140	190	190	20375	21486	20733	728.9	20832	687.32	21469	67.48	21350	307.62	21085
Average					3.63%	1.31%	227.29s	1.39%	242.28s	3.90%	17.82s	3.06%	89.89s	1.99%
Worst					7.35%	3.04%	744.60s	3.21%	706.89s	6.57%	67.48s	5.02%	307.62s	4.65%
Optima					0	2		5		0		1		2

and TSB. The results for *egl* files are given by Table 11. The method reaches one optimum (egl-e1-A) using standard parameters, and achieves another one (egl-e2-A) using different settings. The computing times are still better than those needed by MA (14 times smaller) and TSB (13 times).

5 Conclusions

A GRASP for the very hard CARPTW has been reinforced by using two different randomized heuristics and a PR based on a distance measure in solution space. The best resulting algorithm competes with the best existing heuristic: 17 out of 24 instances are solved to optimality and the gap to lower bounds is less than 1%. Moreover, it is relatively simple and requires only 4 parameters.

This algorithm is still very effective when time windows are removed. It competes with state-of-the-art metaheuristics for the CARP, while being much faster.

The path relinking process considers solutions encoded without trip delimiters and evaluated using a tour splitting procedure. Using any distance for permutations, the exploration of a trajectory consists of generating a sequence of giant tours with a decreasing distance to the giant tour representing the target solution. This technique is very flexible and has the potential to be easily adapted to other capacitated vehicle routing problems, provided the underlying splitting problem stays polynomial.

Like in the VRPTW, the time windows considered concern the service: it is still possible to traverse an edge without servicing it, inside its window. Sometimes, works in a street (e.g. repairing an optic fiber) prevent deadheading traversals too. Our goal is now to tackle such cases, in which the shortest path between two edges is affected by the time windows of traversed edges.

References

1. Beasley, J.E.: Route-first cluster-second methods for vehicle routing. Omega 11, 403–408 (1983)
2. Beullens, P., Muyldermans, L., Cattrysse, D., Van Oudheusden, D.: A guided local search heuristic for the capacitated arc routing problem. European Journal of Operational Research 147, 629–643 (2003)
3. Brandão, J., Eglese, R.: A deterministic tabu search algorithm for the capacitated arc routing problem. Computers & Operations Research 35, 1112–1126 (2008)
4. Cordeau, J.F., Gendreau, M., Hertz, A., Laporte, G., Sormany, J.S.: New heuristics for the vehicle routing problem. In: Langevin, A., Riopel, D. (eds.) Logistic Systems: Design and Optimization, pp. 279–298. Wiley, Chichester (2005)
5. Cormen, T.H., Leiserson, C.E., Rivest, R.L.: Introduction to Algorithms. MIT Press, Cambridge (1990)
6. Feo, T.A., Bard, J.: Flight scheduling and maintenance base planning. Management Science 35, 1415–1432 (1989)
7. Feo, T.A., Resende, M.G.C.: Greedy randomized adaptive search procedures. Journal of Global Optimization 6, 109–133 (1995)
8. Glover, F., Laguna, M.: Tabu search. In: Reeves, C.R. (ed.) Modern Heuristic Techniques for Combinatorial Problems, pp. 70–150. Blackwell, Oxford (1993)

9. Golden, B.L., DeArmon, J.S., Baker, E.K.: Computational experiments with algorithms for a class of routing problems. Computers & Operations Research 10, 47–59 (1983)
10. Guéguen, C.: Exact solution methods for vehicle routing problems. PhD thesis, Central School of Paris, France (1999) (in French)
11. Hertz, A., Laporte, G., Mittaz, M.: A tabu search heuristic for the capacitated arc routing problem. Operations Research 48, 129–135 (2000)
12. Hertz, A., Mittaz, M.: A variable neighborhood descent algorithm for the undirected capacitated arc routing problem. Transportation Science 35, 425–434 (2001)
13. Ho, S.C., Gendreau, M.: Path relinking for the vehicle routing problem. Journal of Heuristics 12, 55–72 (2006)
14. Kindervater, G.A.P., Savelsbergh, M.W.P.: Vehicle routing: Handling edge exchanges. In: Aarts, E.H.L., Lenstra, J.K. (eds.) Local Search in Combinatorial Optimization, pp. 337–360. Wiley, Chichester (1997)
15. Lacomme, P., Prins, C., Ramdane-Chérif, W.: Competitive memetic algorithms for arc routing problems. Annals of Operations Research 131, 159–185 (2004)
16. Martí, R., Laguna, M., Campos, V.: Scatter search vs. genetic algorithms: An experimental evaluation with permutation problems. In: Rego, C., Alidaee, B. (eds.) Metaheuristic Optimization via Memory and Evolution: Tabu Search and Scatter Search, pp. 263–283. Kluwer, Boston (2005)
17. Mester, D., Bräysy, O.: Active guided evolution strategies for large-scale vehicle routing problems with time windows. Computers & Operations Research 32, 1593–1614 (2005)
18. Mullaseril, P.A.: Capacitated rural postman problem with time windows and split delivery. PhD thesis, MIS Department, University of Arizona (1996)
19. Pacheco, J., Martí, R.: Tabu search for a multi-objective routing problem. Journal of the Operational Research Society 57, 29–37 (2006)
20. Pearn, W.L.: Augment-insert algorithms for the capacitated arc routing problem. Computers & Operations Research 18, 189–198 (1991)
21. Resende, M.G.C., Ribeiro, C.C.: GRASP with path-relinking: recent advances and applications. In: Ibaraki, T., Nonobe, K., Yagiura, M. (eds.) Metaheuristics: Progress as Real Problem Solvers, pp. 29–63. Springer, Berlin (2005)
22. Tagmouti, M., Gendreau, M., Potvin, J.-Y.: Arc routing problems with time-dependent service costs. European Journal of Operational Research 181, 30–39 (2007)
23. Wøhlk, S.: Contributions to Arc Routing. PhD thesis, Faculty of Social Sciences, University of Southern Denmark (2005)

A Scatter Search Algorithm for the Split Delivery Vehicle Routing Problem*

Vicente Campos, Angel Corberán, and Enrique Mota

Dept. Estadística i Investigació Operativa, Universitat de Valéncia, Dr. Moliner 50, 46100 Valéncia, Spain
vicente.campos@uv.es, angel.corberan@uv.es, enrique.mota@uv.es

Summary. In this chapter we present a metaheuristic procedure constructed for the special case of the Vehicle Routing Problem in which the demands of clients can be split, i.e., any client can be serviced by more than one vehicle. The proposed algorithm, based on the scatter search methodology, produces a feasible solution using the minimum number of vehicles. The quality of the obtained results is comparable to the best results known up to date on a set of instances previously published in the literature.

Keywords: Scatter search, Vehicle routing, Split delivery.

1 Introduction

In this chapter we consider a variant of the Vehicle Routing Problem (VRP) in which the demand of any client can be serviced by more than one vehicle, the Split Delivery Vehicle Routing Problem (SDVRP). This relaxation of the classical VRP was first proposed by Dror and Trudeau ([9] and [10]), who showed that important savings on the total solution cost could be obtained as well as a reduction in the total number of vehicles used in the solution by allowing clients to be serviced by more than one vehicle. They also showed that this problem is also NP-hard. The SDVRP has received great attention in the last years.

Mullaseril, Dror and Leung [18] studied the problem of distributing feed to cattle at a large livestock ranch in Arizona. They modeled this problem as an arc routing one (in fact, as a Capacitated Rural Postman Problem with split deliveries and time windows). The about 100.000 head of cattle are fed each day, within a specified time window, by six trucks that deliver feed to the large pens connected by a road network. Sometimes, the last pen on a route does not receive its full load and another truck, servicing a different route, has to visit it again in order to complete the load. The computational experiments showed that allowing split deliveries produced a significant reduction in the total distance traveled by the vehicles in most of the considered situations.

* A preliminary version of this work was published in: Cotta, C. and van Hemert, J. (eds.), *Evolutionary Computation in Combinatorial Optimization*, Lecture Notes in Computer Science 4446, pp. 121–129. Springer, Berlin (2007).

A. Fink and F. Rothlauf (Eds.): Advances in Computational Intelligence, SCI 144, pp. 137–152, 2008.
springerlink.com

In 1998, Sierksma and Tijssen [19] presented another application. The problem was to schedule the helicopter flights from an airport near Amsterdam to 51 offshore platforms in the North Sea in order to exchange the employees, which work every other week. A person leaving a platform is exchanged for another person arriving for working at the same platform. The helicopters have a fixed capacity and, because of fuel constraints, a maximum flying distance for each route. The problem was modeled as a Split Delivery Vehicle Routing Problem, in which the total number of exchanges at a given platform could be done by more than one helicopter.

More details about the above applications, as well as some comments on their resolution and results, can be found in the recent paper by Chen, Golden and Wasil [6]. Moreover, that paper also mentions another interesting application by Song, Lee and Kim [20] related to the distribution of newspapers from printing plants to agents in Seoul (South Korea).

Dror, Laporte and Trudeau [8] proposed a branch and bound algorithm based on an integer and linear SDVRP formulation, to which several classes of new valid inequalities were added. The procedure was tested on three small instances up to 20 clients and varying client demands. The SDVRP was studied from a polyhedral point of view in [5]. Based on the partial description of the SDVRP polyhedron, the same authors implemented a branch and cut algorithm capable of solving some medium size instances up to 51 clients. The strengthened linear relaxation produces a good lower bound to the optimal solution value.

In their original work, Dror and Trudeau ([9], [10]) propose a two stage procedure that first obtains a feasible VRP solution and then improves it using specific routines such as the route addition and k-split interchanges. The route addition routine consists of creating a new route to service a client whose demand is split in several routes if the total distance is reduced. A k-split interchange considers a client i with demand d_i and removes i from all the routes that service it. Then, the routine considers all subsets of routes having a "residual" capacity greater than d_i and computes the total insertion cost of client i into all the routes of the subset. Finally, the subset leading to the least insertion cost is chosen and the interchange takes place. These basic but important procedures have also been used in successive heuristic and metaheuristic procedures later on.

In [11], Frizzell and Giffin studied the SDVRP with time windows but on a special network (the clients are located on a grid) and proposed a constructive heuristic followed by some improvement procedures (1-0 exchanges and 1-1 interchanges) that will be described later.

A Tabu Search procedure was developed by Archetti, Hertz and Speranza [1]. It produces an initial feasible solution using the GENIUS algorithm for the TSP ([12]). In the Tabu Search phase moves are made according to two procedures: one orders the routes servicing client i according to the saving obtained by removing i, while the other looks for the "best neighbor" solution of the current one. In a final and improvement phase, GENIUS is applied to each individual route. A variant of this algorithm (SPLITABU) consists of applying 2-opt and node interchange procedures each time the best solution encountered so far is improved. This variant produces better results and is denoted as SPLITABU-DT.

Archetti, Savelsbergh and Speranza [4] use the above Tabu Search procedure to identify parts of the solution space that are likely to contain high quality solutions. Once a set of promising routes is selected, an integer program (a route-based formulation for the SDVRP) is run trying to obtain improved feasible solutions.

In a recent paper, Chen, Golden and Wasil [6] propose a procedure that first constructs a feasible VRP solution using the Clarke and Wright [7] savings algorithm. The solution of an integer program provides then the optimal reallocation of the endpoints of a route in order to maximize the total savings. The program is run for a maximum time of T seconds, with a given neighbor list for each endpoint that is a function of the number of endpoints, and the best feasible solution found is saved. Then, using the feasible solution as the initial one, a new program is run, with a larger size for the neighbor list and a smaller limit for the running time. To the final and best solution obtained, a variable length record-to-record travel algorithm [16] that considers 1-0 exchanges, 1-1 interchanges and 2-opt moves, is applied.

Before closing this introduction, we should mention some structural properties of the problem. In [9] Dror and Trudeau showed that "if the cost matrix satisfies the triangle inequality, then there exists an optimal solution to the SDVRP where no two routes have more than one client with a split demand in common". Archetti, Savelsbergh and Speranza ([2]) define by n_i the number of vehicles servicing client i and by $n_i - 1$ the number of splits at client i. Then they show that when the cost matrix satisfies the triangle inequality there is an optimal solution to the SDVRP where the total number of splits (the sum of the number of splits of every client) is less than the number of routes.

Moreover, Archetti, Savelsbergh and Speranza addressed in [3] the question: To split or not to split? They showed first that, assuming that all the distances among clients and the depot satisfy the triangle inequality, the ratio between the minimum number of routes required to satisfy the client demands in a VRP solution over the minimum number in a SDVRP solution is always less than or equal to 2. They also proved that this bound is tight. In what refers to the ratio between optimal solution values, the same authors showed in [2] that the same bound applies. So, allowing splitting the demands may produce important savings both in the total number of vehicles used and in the total solution cost, as already pointed out by Dror and Trudeau. Moreover, Archetti, Savelsbergh and Speranza conducted an empirical study of the last mentioned ratio, as a function of client locations and client demands, concluding: Cost reductions seem to be due to the ability to reduce the number of routes, without depending on client locations, and mainly depend on the relation between mean demand and vehicle capacity and on the variance of the demands. They obtained the largest benefits when the mean demand is greater than half the vehicle capacity but less than three quarters of the vehicle capacity. Our own computational study also points to this direction.

The chapter is organized as follows: In Sect. 2 the problem is defined and some notation is presented. Section 3 describes the main features of the proposed metaheuristic and in Sect. 4 we present the computational results. Conclusions and future work are summarized in Sect. 5.

2 Problem Definition and Notation

The Vehicle Routing Problem with Split Demands is defined on an undirected and complete graph $G = (V, E)$, where $V = \{0, 1, 2, \ldots, n\}$ is the set of vertices (vertex 0 denotes the depot and $1, \ldots, n$ represent the set of clients). Each edge $e = (i, j)$ has an associated cost or distance c_e between clients i and j. Moreover, each vertex has a known demand d_i ($d_0 = 0$) and there is a fleet of identical vehicles of capacity Q located at the depot. A feasible solution consists of a set of routes, each one beginning and ending at the depot, such that:

- The demand of every client is satisfied, and
- The sum of the demands serviced by any vehicle does not exceed its capacity Q

The SDVRP version defined above is a difficult problem that presents an outstanding characteristic that makes it different from the classical VRP: *there is always a feasible solution using the minimum number of vehicles k*. It is easy to see that this minimum number corresponds to the smallest integer greater than or equal to $\sum_i d_i/Q$. This is not always true if the demand of a client can not be split, since in this case the minimum number of vehicles corresponds to the optimal solution of a Bin Packing Problem.

To the usual and explicit objective of minimizing the total solution cost, we add the implicit one of minimizing the number of vehicles used in the solution. We consider that this is a very important objective, since in most of the practical applications involving several vehicles there is a fixed cost associated to each used vehicle. Moreover, the total fixed cost of a fleet is usually greater than the total and variable cost of a feasible solution. This variable cost usually depends on the total distance traveled by the fleet. Note that a term in the objective function penalizing the excess of vehicles could be added, or bicriteria techniques could also be taken into account, since it is possible in some instances to decrease the total cost by increasing the number of vehicles. Instead we propose a Scatter Search procedure, following the framework presented in [14] and [15], which generates a population of feasible solutions with the minimum number of vehicles.

3 A Scatter Search Procedure

In this section we describe the main features of a Scatter Search procedure designed for the SDVRP. This is, as far as we know, the first time that such a technique is applied to this routing problem. The overall procedure is outlined in Fig. 1. Scatter Search is a population-based method that has been shown to

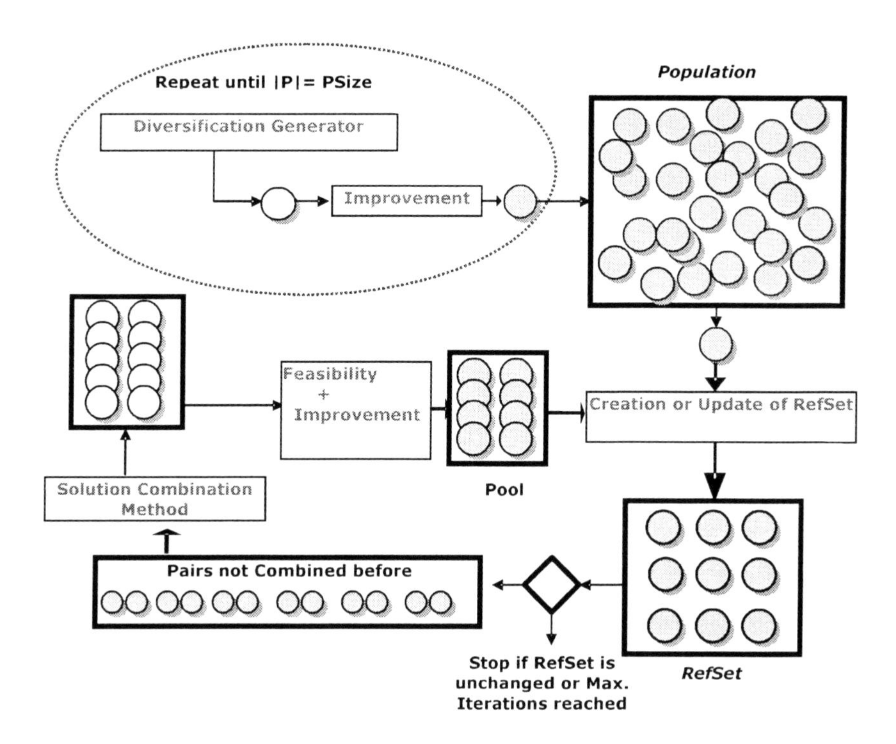

Fig. 1. Scheme of a Scatter Search procedure

yield promising outcomes for solving hard combinatorial optimization problems. The Scatter Search process tries to capture information not contained separately in the original solutions, takes advantage of improving methods, and makes use of strategy instead of randomization to carry out component steps. Our implementation is based on the description given in Glover [14].

The procedure starts with the construction of an initial reference set of solutions (*RefSet*) from a large set P of diverse solutions. The solutions in *RefSet* are ordered according to quality, where the best solution is the first one in the list. The search is then initiated by combining solutions in *RefSet*. Although the method can use combinations of any number of solutions, we have restricted our attention to combinations of two solutions. *RefSet* is updated with the newly created solutions. If the new solution is good enough, it can immediately replace the worst solution in *RefSet* (dynamic strategy) or be stored in a secondary set that will be used to update *RefSet* when all possible combinations of solutions have been performed (static strategy). In the last case, the new *RefSet* will contain the best solutions among those in the old *RefSet* and in the secondary set. The dynamic strategy changes *RefSet* very quickly and usually produces good solutions in short times, while the static strategy can be slower but usually produces better solutions. In our implementation we have used the dynamic strategy. The procedure stops when, after combining all the solutions in *RefSet*, it remains unchanged or a maximum number of iterations is reached. The main characteristics and particularities introduced by the authors are briefly described in the next subsections.

3.1 Creating a Population

We have adapted two standard VRP heuristic procedures to the split demands case in order to obtain SDVRP feasible solutions. The first one, called here Big Tour, uses the Lin and Kernighan [17] heuristic to build a giant tour through the n clients and the depot. From this tour it is always possible to obtain k routes and, thus, a feasible SDVRP solution: Let us first renumber the clients so that the TSP tour is $0-1-2-\ldots-(n-1)-n-0$. The first vehicle leaves the depot using edge $(0,1)$ and services, successively, clients $1,2,\ldots$ up to client i_1 for which the total demand serviced by this first route is at least equal to the vehicle capacity. Client i_1 is either completely serviced or its demand is split between routes 1 and 2. In the first case, edges $(i_1,0)$ and $(0,i_1+1)$ are added, corresponding to the last edge in route 1 and the first one in route 2. In the second case edge $(i_1,0)$ is added twice and route 2 continues using edge (i_1,i_1+1). The feasible solution finally obtained satisfies the two structural properties mentioned in the introduction, i.e., the total number of splits is less than the number of routes and any two routes have, at most, one client in common. We can take into account the difference between the total capacity kQ of the vehicles fleet and the total demand of the clients and adjust the load in each vehicle to an average load so that the solution finally obtained uses k balanced (in terms of load) routes. In this way, we avoid obtaining a solution having $k-1$ routes with load Q and a last route with possibly a very small load.

The same Big Tour is used to generate additional solutions, all of them following the same sequence of clients but starting each one at a different client, i.e., the second feasible solution (for instance) could use as the first edge in the first route edge (0,2) and then edge (2,3) and so on. The last edge in route k would then be edge (1,0). However, in order to obtain solutions that differ substantially, the starting clients are selected in a nonconsecutive order.

The second procedure is a modified and accelerated version of the classical Clarke and Wright parallel savings algorithm [7]. According to this procedure, from an initial solution consisting of n return trips to each client, the best available saving, computed as $s_{ij}=c_{0i}+c_{0j}-\lambda c_{ij}$, is used to merge the single routes $(0,i,0)$ and $(0,j,0)$ into a new route $(0,i,j,0)$ and the procedure is repeated until no merge is feasible, in terms of vehicle capacity, or there are no more available savings. For each client, its neighborhood is computed as the subset of its closest clients, and only these savings are calculated. We allow splitting the demand of a client l only when the best available saving corresponds to merging a given route r with a return trip from client l and the total demand exceeds the vehicle capacity Q; in this case, part of the demand of client l is serviced in route r and we maintain a return trip from client l with the unsatisfied demand. In order to limit the complexity of the procedure, any client is serviced by, at most, two vehicles. This procedure does not guarantee a feasible solution using the minimum number of vehicles (k) but in all our computational experiences, feasible solutions using k vehicles are obtained. In order to generate more than one solution, we prohibit half of the savings used in a solution when computing the next one. Savings are prohibited with probabilities directly proportional to the frequency of use of each saving in the previously generated solutions.

Finally, we have also implemented a sequential version of the Clarke and Wright savings algorithm [7] in which routes are generated one after the other. Thus, the procedure has always just one *active route* which grows by adding new clients either to the first serviced client or to the last one (the two "endpoints" of the route). In this case, only the demand of the last client added to the route can be split, but the number of vehicles that may service this client is not limited. Savings are again computed, for a pair of clients i and j as $s_{ij} = c_{0i} + c_{0j} - \lambda c_{ij}$ and, in both versions, different values of parameter λ are considered in order to diversify the solutions. We have chosen $\lambda = 1, 0.6$ and 0.4.

3.2 Improving a Feasible Solution

A local search phase is applied to each solution in the original population in order to reduce its cost, if possible. We have implemented procedures for client moves, such as the 1-0 exchanges, tried first and consisting of shifting one client from one route to another route, and 1-1 interchanges, consisting of interchanging one client from a route with another client in another route. These moves are applied to every non split client. We have also implemented 2-split changes, that take a client out from every route visiting it and look for a pair of routes that, jointly, could service its demand (2-split changes are a particular case of the k-split changes, first introduced by Dror and Trudeau in [9]). Finally, 2-2 interchanges are checked. With them, we try to interchange one edge of a route with another from another route. When such improvements are no longer possible, the routes in the solution are re-optimized using a 2-opt procedure or the more complex Lin and Kernighan algorithm. The same procedures are applied to a feasible solution entering the reference set, as described in the next subsection.

3.3 The Reference Set

The *Psize* feasible solutions in the population are ordered according to the cost and b of them are selected to be in the reference set *RefSet*. One half corresponds to the best feasible solutions and the remaining solutions add the necessary diversity to this set, since they correspond to those solutions in the population that are the most different when compared to the best ones. As a measure of the difference between two solutions we compute the total number of edges in one solution but not in the other. Each pair of solutions in the reference set is combined to produce another solution that enters this set only when its cost, after applying the improvement procedures, is lower than the cost of the worst solution, which is then eliminated. Therefore, the cost is the criterion used in the updating of *RefSet*. The overall procedure stops when, after every possible combination (one iteration), no new feasible solution enters in the reference set or a maximum number of iterations, previously fixed, is reached.

In our computational experiments, we have tested different population sizes combined with the size of the Reference Set. As expected, the quality of the solutions improves as $Psize$ and b increase, although at cost of greater computing times. Accordingly, we have chosen $Psize = 150$ and $b = 25$ as the final values for the computational experiences and comparisons.

3.4 The Combination Method

We have devised a procedure that captures the essential characteristics of a feasible SDVRP solution and tries to maintain those that could be satisfied by the good solutions. In order to do that, for each solution in the reference set we define a set of *critical* clients, consisting of:

1. all its split clients,
2. all the clients in routes with just 1 or 2 clients,
3. the client whose removal from a route produces the greatest saving cost, for each route with at least 3 clients, and finally
4. every client such that at least one among its three closest neighbors belongs to a different route.

When combining feasible solutions A and B in the reference set we consider, in turn, a *critical client* in A, in classes 1 to 3 above, and we move this client, thus modifying solution A, following the recommendation for this same client in solution B. If it is a split client in B, we consider that there is no recommendation and so we take the next *critical client* in A. If it is not, we consider its two adjacent clients, say α and β, in B and we locate these clients in solution A. If clients α and β are two consecutive clients in the same route in solution A, we understand that solution B recommends to insert the critical client between α and β; otherwise, the critical client is moved to the "best recommended" position, i.e., inserted after client α or inserted before client β. In order to simplify the combination procedure and since the moves already performed may produce a route overload, to move a client to an already unfeasible route is prohibited. Note that combining solutions B and A is also possible and produces a different combination.

When all the *critical clients* of A have been considered, the combination method has produced a new and maybe unfeasible solution because the load in some routes can exceed the vehicle capacity Q. A routine is then applied that considers some moves aimed at obtaining a feasible solution. The idea of this routine is looking for clients in an unfeasible route that could be moved completely to (or partially serviced by) another route in order to obtain feasible routes.

Each time a feasible solution is obtained as a combination of two solutions in the Reference Set, the improve procedures described in Sect. 3.2 are applied. Once all the possible comparisons have been considered, if no new solution enters the reference set, instead of finishing the overall procedure at this moment, we augment the set of *critical clients* of each solution in *RefSet* by including those in case 4 above and we run all the possible combinations once again. The new neighborhood created for each solution replaces the usual rebuilding phase. If a feasible solution enters now the Reference Set, the procedure continues, otherwise we stop it. The procedure considers also the combinations already performed so that repetitions are avoided and only pairs of feasible solutions not yet considered are combined.

4 Computational Experiments

In this section we present the results obtained by the Scatter Search procedure on a wide set of instances taken from the literature. They are compared with the results obtained with the more relevant algorithms designed for solving the SDVRP.

4.1 The Instances

In order to test their algorithm, Dror and Trudeau [9] selected three basic instances with 75, 115 and 150 clients, with randomly generated demands from six different scenarios, expressed as a fraction of the vehicle capacity. The same pattern has been used since then, so most of the published and available test instances are generated from known VRP instances, varying the demands of clients. In two cases, as far as we know, other instances are proposed: in [3] some of the Solomon instances are used and in [6] the authors propose new and geometric instances. However the published computational results do not always allow an easy comparison.

To study the computational behavior of our algorithm, we have generated the same set of instances used by Archetti, Hertz and Speranza in [1]. Starting from the VRP problems 1 to 5, 11 and 12 taken from [13], which have between 50 and 199 clients and satisfy the triangle inequality, the demands of customers have been computed in the following way. First, two parameters α and γ ($\alpha \leq \gamma$) are chosen in the interval $[0, 1]$. Then, the demand d_i of customer i is set equal to $d_i = \lfloor \alpha Q + \delta(\gamma - \alpha)Q \rfloor$, where δ is a random number in $[0, 1]$. This procedure is equivalent to generate the demands of an instance in the interval $(\alpha Q, \gamma Q)$. As in [9] and in the above mentioned paper by Archetti, Hertz and Speranza, we have considered the following combinations of parameters (α, γ): $(0.01, 0.1)$, $(0.1, 0.3)$, $(0.1, 0.5)$, $(0.1, 0.9)$, $(0.3, 0.7)$ and $(0.7, 0.9)$. Considering also the case where the original demands are not changed (represented as O.D. in the tables), a total of 49 instances are obtained. Distances among clients have been computed as follows:

$$c_{ij} = round\left(10000\sqrt{(x_i - x_j)^2 + (y_i - y_j)^2}\right) \quad (1)$$

where c_{ij} represents the cost or distance between clients i and j and *round* is the function which gives the integer number closest to its argument.

4.2 Computational Results

We have first tested the quality of the feasible solutions obtained by the three constructive procedures described in Sect. 3.1 on the whole set of test instances. For each instance and each of the three algorithms, we have obtained 50 solutions which have been improved using the procedures described in Sect. 3.2. Table 1 shows the average values for all the instances in each class defined by a given value of parameters (α, γ) and Fig. 2 summarizes the results obtained. These results indicate that the values of the solutions obtained by the sequential version are always worse than the values obtained by the other two procedures, which are similar.

Table 1. Average values for three constructive algorithms for the SDVRP

Instance class	Big Tour	Sequential CW	Parallel CW
OD	10160398,7	12167770,7	9863394,7
0.01-0.1	8196954,3	9405272,8	8160844,3
0.1-0.3	17922662,7	22271986,9	17572631,2
0.1-0.5	24246567,9	29550626,9	24172589,7
0.1-0.9	37063229,9	41902227,9	37405057,9
0.3-0.7	37575097,1	42143493,7	38055871,1
0.7-0.9	58107028,0	66310339,9	60157794,6

We have decided then to discard the sequential version of the Clarke and Wright savings algorithm and use procedure Big Tour to generate half of the population of feasible solutions, of size *Psize*, and generate the remaining feasible solutions using the parallel version of the Clarke and Wright heuristic.

Computational results on the whole set of instances are shown in Tables 4 and 5, which compare the results obtained by the Scatter Search procedure (SS-SDVRP) with those obtained with two Tabu Search (TS) algorithms presented in [1]. Other characteristics of the feasible solutions are also included in the tables. The first two columns show the instance name, which also indicates the number of clients, and the values of α and γ and thus, the corresponding interval where demands have been generated. Instances in the first seven rows have original demands (denoted by O.D.). Column 3 presents the value (z) obtained by our Scatter Search procedure. A value in bold indicates that this value is at least as good as all the ten values obtained by applying the SPLITABU and the SPLITABU-DT procedures ([1]).

Column 4 indicates the number of vehicles in the feasible solution obtained (k), which corresponds always to the minimum number. Total time in seconds is presented in column 5 (t). The procedure was implemented in C and run on a PC Pentium IV, 1GB RAM, CPU 2.40 GHz. The minimum solution value ($zmin$) among the five executions that each instance is run with SPLITABU and SPLITABU-DT is shown in columns 6 and 9, respectively. Similarly,

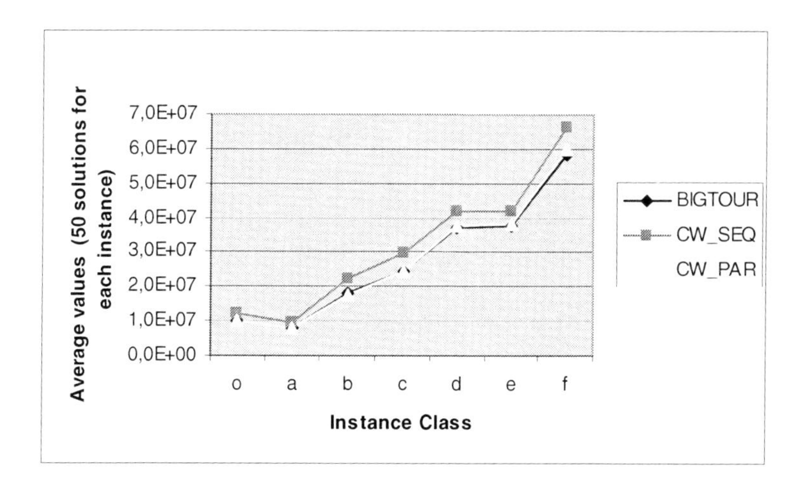

Fig. 2. Quality comparison for the three constructive heuristics

Table 2. Computational results grouped by client demands

	SS-SDVRP			SP-TABU			SP-TABU-DT		
	k	z	t	k	z	t	k	z	t
O.D.	9.7	**916.4**	295.8	9.7	957.1	420.0	9.7	930.4	138.5
0.01-0.1	6.0	**755.9**	517.7	NA	818.1	197.1	NA	786.0	121.1
0.1-0.3	22.5	**1677.5**	208.8	24.7	1759.0	418.7	24.7	1709.2	228.5
0.1-0.5	33.3	**2326.6**	155.2	37.5	2407.4	827.6	37.5	2357.2	607.9
0.1-0.9	55.0	3613.1	112.2	63.5	3674.7	1035.9	63.5	**3548.2**	1087.9
0.3-0.7	56.3	3704.0	72.5	61.0	3830.3	826.4	61.0	**3621.1**	1285.3
0.7-0.9	92.0	5757.6	47.9	94.8	5837.2	4108.3	94.8	**5565**.9	5325.2

columns 7 and 10 present the average solution value for the 5 runs (*zmean*). Columns 8 and 11 give the average times, in seconds of a PC Pentium IV, 256 MB RAM, CPU 2.40 GHz. Finally, column 12 gives the number of vehicles in the feasible solution as presented in a preliminary version of [4].

Table 2 summarizes the results shown in Table 4. It gives the average number of vehicles used (k), the average solution cost (z) and the average computing time in seconds (t) on each subset of 7 instances grouped by a given class of client demands. Best average values are denoted in bold. Figure 3 illustrates the behavior of the three metaheuristics.

Considering the original demands, the quality of the solutions obtained with the SS algorithm is better, although at a greater computational effort. Every solution uses the minimum number of vehicles. Note that the values obtained by the Scatter Search algorithm were produced maintaining all the parameters unchanged for all the instances and with only one execution per instance. The

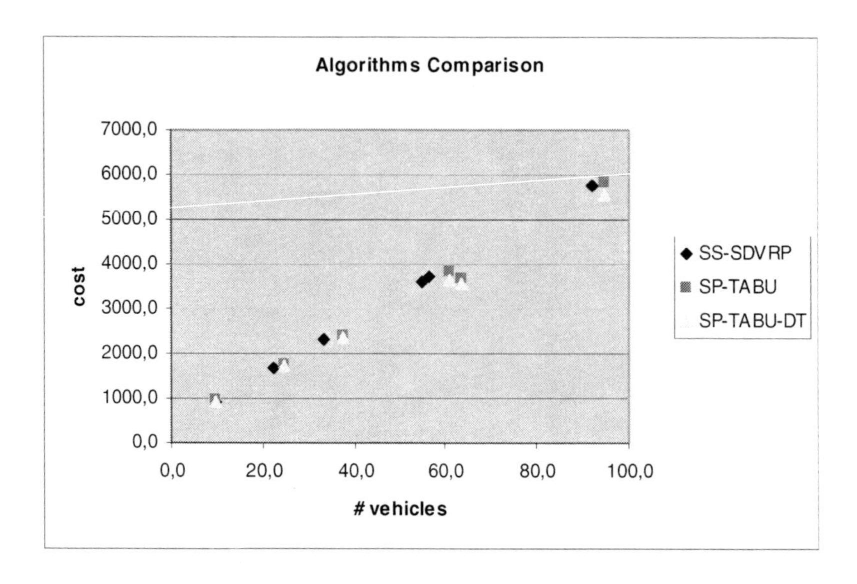

Fig. 3. Number of vehicles used by the SS and TS metaheuristics

number of vehicles used is not available, (NA), for both TS methods on the second group of instances and the solutions obtained by the SS are also better than the best solutions obtained with the TS procedures. When the demand is generated in the interval $(0.1Q, 0.3Q)$ the solution quality is also better in the case of the SS algorithm. The number of vehicles in the solutions obtained with the Tabu Search procedures is no longer the minimum one and the difference reaches 3 vehicles in 2 out of 6 instances. On the instances with demands in $(0.1Q, 0.5Q)$ the SS produces worse solutions than the best ones obtained in the 5 runs of the two TS procedures, although on average the behavior of the SS algorithm is still slightly better (see Table 2) and it uses 4.2 vehicles less on average. On the remaining instances, the SS solutions are worse than the ones obtained with the TS procedures. Clearly our algorithm does not perform well on this kind of instances with big demands. This could be explained by the fact

Table 3. Comparison on the instances with original demands

Number Clients	SS-SDVRP		SP-TABU-DT		Chen et al. procedure	
	z	t	z	t	z	t
50	**524.61**	49.7	530.65	13.0	**524.61**	1.8 (3.4)
75	**829.01**	166.6	845.82	36.0	840.18	4.0 (57.0)
100	**819.56**	192.4	833.35	58.0	**819.56**	3.7 (126.5)
120	**1042.11**	270.3	1053.54	38.0	1043.18	5.6 (136.4)
150	1045.22	527.1	1064.38	389.0	**1041.99**	10.0 (308.0)
199	1324.73	588.3	1339.98	386.0	**1307.40**	18.1 (618.5)

Table 4. Computational results on the whole set of instances (I)

Inst.	Dem.	SS-SDVRP			SP-TABU			SP-TABU-DT			
		z	k	t	zmin	zmean	t	zmin	zmean	t	k
p1-50	O.D.	**5246113**	5	49.7	5276751	5300570	17	5306520	5335535	13	5
p2-75	O.D.	**8290121**	10	166.5	8404529	8516729	64	8458221	8495410	36	10
p3-100	O.D.	**8294477**	8	276.1	8341357	8461844	60	8333566	8356191	58	8
p4-150	O.D.	**10452237**	12	527.1	10570279	10621988	440	10643847	10698369	389	12
p5-199	O.D.	**13247325**	16	588.3	13573455	13678177	1900	13399777	13428515	386	16
p6-120	O.D.	**10421158**	7	270.3	10763753	10847331	40	10535425	10560148	38	7
p7-100	O.D.	**8195581**	10	192.4	**8195581**	8226045	86	**8195581**	8253184	49	NA
p1-50	.01-0.1	**4607896**	3	51.8	**4607896**	4638532	9	**4607896**	4637571	5	NA
p2-75	.01-0.1	**5969872**	4	144.0	6041186	6076579	42	6016174	6052376	13	NA
p3-100	.01-0.1	**7268078**	5	272.1	7572459	7727881	59	7374514	7522012	31	NA
p4-150	.01-0.1	**8712624**	8	743.3	8864415	8949843	258	8819995	8909533	173	NA
p5-199	.01-0.1	**10231356**	10	1874.8	10452620	10735985	754	10478739	10562679	526	NA
p6-120	.01-0.1	**9765696**	6	370.9	10641948	10957537	61	10760897	10846959	42	NA
p7-100	.01-0.1	6359953	5	166.5	6407126	6628037	71	6358865	6487359	58	NA
p1-50	0.1-0.3	**7410565**	10	66.4	7507084	7644041	27	7515968	7614021	22	11
p2-75	0.1-0.3	**10718694**	15	143.8	10835052	10990297	78	10879305	10953225	45	16
p3-100	0.1-0.3	**13975030**	20	305.1	14189700	14288683	122	14193762	14248114	96	22
p4-150	0.1-0.3	19372005	29	326.6	19284742	19406720	545	19079235	19182459	393	32
p5-199	0.1-0.3	24331737	38	32.1	24120235	24199773	1224	23780537	23841545	755	41
p6-120	0.1-0.3	**27425985**	23	380.8	28520739	29009898	516	29145714	29187092	143	26
p7-100	0.1-0.3	**14188103**	20	206.3	14531928	14709592	85	14379543	14620077	146	v
p1-50	0.1-0.5	9978334	15	87.1	9940561	10076838	56	9972128	10086663	28	16
p2-75	0.1-0.5	14635982	22	126.8	14407823	14501086	71	14321606	14436243	123	24
p3-100	0.1-0.5	19080227	29	225.2	18788510	18878331	206	18857414	18947210	136	33
p4-150	0.1-0.5	26499703	43	21.3	26168532	26340901	564	26089113	26327126	739	49
p5-199	0.1-0.5	32919600	56	31.2	32765665	32981871	3811	32473125	32844723	2668	63
p6-120	0.1-0.5	**39796717**	34	329.0	41231758	41667801	259	41311259	42061210	268	40
p7-100	0.1-0.5	19953407	29	266.5	19924980	20300366	188	19815453	20299948	293	NA

Table 5. Computational results on the whole set of instances (II)

Inst.	Dem.	SS-SDVRP			SP-TABU			SP-TABU-DT			
		z	**k**	**t**	**zmin**	**zmean**	**t**	**zmin**	**zmean**	**t**	**k**
p1-50	0.1-0.9	15543768	25	92.6	14815710	14939233	34	14438367	14699221	61	26
p2-75	0.1-0.9	21823368	37	119.9	21131636	21212778	311	21072124	21244269	193	41
p3-100	0.1-0.9	28942137	48	177.9	28116117	28266122	412	27467515	27940774	649	56
p4-150	0.1-0.9	40628821	71	50.4	39785729	40062807	1822	38497320	39097249	2278	84
p5-199	0.1-0.9	50745658	93	50.7	50058905	50396524	2598	47374671	48538254	3297	107
p6-120	0.1-0.9	63573283	56	20.6	65072461	65603347	1037	62596720	65839735	878	67
p7-100	0.1-0.9	31663064	48	272.7	31192745	31580865	523	30105041	31015273	260	NA
p1-50	0.3-0.7	15321944	25	92.4	14941462	15093118	52	14870198	14969009	49	26
p2-75	0.3-0.7	22288975	37	11.1	21666219	21759879	184	21497382	21605050	129	39
p3-100	0.3-0.7	29863298	49	17.0	28958054	29147859	454	27642538	28704954	810	53
p4-150	0.3-0.7	41856832	73	23.0	41228032	41467544	1512	39671062	40396994	3008	80
p5-199	0.3-0.7	52650121	96	327.3	53329707	53689192	2279	50014512	51028379	3566	103
p6-120	0.3-0.7	64810943	58	20.5	67207650	68663390	477	64330110	66395522	659	65
p7-100	0.3-0.7	32487607	49	16.0	31631372	32220334	411	28821235	30380225	778	NA
p1-50	0.7-0.9	23124751	40	5.8	21733326	21763923	160	21483778	21652085	106	42
p2-75	0.7-0.9	33878605	60	10.5	32184116	32294627	437	31381780	31806415	869	61
p3-100	0.7-0.9	45761339	80	38.3	43619465	43687723	1891	42788332	43023114	1398	82
p4-150	0.7-0.9	64794550	119	30.5	63345083	63542058	8783	60998678	61963577	10223	123
p5-199	0.7-0.9	83237230	158	215.0	82071543	83439543	11347	76761141	79446339	21849	162
p6-120	0.7-0.9	101583160	95	20.4	103067404	105505745	2033	100726022	103040778	1826	99
p7-100	0.7-0.9	50652558	80	13.8	49334893	49607513	1865	47735921	48677857	1004	NA

that the SS algorithm is designed to find solutions with the minimum number of vehicles while the TS algorithms minimize the total distance traveled. Note that the solutions obtained with the TS methods always use a larger number of vehicles that in some cases increases by up to 14 vehicles, as in instance p5-199 with demands in $(0.1Q, 0.9Q)$.

Finally, comparisons on all the instances with the results presented in [4] and [6] were not possible since the instances and other significant details were not available to the authors. Though in [6] the general characteristics of the instances are maintained as in [1], the instances themselves are different. However a comparison is possible among the SS, Tabu-DT and the Chen et al. algorithm on 6 of the instances with original demands. Results are summarized in Table 3. Columns z and t show the solution value and the time used to obtain it in the case of the SS and the Chen et al. procedure. In the case of the Tabu-DT algorithm, column z shows the value of the best solution obtained in 5 executions of the algorithm, while t shows the average computing time. Times shown for the Chen et al. procedure are obtained in a slower machine (PC Pentium IV, 512 MB RAM, CPU 1.70 GHz). As it can be seen, the best results are obtained by the SS and the Chen et al. procedure, although this last method is faster than ours.

Note however that a time limit is a parameter in the Chen et al. procedure and that greater CPU times were given to the algorithm in order to compare in [6] its results with the Tabu-DT method. Therefore, we have included in brackets the average CPU times used in [6] for solving instances of similar sizes when the client demands are generated in the range *(0.1Q, 0.3Q)*. These average CPU times are the second best times given in [6] for the six scenarios considered.

5 Conclusions and Further Research

The first results obtained with the Scatter Search procedure indicate that it is able to obtain very good feasible solutions *with the minimum number of vehicles* within reasonable computing times. When the demands are well over half the capacity of the vehicle, the values of the solutions are not so good, because our procedure was not initially designed for these situations. The set of published and available instances is limited and quite small. In the future, we want to work on the elaboration of bigger test instances that will be publicly available and include some other refinements to the overall algorithm like another generator of feasible solutions and more procedures to be applied to the unfeasible solutions produced in the combination phase.

Acknowledgments. The authors want to thank the support of the Spanish Ministerio de Educación y Ciencia, through grants MTM2006-14961-C05-02 and TIN2006-02696. We are also grateful to C. Archetti and M.G. Speranza that kindly helped us to generate the instances used in the computational experiences and to an anonymous referee for his/her careful reading of the manuscript.

References

1. Archetti, C., Hertz, A., Speranza, M.G.: A tabu search algorithm for the split delivery vehicle routing problem. Transportation Science 40, 64–73 (2006)
2. Archetti, C., Savelsbergh, M.W.P., Speranza, M.G.: Worst-case analysis for split delivery vehicle routing problems. Transportation Science 40, 226–234 (2006)
3. Archetti, C., Savelsbergh, M.W.P., Speranza, M.G.: To split or not to split: That is the question. Transportation Research E: Logistics and Transportation Review 44, 114–123 (2008)
4. Archetti, C., Savelsbergh, M.W.P., Speranza, M.G.: An optimization-based heuristic for the split delivery vehicle routing problem. Transportation Science (to appear, 2008)
5. Belenguer, J.M., Martínez, M.C., Mota, E.: A lower bound for the split delivery vehicle routing problem. Operations Research 48, 801–810 (2000)
6. Chen, S., Golden, B., Wasil, E.: The split delivery vehicle routing problem: Applications, algorithms, test problems and computational results. Networks 49, 318–329 (2007)
7. Clarke, G., Wright, J.V.: Scheduling of vehicles from a central depot to a number of delivery points. Operations Research 12, 568–581 (1964)
8. Dror, M., Laporte, G., Trudeau, P.: Vehicle routing with split deliveries. Discrete Applied Mathematics 50, 239–254 (1994)
9. Dror, M., Trudeau, P.: Savings by split delivery routing. Transportation Science 23, 141–145 (1989)
10. Dror, M., Trudeau, P.: Split delivery routing. Naval Research Logistics 37, 383–402 (1990)
11. Frizzell, P.W., Giffin, J.W.: The split delivery vehicle routing problem with time windows and grid network distances. Computers and Operations Research 22, 655–667 (1995)
12. Gendreau, M., Hertz, A., Laporte, G.: New insertion and postoptimization procedures for the travelling salesman problem. Operations Research 40, 1086–1094 (1992)
13. Gendreau, M., Hertz, A., Laporte, G.: A tabu search heuristic for the vehicle routing problem. Management Science 40, 1276–1290 (1994)
14. Glover, F.: A template for scatter search and path relinking. In: Hao, J.-K., Lutton, E., Ronald, E., Schoenauer, M., Snyers, D. (eds.) AE 1997. LNCS, vol. 1363, pp. 13–54. Springer, Berlin (1998)
15. Laguna, M., Martí, R.: Scatter Search – Methodology and Implementations in C. Kluwer Academic Publishers, Boston (2003)
16. Li, F., Golden, B., Wasil, E.: Very large-scale vehicle routing: New test problems, algorithms and results. Computers and Operations Research 32, 1197–1212 (2005)
17. Lin, S., Kernighan, B.W.: An effective heuristic algorithm for the travelling salesman problem. Operations Research 21, 498–516 (1973)
18. Mullaseril, P.A., Dror, M., Leung, J.: Split-delivery routing in livestock feed distribution. Journal of the Operational Research Society 48, 107–116 (1997)
19. Sierksma, G., Tijssen, G.A.: Routing helicopters for crew exchanges on off-shore locations. Annals of Operations Research 76, 261–286 (1998)
20. Song, S., Lee, K., Kim, G.: A practical approach to solving a newspaper logistic problem using a digital map. Computers and Industrial Engineering 43, 315–330 (2002)

Stochastic Local Search Procedures for the Probabilistic Two-Day Vehicle Routing Problem

Karl F. Doerner[1], Walter J. Gutjahr[2], Richard F. Hartl[1], and Guglielmo Lulli[3]

[1] Department of Business Administration, University of Vienna, Bruenner Strasse 72, 1210 Vienna, Austria
{karl.doerner,richard.hartl}@univie.ac.at

[2] Department of Statistics and Decision Support Systems, University of Vienna, Universitätsstraße 5, 1010 Vienna, Austria
walter.gutjahr@univie.ac.at

[3] Department of Computer Science, University of Milano Bicocca, Via Bicocca degli Arcimboldi 8, 20126 Milano, Italy
guglielmo.lulli@disco.unimib.it

Summary. This chapter is motivated by the study of a real-world application on blood delivery. The Austrian Red Cross (ARC), a non-profit organization, is in charge of delivering blood to hospitals on their request. To reduce their operating costs through higher flexibility, the ARC is interested in changing the policy of delivering blood products. Therefore it wants to provide two different types of service: an urgent service which delivers the blood within one day and the other, regular service, within two days. Obviously the two services come at different prices.

We formalize this problem as a stochastic problem, with the objective to minimize the average long-run delivery costs, knowing the probabilities governing the requests of service. To solve real instances of our problem in a reasonable time, we propose three heuristic procedures whose core routine is an Ant Colony Optimization (ACO) algorithm, which differ from each other by the rule implemented to select the regular blood orders to serve immediately. We compare the three heuristics on both a set of real-world data and on a set of randomly generated synthetic data.

Computational results show the viability of our approach.

Keywords: Stochastic local search, Vehicle routing, Blood delivery.

1 Introduction

The Austrian Red Cross (ARC), a non-profit organization, is in charge of delivering blood to hospitals on their request any time that they put an order. In current operations, the ARC is obliged to fulfill any order within the following day. This policy leads to high delivery costs. Quite often, the ARC has to pay extra working hours to their drivers in order to fulfill all the orders. Even solving a Vehicle Routing Problem (VRP) every day of operations will not ameliorate the current situation. As an alternative, the ARC is interested in changing policy in order to acquire higher flexibility. More specifically the ARC is investigating

A. Fink and F. Rothlauf (Eds.): Advances in Computational Intelligence, SCI 144, pp. 153–168, 2008.
springerlink.com

the possibility to provide two different types of service: one which delivers the blood within one day and the other within two days. Obviously, these services must have different prices.

If this policy is implemented, the ARC will be confronted every day with two different types of blood orders: *urgent* blood orders to be delivered within the current day, and *regular* blood orders which allow the delivering within the next day. The ARC has both to decide when to serve hospitals with regular blood orders and solve a VRP problem to find the optimal sequence for serving the selected hospitals (including of course all urgent services). Decisions on each single day have a downstream effect on decisions and their corresponding costs for the following days. Hence, these decisions should depend on the foreseen blood orders for the following days and are taken with the goal to minimize the long-run total expected delivery costs. The blood is delivered from one central depot.

The main feature of the problem, together with the two different types of blood orders and the decisions involved, is blood orders uncertainty represented, in our modeling approach, with random variables. This leads to a stochastic problem which differs from its deterministic counterpart in several fundamental respects. We capture it in the formal framework of a stochastic two-day delivery VRP (see the Appendix for a mathematical description). The problem was introduced in [7].

Network routing problems have been deeply investigated both in a deterministic [15] and in a stochastic [9, 11] setting. Nevertheless, at the best of our knowledge, our specific problem is quite new in the routing problem arena. The two-period symmetric TSP studied by Butler et al. [5] is a work which resembles our problem in some respects, but it remains within a deterministic framework. Under different model assumptions, several approaches and methodologies have been developed to handle stochastic routing problems, which are, however, not easily transferable to our situation.

Recently Angelleli et al. [2, 3] analyzed the competitive ratio for different dispatch strategies for a very similar problem. For instance, the problem under consideration could be viewed as a Markov decision process (MDP) (see [12]), but the corresponding solution techniques fail for real instances of our problem since the number of states of the MDP explodes with a growing number of hospitals. An alternative would consist in the application of two-stage stochastic programs where a planned or an "a priori" solution is determined at the first stage and then at the second stage, once the realizations of the random variables are disclosed, a recourse or corrective action is applied to the first stage decision. Also this consideration, however, would lead to exploding computation times: Note that both in the MDP model and in the two-stage stochastic program, the evaluation of the solution cost or of the recourse action, which require the solution of a VRP, are themselves NP-hard problems. Andreatta and Lulli considered a TSP variant of the problem as a Markov decision process with an aggregate model [1].

To solve real instances of our problem in a reasonable time, we propose three heuristic procedures which differ from each other by the rule implemented to select the regular blood orders to serve immediately. The first is an advanced heuristic (Sampling-Based Local Search) which incorporates Iterative Local Search and Ant Colony Optimization (ACO) as well as bootstrap and sampling procedures. The other two algorithms (Fixed Percentage Selection Procedure and Estimated Savings Selection Procedure) implement a simple postponement procedure, and a selection algorithm based on preprocessing results, respectively.

The core routine for the three algorithms is the Savings-Based Ant Colony Optimization (SBACO) algorithm used to solve the VRPs. SBACO builds on the Ant Colony Optimization (ACO) metaheuristic and combines it with the well-known Savings algorithm [6] for VRPs. The success principles of ACO (which is, as other metaheuristics, derived from analogies with processes in nature) consist in an intelligent exploitation of the problem structure and in an effective interplay between intensification (concentrating the search to a specific region of the search space) and diversification (elaborating various diverse regions in the solution space). These features are extended in SBACO by suitable memory structures and an efficient use of the "savings" values as a valuable guide for tour determination.

2 Solution Procedures

In the description of the solution procedures, we use the following notation:

$H_0^{(t)}$ = the set of urgent orders, placed on day t (to be served during day t),

$H_1^{(t)}$ = the set of regular orders, placed on day t (to be served within two days).

For each day t, the task is to select the set $B^{(t)}$, subset of $H_1^{(t)}$, containing the regular orders to be served during the current day t. $c(A)$ denotes the delivery cost for serving the orders in the set A.

2.1 Sampling-Based Local Search

Sampling-Based Local Search (SBLS) consists of three nested iterations (Loop 1, Loop 2 and Loop 3).

1. An execution of Loop 1 evaluates a possible solution, that is, a subset $B^{(t)}$ of customers to be serviced already within 24 hours, out of the set $H_1^{(t)}$ of customers who expect their service within 48 hours. The VRP for the set consisting of these selected customers plus the customers who *must* be serviced on the current day t (including those "inherited" from the previous day $t-1$) is solved. Then, the procedure goes into Loop 2, where the costs for the customers postponed to the next day are estimated by sampling. Finally, the set $B^{(t)}$ is updated by a local search move.
2. In Loop 2, we take a sample of the possible scenarios for the demand on day $t+1$ and apply local search to find a set of customers $B^{(t+1)} \subseteq H_1^{(t+1)}$ to update the set $B^{(t+1)}$, the procedure goes into Loop 3.

3. In Loop 3, the cost occurring on day $t+1$ is determined by solving the corresponding VRP by SBACO. To this delivery cost, a penalty term for the customers postponed to day $t+2$ is added (otherwise, the local search procedure would always serve only the minimum possible number of customers on day $t+1$). This penalty term is computed to estimate the *additional* costs on day $t+2$ caused by the postponed customers. The simplest way to do this is to multiply the number of postponed customers by an estimate of the average additional costs incurred by such a customer depending on the distance to the depot. This estimate can be based on previous days for which costs are already known (for details, see below). We also use a second estimate based on distances and take a weighted mean.

Let us now outline how, for the penalty term computation, a sequence of improving estimates c_t $(t = 0, 1, \ldots)$ of the additional costs incurred by a postponed customer is computed. We start with an arbitrary initial value c_0 on day 0 and modify c_t in each iteration to a value c_{t+1} by taking a weighted mean between the current value c_t and the average (additional) cost on day t for the service of a customer postponed from day $t-1$ to day t.

When the process approaches its steady state, it can be expected that c_t approaches the true value of the cost parameter under consideration.

In the second estimation approach, we consider the sum of distances from the depot to the postponed customers instead of the number of these customers. Now, we estimate the average (additional) cost *per distance unit* for a postponed customer. Similarly as before, the estimate $\bar{c}_t$ for this value is updated by taking a weighted mean between the current value and the average (additional) cost per distance unit on day t for the service of a customer postponed from day $t-1$ to day t.

The pseudocode of SBLS is given in the Appendix.

2.2 Fixed Percentage Selection Procedure

The Fixed Percentage Selection Procedure (FPSP) uses a very simple rule to select the regular blood orders to be served on the same day. For the selection decision, the percentage p of regular customers to be served immediately is fixed in advance. To select the orders, we use the following rule: For each customer i with a regular order, we solve the VRP for the set of all urgent or regular orders (including those "inherited" from the last day) *except* the order of customer i. The lower the resulting cost C_i^-, the higher is the immediate cost reduction on the current day achieved by postponing customer i to the next day. Therefore, the p percent of customers with the lowest values of C_i^- are postponed to the next day.

After some pre-tests, we chose $p = 50$ percent for our real-world application. Both values between 30 and 40 percent and values between 60 and 70 percent yielded distinctly worse results in our application case. It is clear that the optimal value for p heavily depends on the special structure of order frequencies and their oscillations.

Once the selection decision is taken, we compute the new VRP solution by the SBACO algorithm. Note that in this procedure we do not use any probabilistic information on blood orders for the following day. The pseudocode of FPSP is given in the Appendix.

2.3 Estimated Savings Selection Procedure

In the Estimated Savings Selection Procedure (ESSP), we postpone a regular blood order to be served on the next day if the extra delivery cost for serving this customer in addition to those with urgent orders and those "inherited" from the previous day exceeds the estimated additional cost for serving this customer on the next day. This is obviously a simple approximation first because each single regular customer is considered independently from the others and dependencies between regular orders are not considered and second because no downstream effects of current decisions in the following stages of the problem are taken into account.

For the described decision rule, estimates of the additional costs incurred by each postponed customer are required. In a preprocessing phase, we compute these estimates for each day d of week (Monday to Sunday) separately, taking varying frequencies of demands during the week into account. Estimates are obtained by sampling, based on historical data on request frequencies. As a sample size for the demand structure on a specific day of week, we took $N = 100$ samples in our computational experiments. The pseudocode of ESSB is given in the Appendix.

2.4 Savings-Based ACO Algorithms for the VRP

The SBACO procedure is the core subroutine of all three described heuristics. SBACO heuristically solves a given VRP applying the ACO metaheuristic.

The Ant Colony Optimization (ACO) metaheuristic introduced by Colorni, Dorigo and Maniezzo [8] imitates the behavior shown by real ants when searching for food. Ants exchange information about food sources via pheromone, a chemical substance which they deposit on the ground as they move along. Short paths from the nest to a food source obtain a larger amount of pheromone per time unit than longer paths. As a result, an ever increasing number of ants is attracted to follow such routes, which in turn reinforces the corresponding pheromone trails. Artificial ants not only imitate the learning behavior described above, but also apply additional, problem specific heuristic information. ACO has been successfully applied to various hard combinatorial optimization problems [8].

The Savings-Based ACO Algorithm [13, 14] mainly consists of the iteration of three steps: (1) generation of solutions by ants according to pheromone information; (2) application of a local search to the ants' solutions, and (3) update of the pheromone information.

Solutions are constructed based on the well known Savings Algorithm due to Clarke and Wright [6]. In this algorithm, the initial solution consists of the assignment of each customer to a separate tour. After that, for each pair of customers i and j, the following savings values are calculated:

$$s_{ij} = d_{i0} + d_{0j} - d_{ij}, \tag{1}$$

where d_{ij} denotes the distance between locations i and j, the index 0 denotes the depot, and s_{ij} represent the savings of combining two customers i and j on one tour contrary to serving them on two different tours. In the iterative phase, customers or partial tours are combined by sequentially choosing feasible entries from the list of saving values. A combination is infeasible if it violates either the capacity or the tour length constraints. The decision making about combining customers is based on a probabilistic rule taking into account both savings values and the pheromone information. Let τ_{ij} denote the pheromone concentration on the arc connecting customers i and j telling us how good the combination of these two customers i and j was in previous iterations. In each decision step of an ant, we consider the k best combinations still available, where k is a parameter of the algorithm. Let Ω_k denote the set of k "neighbors", that is, the k feasible combinations (i,j) yielding the largest savings, considered in a given decision step, then the decision rule is given by equation (2).

$$\mathcal{P}_{ij} = \begin{cases} \frac{s_{ij}^{\beta}\tau_{ij}^{\alpha}}{\sum_{(h,l)\in\Omega_k} s_{hl}^{\beta}\tau_{hl}^{\alpha}} & \text{if } (i,j) \in \Omega_k \\ 0 & \text{otherwise.} \end{cases} \tag{2}$$

In (2), $\mathcal{P}_{ij}$ is the probability of choosing to combine customers i and j on one tour, while α and β bias the relative influence of the pheromone trails and the savings values, respectively. This algorithm results in a (sub-)optimal set of tours through all customers, once no more feasible savings values are available.

The used pheromone update rule was proposed in [4] and its pheromone management centers around two concepts borrowed from Genetic Algorithms, namely ranking and elitism to deal with the trade-off between exploration and exploitation. This paradigm was used for solving the VRP. We will briefly depict the pheromone update scheme here. Let $0 \le \sigma \le 1$ denote the trail persistence and e the number of elitists. Then, the pheromone update scheme can formally be written as

$$\tau_{ij} := \sigma\tau_{ij} + \sum_{r=1}^{e-1} \Delta\tau_{ij}^{r} + e\Delta\tau_{ij}^{*} \tag{3}$$

First, the best solution found by the ants up to the current iteration is updated as if e ants had traversed it. The amount of pheromone laid by the elitists is $\Delta\tau_{ij}^{*} = 1/L^{*}$ if (i,j) belongs to the best solution so far, and $\Delta\tau_{ij}^{*} = 0$ otherwise, where L^{*} is the objective value of the best solution found so far. Secondly, the $e-1$ best ants of the current iteration are allowed to lay pheromone on the arcs they traversed. The quantity laid by these ants depends on their rank r as well as their solution quality L^{r}: the r-th best ant lays $\Delta\tau_{ij}^{r}$ on each arc (i,j), where $\Delta\tau_{ij}^{r} = (e-r)/L^{r}$ if arc (i,j) is traversed by this ant, and $\Delta\tau_{ij} = 0$ otherwise.

Arcs belonging to neither of those solutions just face a pheromone decay at the rate $(1 - \rho)$, which constitutes the trail evaporation.

A solution obtained by the above mentioned procedure can then be subjected to a local search in order to ensure local optimality. In our algorithms we sequentially apply the *swap* neighborhood between tours to improve the clustering, and the 2-opt algorithm within tours to improve the routing.

3 Procedures Comparison and Simulation

The developed solution procedures use information on the probability distribution of blood orders for each hospital on any given day. To estimate such probability distributions of the demand, we used one-year historical real data provided by the ARC. We analyzed these data with the statistical package SPSS. Moreover, we had to specify (estimate) the (hospital-specific) ratio between urgent and regular orders. We predicted the ratio for each hospital and each day of week by fitting a reasonable model of customer behavior to the real data, since the policy allowing two types of orders is not yet implemented and therefore no data is available.

Ratio prediction was done by clustering hospitals into groups according to their size, by carefully analyzing their past order record, and by making suitable assumptions on inventory strategies: it is assumed that hospitals will send an order as soon as their blood store falls below a certain lower bound (urgent order) or will presumably fall below this bound the next day (regular order), and they will fill the store to a certain upper bound. We verified the consistency of estimates obtained in this way by using data provided directly by hospitals via questionnaires. We do not go further into details for the sake of brevity.

We compare the three algorithms by computing their performance over a time horizon of a half year (180 days). For the comparison, two data sets were used:

1. A set of real data with the orders of 55 hospitals (see Fig. 1) within the indicated 180-days period. Only the ratio between urgent and regular orders is estimated by the inventory model mentioned above.
2. A randomly generated synthetic data set obtained directly by simulation from the inventory model. The inventory model makes certain assumptions on the customer behavior that also influence the total number of orders (urgent plus regular orders) on each day. This leads to a deviation from the ordering behavior observed in practice. In particular, the empirical data show that hospitals are rather reluctant to send orders on weekends; they seem to do it in urgent cases, but otherwise they prefer to delay an order to the beginning of the next week. As a consequence, our synthetic data set contains a distinctly higher number of orders on weekends than the real data set, and most of them are regular orders which would presumably have been delayed in the real customer behavior.

   We generated ten different instances (based on the same basic parameters) for the tests.

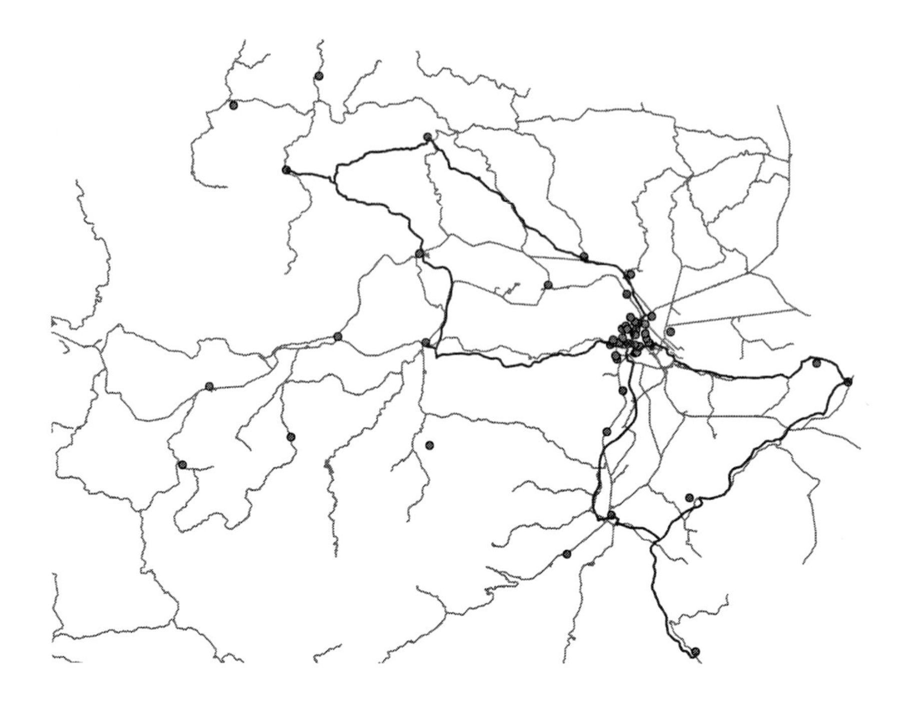

Fig. 1. Distribution of the hospitals

Table 1. Comparison of the results (in hours)

method	runtime	avg. sol. quality real-world data	avg. sol. qual. generated data	% of postp. customers
no postp.	–	7374.8	7792.1	0
SBLS	41 sec/day	7188.3	7537.5	54
FPSP	2.9 sec/day	7185.2	7538.8	50
ESSP	12 hours preproc. + 2.5 sec/day	7061.6	7490.3	44

The three algorithms were implemented in C and run on a Pentium PC with 3.6 GHz. Each algorithm was run ten times in order to take into account the probabilistic nature of all the three heuristic approaches. Note that even the FPSP heuristic which does not use any sampling, has a probabilistic nature since it uses SBACO as a subprocedure for solving VRPs, which is a probabilistic algorithm. In the case of the real data set, all ten runs were applied to the same problem instance. In the case of the synthetic set containing ten particular instances, run 1 of each algorithm was applied to instance 1, etc. In Table 1, we report average values over these ten runs. Delivery costs are indicated in hours.

In the first row of Table 1, we report the delivery costs of solution which corresponds to the current blood delivery policy of the ARC, i.e., the status-quo situation. Rows two through to four contain the results for the algorithms SBLS, FPSP and ESSP, respectively.

The SBLS and FPSP algorithms perform equally good on the real data, achieving driving time reductions by 187 resp. 190 hours in average, which are 2.5 resp. 2.6 percent of total driving time. ESSP even reduces the driving time by 313 hours, which are 4.2 percent of total driving time. In the synthetically generated data set, the reductions amount to 3.3 percent for SBLS and FPSP and to 3.9 percent for ESSP.

It is interesting to see that the best-performing algorithm, ESSP, only postpones 44 percent of orders in average, whereas the other two algorithms postpone a larger number (50 percent – which is the percentage chosen in advance – for FPSP, and even 54 percent for SBLS).

Comparing computation times, both FPSP and ESSP are very fast when applied to the planning problem of a single day, while SBLS needs a computation time that is more than ten times higher. However, for our hospital compound of 55 hospitals, the runtime is still less than one minute per day, which should not cause any practical problems.

ESSP, although very fast in the daily runs, requires a preprocessing run where estimates for the costs of postponed customers are determined. This run took about twelve hours in our experiments. If it is carried out once a year (which is reasonable since order frequencies and other conditions may change), ESSP is the slowest of the three algorithms with about two minutes runtime per day, but even this runtime is easily feasible in practice.

The SBLS algorithm might seem to be dominated by FPSP which produces results of a comparable quality in less computation time, but it should be emphasized that SBLS has its own advantages: it requires less parameters to be tuned than FPSP and ESSP, and it is a *self-adjusting* procedure that reacts to changes in the order behavior of the customers by automatically adapting the values c_t and $\bar{c}_t$. The application of one of the two other algorithms runs the risk that parameter settings can become ineffective if changes in the customer behavior are not responded in time by a suitable re-tuning of the percentage p of postponed customers in the case of FPSP, or by a new run of the preprocessing procedure in the case of the ESSP.

Figures 2 to 5 show the delivery costs for certain days of the week in their development over the 180-days-period. The demand on a working day is usually higher than on a weekend day. Therefore, we have selected Tuesday as a typical working day and Saturday as a typical weekend day for a closer analysis. For both days of week, we depict the behavior of the algorithms both for the synthetic and the real data set. Let us start with looking at Figs. 2 and 3 referring to the synthetically generated data. It is immediately seen that all three algorithmic techniques lead to consistently lower driving times than the baseline policy where no orders are delayed. The gain is particularly impressive on Saturday. For Tuesday, no consistent advantage of one of the algorithms SBLS, FPSP

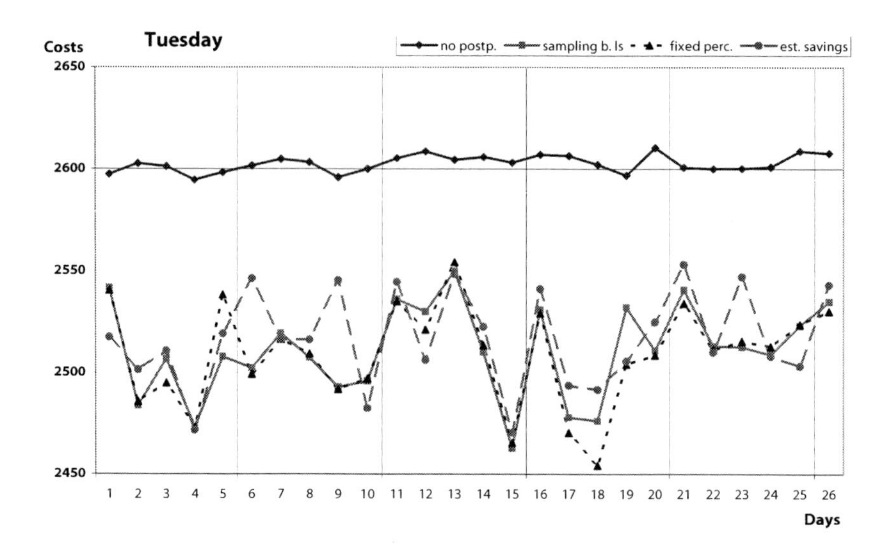

Fig. 2. Delivery costs on Tuesday (in minutes) – synthetic data

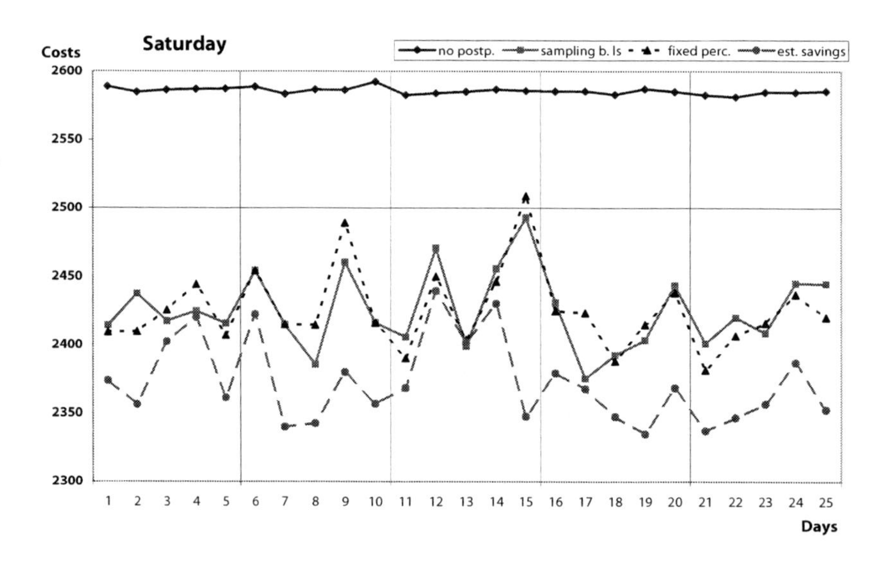

Fig. 3. Delivery costs on Saturday (in minutes) – synthetic data

and ESSP over another can be stated. On Saturday, however, ESSP performs distinctly better than SBLS and FPSP.

Figures 4 and 5 show the corresponding effects for the real data set. Comparing Fig. 4 with Fig. 2 (both for Tuesday) reveals that the curves for SBLS and

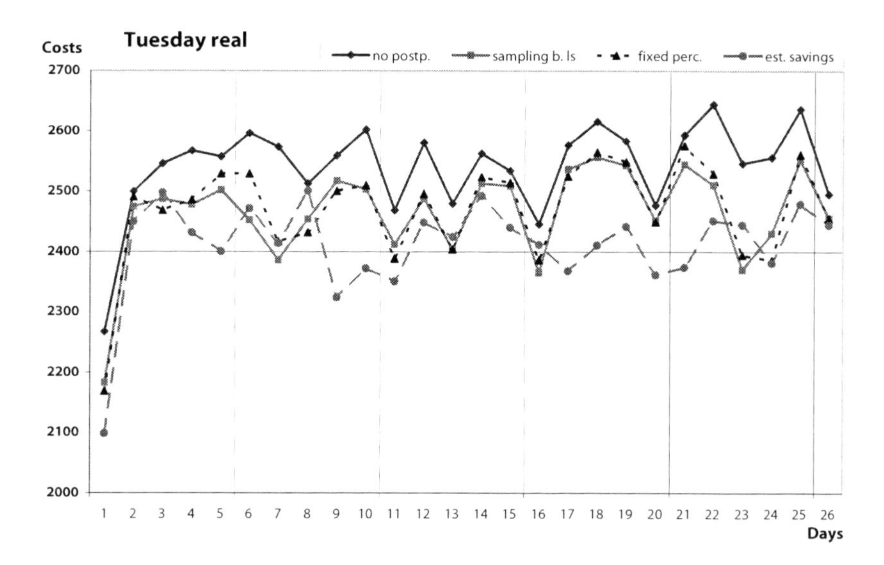

Fig. 4. Delivery costs on Tuesday (in minutes) – real-world data

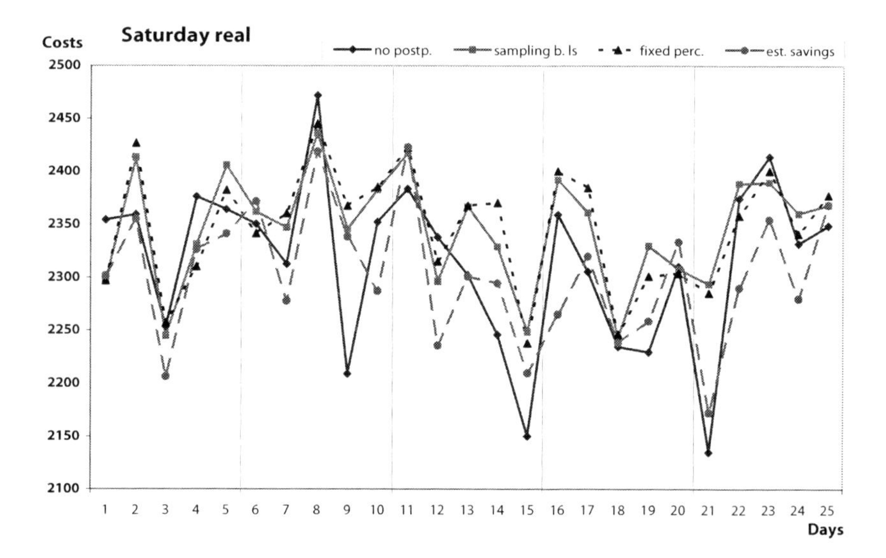

Fig. 5. Delivery costs on Saturday (in minutes) – real-world data

FPSP are now closer to the baseline curve for "no postponement", and a clear advantage of ESSP over the other algorithms can be recognized.

Surprisingly, Fig. 5 for Saturday shows another picture. Here, the baseline policy of "no postponement" is astonishingly good, outperforming in some weeks even the ESSP results. An explanation of this effect can be the following: As

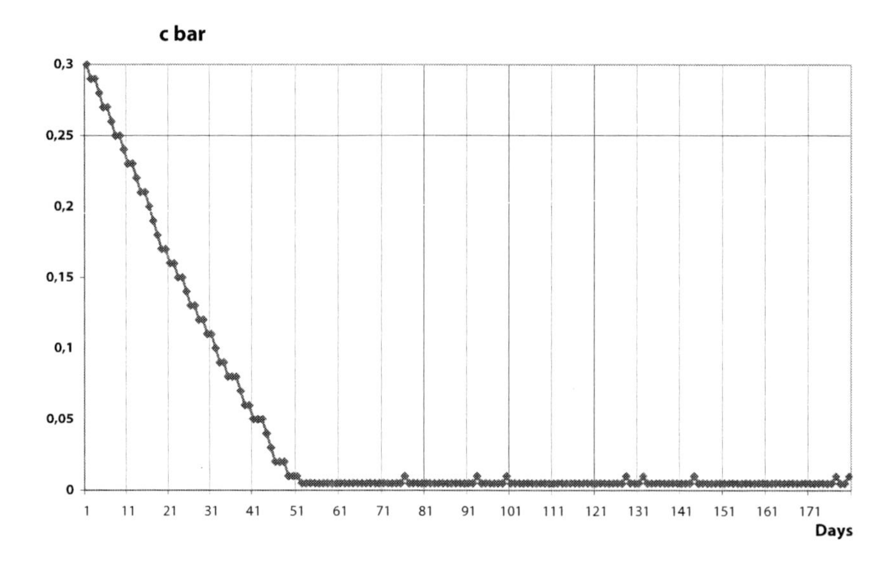

Fig. 6. The development of the parameter $\bar{c}$

mentioned above, the generation mechanism for the synthetic data distinctly overestimates the number of orders on weekend days. A lot of orders, in particular regular orders, are created that would not occur in practice. Dealing with many regular orders favors our three postponement algorithms, since they are able to cope with a mix of regular and urgent orders. On the other hand, they have no advantage over the baseline algorithm if there are only (or nearly only) urgent orders present, since these must be satisfied on the same day anyway. This effect is enhanced by the specific situation on Saturday: in the real-data situation, SBLS, FPSP and ESSP can be expected to have postponed a certain amount of orders from Friday to Saturday, but due to the reduced order frequency on Saturday, a much smaller amount of orders will be further postponed to Saturday. Thus, these algorithms use a good deal of their effort on Saturday for working off the remaining load from Friday. The baseline algorithm, however, has satisfied all Friday orders already on Friday, so the cost registered on Saturday is only due to the relatively low load of Saturday orders.

To give an example for the development of the values c_t or $\bar{c}_t$ in the SBLS solution procedure, we present in Fig. 6 the curve for the $\bar{c}_t$ case. $\bar{c}_t$ expresses an estimate of the additional cost per distance unit for a postponed customer. Since we measure both costs and distances in time units, $\bar{c}_t$ is a dimensionless number, a value x meaning that servicing an additional customer in one hour distance from the depot (by including this customer into a present tour) requires an additional travel time of x hours. As observed from Fig. 6, the initial value for $\bar{c}_t$ has been chosen much too high. The procedure corrects this by successively reducing $\bar{c}_t$, until the value oscillates around 0.008.

4 Conclusions and Directions of Future Research

We have investigated the potential impact of introducing different order types in the blood delivery strategies of the Austrian Red Cross. The reduction in delivery costs were analyzed when offering the customers different prices for different delivery time windows. We have designed and implemented three alternative solutions approaches. The obtained results show the viability of the proposed approach. The algorithms perform very well on the real data, achieving driving time reductions by up to 313 hours for 180 delivery days, which is a reduction in driving time by 4.2 percent of total driving time. The implication of our study is that it is worthwhile for the ARC to offer different order types to the hospitals.

In a further work [10] we investigate the potential value of switching from the current vendee managed inventory set up to a vendor managed inventory system. We develop and evaluate two alternative delivery strategies. The first strategy retains the concept of regions and the use of fixed routes. The second strategy combines more flexible routing decisions with a focus on delivery regularity for each hospital. For an initial assessment of the potential benefits of these delivery strategies, we investigate a simplified setting: we consider a single blood product and we assume known and constant daily demand for each hospital.

Acknowledgment

We want to thank Franz Jelinek from the Austrian Red Cross for supporting this project. Any opinions expressed herein do not necessarily reflect those of the ARC. Furthermore, we are grateful to Ortrun Schandl for collecting the data from the hospitals and analyzing the questionnaire. Financial support from the Oesterreichische Nationalbank under grant #11187 is gratefully acknowledged.

References

1. Andreatta, G., Lulli, G.: A multi-period TSP with stochastic regular and urgent demands. European Journal of Operational Research 185, 122–132 (2008)
2. Angelelli, E., Savelsbergh, M.W.P., Speranza, M.G.: Competitive analysis of a dispatch policy for a dynamic multi-period routing problem. Operations Research Letters 35, 713–721 (2007)
3. Angelelli, E., Savelsbergh, M.W.P., Speranza, M.G.: Competitive analysis for dynamic multi-period uncapacitated routing problems. Networks 49, 308–317 (2005)
4. Bullnheimer, B., Hartl, R.F., Strauss, Ch.: A new rank based version of the ant system: A computational study. Central European Journal of Operations Research 7, 25–38 (1999)
5. Butler, M., Williams, H.P., Yarrow, L.A.: The two-period travelling salesman problem applied to milk collection in Ireland. Computational Optimization and Applications 7, 291–306 (1997)
6. Clarke, G., Wright, J.W.: Scheduling of vehicles from a central depot to a number of delivery points. Operations Research 12, 568–581 (1964)

7. Doerner, K.F., Gutjahr, W.J., Hartl, R.F., Lulli, G.: A probabilistic two-day delivery vehicle routing problem. In: The Fifth Symposium on Transportation Analysis (TRISTAN V), Preprints, Le Gosier, Guadeloupe, French West Indies, June 13–18 (2004)
8. Dorigo, M., Stützle, T.: Ant Colony Optimization. MIT Press, Cambridge (2004)
9. Gendreau, M., Laporte, G., Seguin, R.: Stochastic vehicle routing. European Journal of Operational Research 88, 3–12 (1996)
10. Hemmelmayr, V., Doerner, K.F., Hartl, R.F., Savelsbergh, M.W.P.: Delivery strategies for blood products supplies. OR Spectrum (to appear, 2008)
11. Powell, W.B., Jaillet, P., Odoni, A.R.: Stochastic and dynamic networks and routing. In: Ball, M.O., et al. (eds.) Handbook in Operations Research and Management Science, vol. 8: Network Routing, pp. 141–296. Elsevier, Amsterdam (1995)
12. Puterman, M.L.: Markov Decision Processes. Wiley, Chichester (1994)
13. Reimann, M., Stummer, M., Doerner, K.F.: A savings based ant system for the vehicle routing problem. In: Proceedings of the Genetic and Evolutionary Computation Conference (GECCO 2002), July 2002, pp. 1317–1325. Morgan Kaufmann, San Francisco (2002)
14. Reimann, M., Doerner, K.F., Hartl, R.F.: D-Ants: Savings based ants divide and conquer the vehicle routing problem. Computers & Operations Research 31, 563–591 (2004)
15. Toth, P., Vigo, D. (eds.): The Vehicle Routing Problem. SIAM Monographs on Discrete Mathematics and Applications, Philadelphia (2002)

Appendix

VRPs are well known combinatorial optimization problems that involve the construction of a set of vehicle tours starting and ending at a single depot and satisfying the demands of a set of customers, where each customer is served by exactly one vehicle and neither vehicle capacities nor maximum tour lengths are violated.

We extend the general VRP modelling framework to our problem situation as follows: As stated before, we distinguish *urgent* requests from a set $H_0^{(t)}$ of customers requiring delivery within the first day, and *regular* requests from a set $H_1^{(t)}$ of customers allowing delivery within two days. A decision has to be made on which day to serve customers belonging to the set $H_1^{(t)}$.

The features of our real application suggest the following hypotheses on the model:

- orders are random variables,
- delivery can take place any time during the day, that is, there are no time windows,
- there are no capacity constraints on vehicles.

Under mild assumptions, the described problem can be considered as a Markov decision process (see, e.g., Puterman [12]). Indeed, the state of the system depends exclusively on the state and the decision taken at the previous stage. However, for computational reasons, optimization methods for Markov decision processes cannot be applied anymore to instances of our problem of a realistic size.

With the notation introduced at the beginning of Sect. 2, the problem can be formalized as follows:

$$\text{Min } c(H(0) \cup B^{(0)}) + E[\sum_{t=1}^{T} c(H(t) \cup B^{(t)})]$$

$$\text{s. t.} \quad H(t) = (H_1^{(t-1)} \backslash B^{(t-1)}) \cup H_0^{(t)} \qquad \forall t = 1, \ldots, T. \qquad (4)$$

$$B^{(t)} \subseteq H_1^{(t)} \qquad \forall t = 1, \ldots, T.$$

where $H(0) = H_0^{(0)}$, $c(A)$ denotes the delivery costs for serving the customers belonging to the set A (optimal solution of the VRP), E denotes the mathematical expectation, and T is the time horizon. Constraints (4) represent the evolution of the system from one stage to the following one.

Procedure. Sampling-Based Local Search

Initialization: $t = 0$; initialize c_0 and $\bar{c}_0$; $H_1^{(0)} = B^{(0)} = \emptyset$;
on each day $t = 1, 2, \ldots$ {
  Step 0: select $B^{(t)} \subseteq H_1^{(t)}$;
  while termination criterion is not met { // *Loop 1*
    Step 1: solve the VRP for the customers of the set
    $(H_1^{(t-1)} \setminus B^{(t-1)}) \cup H_0^{(t)} \cup B^{(t)}$ by SBACO;
    for a certain number of iterations {// *Loop 2*
      Step 2: sample one instance for the sets
        $H_0^{(t+1)}$ and $H_1^{(t+1)}$ by using the estimated distribution of the demand;
      Step 3: select $B^{(t+1)} \subseteq H_1^{(t+1)}$;
      while termination criterion is not met { // *Loop 3*
        $G = (H_1^{(t)} \setminus B^{(t)}) \cup H_0^{(t+1)} \cup B^{(t+1)}$; // customers chosen for day $t+1$
        compute the tours and delivery costs $f(G)$ by SBACO;
        with $\varphi(A)$ = sum of distances from the depot to the customers in A,
        $totalCosts = f(G) + \lambda \cdot c_t \cdot |H_1^{(t+1)} \setminus B^{(t+1)}| +$
        $+(1-\lambda) \cdot \bar{c}_t \cdot \varphi(H_1^{(t+1)} \setminus B^{(t+1)})$;
        update best value $minTotalCosts$ of $totalCosts$;
        update $B^{(t+1)}$ by a local search move;
      } // close Loop 3
    } // close Loop 2
    Step 4: compute the average value of $minTotalCosts$ for $B^{(t)}$ over the samples;
    chosen in Loop 2;
    $overallCosts$ = costs of Step 1, plus average costs of Step 4;
    Step 5: update $B^{(t)}$ by a local search move;
  } // close Loop 1
  $c_{t+1} = \rho \cdot c_t + (1-\rho) \cdot \frac{c((H_1^{(t-1)} \backslash B^{(t-1)}) \cup H_0^{(t)} \cup B^{(t)}) - c(H_0^{(t)} \cup B^{(t)})}{|H_1^{(t-1)} \backslash B^{(t-1)}|}$;
  with $\varphi(A)$ = sum of distances from the depot to the customers in A,
  $\bar{c}_{t+1} = \rho \cdot \bar{c}_t + (1-\rho) \cdot \frac{c((H_1^{(t-1)} \backslash B^{(t-1)}) \cup H_0^{(t)} \cup B^{(t)}) - c(H_0^{(t)} \cup B^{(t)})}{\varphi(H_1^{(t-1)} \backslash B^{(t-1)})}$;
} // close loop over days t

Procedure. Fixed Percentage Selection Procedure

```
on each day t = 1, 2, ... {
  set L = (H_1^(t-1) \ B^(t-1)) ∪ H_0^(t) ∪ H_1^(t);
  for each customer i ∈ H_1^(t) {
    solve the VRP for the customers of the set L \ {i} and store the
    resulting cost in C_i^-;
  } // close loop over customers i
  sort the customers i according to their values C_i^- in ascending order;
  postpone the first p percent of customers on the resulting list by putting
    them into the set H_1^(t) \ B^(t);
  put the remaining customers into the set B^(t);
} // close loop over days t
```

Procedure. Estimated Savings Selection Procedure

```
(1) Preprocessing:
for day d = 1, ..., 7 of week {
  for ν = 1, ..., N {
    sample one instance for the set H_0^(t) by using the distribution of the demand
    on day d of week;
    solve the VRP for the customers of the set H_0^(t)
    and store the resulting cost in C;
    for each customer i {
      solve the VRP for the customers of the set H_0^(t) ∪ {i} and store the
       resulting cost in C_i;
      compute the additional cost ξ(i, ν, d) = C_i − C for serving customer i;
    } // close loop over customers
  } // close loop over samples ν
  for each customer i
    set the customer-specific cost estimate η(i, d) equal to the average value
    of ξ(i, ν, d) over all ν;
} // close loop over days of week
(2) Planning for consecutive days:
on each day t = 1, 2, ... {
  determine d as the day of week of day t + 1;
  set L = (H_1^(t-1) \ B^(t-1)) ∪ H_0^(t);
  solve the VRP for L and store the resulting cost in Γ;
  for each customer i ∈ H_1^(t)
    solve the VRP for L ∪ {i} and store the resulting cost in Γ_i;
  postpone all customers i with Γ_i − Γ > η_i(i, d) by putting them into the
  set H_1^(t) \ B^(t);
  put the remaining customers into the set B^(t);
} // close loop over days
```

The Oil Drilling Model and Iterative Deepening Genetic Annealing Algorithm for the Traveling Salesman Problem

Hoong Chuin Lau and Fei Xiao

School of Information Systems, Singapore Management University,
80 Stamford Road, Singapore 178902
hclau@smu.edu.sg, feixiao@smu.edu.sg

Summary. In this work, we liken the solving of combinatorial optimization problems under a prescribed computational budget as hunting for oil in an unexplored ground. Using this generic model, we instantiate an iterative deepening genetic annealing (IDGA) algorithm, which is a variant of memetic algorithms. Computational results on the traveling salesman problem show that IDGA is more effective than standard genetic algorithms or simulated annealing algorithms or a straightforward hybrid of them. Our model is readily applicable to solve other combinatorial optimization problems.

Keywords: Oil drilling model, Iterative deepening genetic annealing algorithm, Memetic algorithms, Traveling salesman problem.

1 Introduction

Solving NP-hard optimization problems efficiently still remains as one of ultimate challenges for computer scientists. Most often, rather than seeking optimal solutions, near optimal solutions are acceptable in industry considering limited computational resources and quick response requirements. In the past two decades, a large number of metaheuristic approaches have been proposed to obtain reasonably good solutions for NP-hard problems, including more popular ones such as genetic algorithms (GA) [11], simulated annealing (SA) [12], and tabu search [7, 8]. A large number of hybrid algorithms has also been proposed, such as memetic algorithms [19].

It is well-known that metaheuristics generally require proper tuning of key parameters, and they perform differently on various optimization problems, or even different instances of the same problem. In particular, when it comes to hybrid local search strategies (such as a combination of GA and SA), the performance often hinges on how well the algorithm explores the search space through an (iterated) process of diversification and intensification. Recently in [3], in an attempt to unify different metaheuristics conceptually, a framework called the I&D frame was introduced to put different intensification and diversification components in relation with one another.

A. Fink and F. Rothlauf (Eds.): Advances in Computational Intelligence, SCI 144, pp. 169–184, 2008.
springerlink.com

In this work, we draw a close analogy of this iterated search process to oil drilling, and propose an easily implementable and efficient scheme called the oil drilling model (ODM) to hybridize two different heuristics, where one serves as an intensifier while the other as an diversifier. The goal for our research is to achieve balanced diversification and intensification for a specified problem under a given computational budget. More specifically, by following the ODM, we develop an iterative deepening genetic annealing (IDGA) algorithm, which is composed of a standard genetic algorithm and simulated annealing. An intensification-diversification pyramid is proposed to illustrate the strength of diversification and intensification with different parameter settings in IDGA.

The traveling salesman problem (TSP) is arguably one of the most fundamental and important combinatorial optimization problems in transportation and logistics planning, along with other problems such as the quadratic assignment problem. In this work, we choose TSP as our case study problem.

This chapter proceeds as follows. We present the details of the oil drilling model in the next section. In Sect. 3.1, the iterative deepening genetic annealing algorithm is developed based on the ODM. The intensification-diversification pyramid is presented in Sect. 3.2. In Sect. 4, we discuss details of IDGA to solve the TSP. In Sect. 5, we present our experimental results, comparing IDGA with GA, SA, and a hybrid of GA and SA. In Sect. 6, we present some concluding remarks.

2 The Oil Drilling Model

Finding high-quality solutions for an optimization problem can be viewed as hunting for oil in an unexplored ground. The goal for an oil hunting firm is to find oil in the ground within the cost limits. Granted that there may be some more advanced technology in finding oil today such as using magnetometers and shock waves, we consider a traditional process where drill is the only means to find out whether oil is present. The process is as follows. First, one has to identify *where* to drill. We also have to decide how *deep* to drill for each identified well. The cost for oil hunting is proportional to the number of locations to drill and the depth of each drilled location.

Analogously, in solving combinatorial optimization problems with heuristics, our goal is to hunt for high-quality solutions within prescribed computational cost limits. The ground can be viewed as the solution space and computational limits can be viewed as cost limits in the oil hunting process. We consider local search algorithms (or heuristics with strength in intensification) as the drilling process, which incurs most of the computational cost. To decide where we should start the local search is similar as to decide where to drill. Moreover, to decide about the maximum number of steps in a local search is analogous to deciding how deep to drill.

Furthermore, in an oil hunting process, before drilling a very deep well, one would like to drill some shallow wells to get some samples of the rocks. By analyzing the structure of the rocks, one will narrow the scope to a smaller number of places to drill deeper therein. This process is repeated until finally one can decide to drill the deepest well at a particular location with the highest chance for finding oil. We see oil hunting as a promising model to model the general notion of iterative diversification and intensification in metaheuristic search. Hence, we name our approach conveniently as the oil drilling model, which is proposed and given algorithmically as follows.

Algorithm 1. The Oil Drilling Model (ODM)

1: Initialize search depth, population size, and initial solutions in the population
2: Sieve the solutions according to the population size
3: Perform local search on the remaining solutions with current depth
4: Increase the depth and reduce the population size
5: If population size is greater than 1, goto 2:

As shown in Algorithm 1, ODM is an iterative deepening process, where there are two basic steps for each iteration. The first step is called the *sieving* step. Only the best solutions obtained in the preceding iteration are kept. The second step is called the *search* step. A local search algorithm will be applied on all remaining solutions. After that the depth of the local search will be increased and the population will be decreased. When the population reaches 1, the deepest local search will be performed and the oil drilling process ends. A similar iterative deepening process can be found in iterative deepening A* search [13], which is an efficient exhaustive search algorithm.

3 Iterative Deepening Genetic Annealing (IDGA) Algorithm

3.1 Structure of IDGA

Genetic algorithms [11] have been widely used in solving combinatorial optimization problems. Crossover, as one of the main operators of GA, is very useful in evolving new high-quality solutions which inherit the useful patterns of the parent solutions. As GA maintains a population of solutions, it exhibits good diversification properties. Hence, for this work, GA is selected as the diversification algorithm for ODM.

On the other hand, simulated annealing [12] is notable not only for its convergence towards optimality, but also for simplicity of implementation. Like many other local search algorithms, SA has strengths in intensification. Furthermore SA is not easily trapped in local optimality. The desirable properties of SA inspire us to select it as the intensification algorithm for ODM.

We introduce the iterative deepening genetic annealing (IDGA) algorithm, which results from applying GA as the sieving step and SA as the local search step in ODM. It is different from the traditional hybrid of GA and SA such as [5], where the population of GA as well as the step limits of SA are fixed.

It is worthwhile to provide a brief discussion relating our approach with memetic algorithms. In [14], a memetic algorithm is defined as an evolutionary algorithm that includes individual learning and optimization, so for example, a genetic algorithm with local search [19] is a memetic algorithm. Memetic algorithms have been deployed to solve combinatorial optimization problems; for example, [21] provides a survey for different memetic algorithms on the Euclidean TSP. Recently, [20] studied important design issues of memetic algorithms. In our ODM approach, the sieving step can be a pure selection procedure without applying a crossover operator, and hence it may not be an evolutionary algorithm. Hence, strictly speaking, ODM is different philosophically from a memetic algorithm. However, IDGA, which is an instantiation of ODM comprising a hybrid of GA and SA, can be classified as a memetic algorithm, with the special condition that the population of IDGA is gradually reducing, as we will present below.

The details of IDGA are given as follows. The population of GA will be reduced and the step limits of SA will be increased in each iteration until the population size of GA reaches 1 and the step limits of SA reaches its largest value. Assuming the decreasing rate of the population is $\frac{1}{\alpha}$, and the increasing rate of the step limits is β, the population size POP_{k+1} and step limit $STEP_{k+1}$ of the $(k+1)$-th iteration of IDGA is calculated as

$$POP_{k+1} = POP_k/\alpha, \tag{1}$$

$$STEP_{k+1} = \beta \cdot STEP_k. \tag{2}$$

As the drilling cost is the main cost of the oil drilling process, we will only consider the cost of SA in ODM. Therefore, the total number of steps of SA is calculated as the computational cost, the cost for the k-th iteration of IDGA will be $POP_k \cdot STEP_k$. Assuming the total computational budget is C, and the total number of iterations of IDGA is K, let the ending population of the GA (i.e., final drilling of ODM) be 1 and let the start step limit of SA be 1, we have

$$\alpha^{K-1} + \beta \cdot \alpha^{K-2} + \cdots + \beta^{K-2} \cdot \alpha + \beta^{K-1} \leq C, \tag{3}$$

where $\beta^{k-1} \cdot \alpha^{K-k}$ is the computational cost for the k-th iteration of IDGA. The above inequality can be transformed to the following form:

$$\begin{cases} (\alpha^K - \beta^K)/(\alpha - \beta) \leq C & \text{when } \alpha \neq \beta, \\ K \cdot \alpha^{K-1} \leq C & \text{when } \alpha = \beta. \end{cases} \tag{4}$$

Note that the cost for each iteration of IDGA will increase, so we have $\alpha < \beta$. Therefore, given a computational cost limit C, the maximum possible iteration K of IDGA can be found by binary search, where an upper bound can be calculated as

$$K_{max} = \min\{\log_\alpha C, \log_\alpha C\} + 1. \tag{5}$$

After finding K with binary search, the cost $\overline{C}$ can be calculated as

$$\overline{C} = \alpha^{K-1} + \beta \cdot \alpha^{K-2} + \cdots + \beta^{K-2} \cdot \alpha + \beta^{K-1}. \tag{6}$$

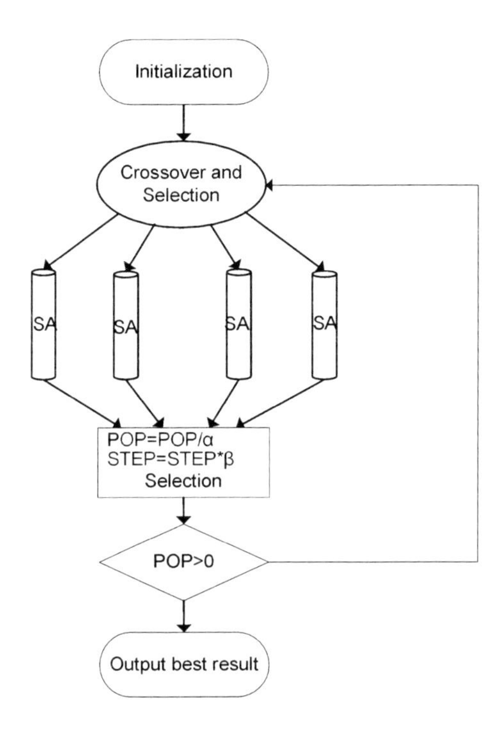

Fig. 1. Iterative deepening genetic annealing algorithm

The extra computational cost, $C - \overline{C}$, will be consumed in the last iteration of IDGA to extend the depth of the last drilling step. A flowchart of IDGA is given in Fig. 1.

In IDGA, a POP number of initial solutions will be randomly generated or generated according to some rules (as the initial spots to drill) and $STEP$, which is the step limit of SA, is initialized to 1. The crossover operator is then used to generate $t \cdot POP$ more solutions and the POP best solutions will be selected according to the selection scheme of GA. Then, the SA is performed on all POP number of solutions with the step limit as $STEP$. After SA has been performed, the population will be reduced to POP/α, while the step limit is increased to $\beta \cdot STEP$, where the POP/α solutions are selected according to the selection scheme of GA. The whole process will be repeated so long as $POP > 1$. The step or depth to drill become largest when POP=1, and the final drilling will be performed at the only place selected, where all the remaining computational budget will be consumed. Here we neglect the details of generating the initial population, the crossover operator, the selection scheme of GA, and the annealing schedule of SA. In Sect. 4, we will provide details for IDGA on the traveling salesman problem.

According to Equation 3, different α and β values will result in different total numbers of iterations in IDGA. Moreover, since the diversification property in

IDGA is mainly determined by size of the population, and the intensification property is determined by depth of SA, the parameters α and β can be viewed as the parameters for diversification and intensification, respectively. Different α and β values will result in different intensification-diversification schemes. The intensification-diversification pyramid given in the next section will clearly illustrate the strength of intensification and diversification under different schemes.

3.2 Intensification-Diversification Pyramid

The depth of SA typically remains constant in a standard hybrid of GA and SA, which implies that the strength of diversification and intensification remains the same throughout the search process. In ODM, more strength is put into the diversification at the start of the search while the focus shifts to intensification at the end of the search. We argue that ODM is better than a standard hybrid scheme, since a greater amount of computational cost should be devoted to exploit the solution structure in order to have the highest chance in obtaining best solutions. Therefore, the starting stage of ODM can be viewed as the selection phase, where different solutions are examined and sieved. The ending stage of ODM is to perform the final drilling, which can be viewed as the exploring phase, where the most computational cost is used to explore the neighbors of the last solution.

As mentioned in the previous section, the parameters α and β determine the strength of intensification and diversification in ODM. To illustrate the different intensification-diversification schemes according to different α and β values, we develop the intensification-diversification pyramid (ID-pyramid). The number of levels of the ID-pyramid is the total number of iterations K in Equation 3, where each level represents an iteration of IDGA. The width of each level is the size of the population in that iteration, and the height of the level is the depth of the SA. With total cost set to 120, we present different ID-pyramids with (α, β) values set to $(1.2, 1.2)$, $(1.5, 1.5)$, $(2, 2)$, $(2, 4)$, $(4, 2)$, and $(4, 4)$, as shown in Fig. 2.

Clearly, the area of each pyramid is equal to the cost limit 120. We also observe that the larger the value of α, the bigger the difference between the width of the neighbor levels in ID-pyramids, and the larger the value of β, the bigger the difference between the height of the neighbor levels in ID-pyramids. The heights of the last level of the ID-pyramids are the highest; this is clearly illustrated in Fig. 2. In fact, as shown in the computational study later, to achieve best performance of ODM, the area of the last level of an ID-pyramid should be larger than the total areas of the rest of the levels. In Fig. 2, the last level ID-pyramid $(4, 2)$ is much lower than the last level of other pyramids, which illustrates that it is not good to let α be larger than β. Detailed experimental results for different ID-pyramids can be found in Sect. 5. In the next section, we will introduce the traveling salesman problem and survey some of the past research, followed by the details of applying IDGA on the TSP.

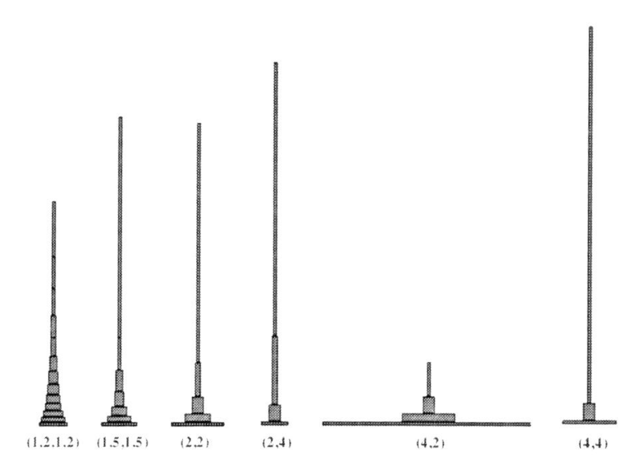

Fig. 2. Intensification-diversification pyramids with cost limit as 120

4 Solving TSP with IDGA

In the traveling salesman problem (TSP), we are given a complete undirected graph $G = (V, E)$, each edge has a nonnegative weight $w(u, v)$, and the objective is to find a cycle of minimum total weight that visits each vertex exactly once and also returns to the starting vertex.

One may model the TSP as an integer programming problem and apply the cutting plane method proposed by Gomory [9]. Further work developed by Applegate et al. [1] has solved a TSP instance with 24,978 cities using approximately 84.8 CPU years on a single Intel Xeon 2.8 GHz processor. It remains highly improbable to solve very large instances (like the World TSP with more than 1 million cities) using the cutting plane method. The best reported tour for the World TSP was found by Helsgaun using the LKH heuristic algorithm [10] (an improved effective implementation of Lin-Kernighan search [15]), which is 0.04996% greater than the lower bound. The cutting plane method developed by Applegate et al. and the LKH heuristics by Helsgaun are state-of-the-art algorithms for solving the TSP. A large number of heuristic algorithms have been developed to obtain reasonably good solutions under a limited computational cost, such as genetic algorithms [18], simulated annealing [22], tabu search [24], and ant colony optimization [6]. There are also hybrid algorithms for the TSP which combine heuristic algorithms like GA and SA with local search algorithms like 2-opt, 3-opt, or Lin-Kernighan search [17, 16]. A recent review on the state-of-art algorithms for the TSP can be found in [2].

In this section, we present results on the performance of IDGA for solving the TSP. Our intention here is not to achieve best results for the TSP, but to demonstrate that applying IDGA to solve the TSP is more effective than applying SA and GA alone or GASA (a simple hybrid GA and SA with fixed population size and search depth). With extensive research done on the TSP, there are numerous techniques we can adopt. Here, we will present details of IDGA embedding some popular techniques such as candidate sets and 2-opt.

Initial Population

To reduce the computational effort and improve the quality of the solutions, candidate sets have been widely used in the approaches for the TSP. The candidate set of each vertex contains the best choices for the neighbors of the vertex when forming the TSP tour. In IDGA, we maintain a candidate set with 6 nearest neighbors for each vertex to balance the performance and computational time. One of the initial solutions for IDGA will be selected as the solution from a classic greedy algorithm. The rest of the initial solutions are generated randomly, where, in building the tour, the current vertex will try to pick up the next vertex in the tour from its candidate set randomly; if all the neighbors in the candidate set of the current vertex have been picked up, it will select the nearest neighbor which has not yet been included in the tour as the next vertex.

All the settings of IDGA are also used in the GA, SA, and a standard combination of GA and SA (GASA) which we benchmark against. Since the SA only has one initial solution, the initial solution of the SA is just the solution from the greedy algorithm.

Crossover

A classic two point crossover is used in IDGA. Both the parents are selected according to the 2-tournament rule. Assuming there are N cities for the TSP, point a and point b is selected on the father's sequence, such that $|b-a|/N \in [0.4, 0.6]$. To have a good diversification in the offspring, two parents will only have one child instead of two, since the siblings from the same parents tend to have similar genes. For the child sequence, the first part from 1 to $b-1$ and the third part from $a+1$ to N are inherited from the father's sequence directly. The second part of the child is the compose of the cities in the second part from a to b of the father's sequence, however, the order of them is according to their order in the mother's sequence. As an example, for the TSP with 6 cities, assuming the father sequence is $[1, 2, 3, 4, 5, 6]$ and the mother sequence is $[4, 1, 2, 5, 6, 3]$, also assuming a is 3, and b is 5, the child sequence will be $[1, 2, 4, 5, 3, 6]$, in which the first part $[1, 2]$, and the third part $[6]$ is from the father directly, the second part is from the second part $[4, 5, 3]$ of the father however in the order of the mother's sequence.

Selection

The fitness function for each sequence is the length of the tour it represents. Each generation of IDGA will generate $4 * POP$ offspring. The ranking method is used to select the sequence for the next generation. POP number of sequences with highest fitness values will be selected among the old POP sequences and their $4 * POP$ offspring.

Annealing Schedule

Instead of providing each SA search in Fig. 1 with a separate annealing schedule, we treat the entire IDGA as a single annealing process to achieve a better convergence. The initial temperature T_0 is set as

$$T_0 = \frac{-L_{best}}{100 \ln(0.01)} \tag{7}$$

so that the probability to accept a solution that is 1/100 worse than L_{best} is 0.01, where L_{best} is the length of the best tour in the initialization procedure. In IDGA, SA mainly plays the role of intensification, and hence the initial temperature is set relatively low.

Accordingly, the final temperature T_e is set as

$$T_e = \frac{-1}{\ln(0.001)} \tag{8}$$

so that the probability to accept a solution which is 1 unit longer than the best length is 0.001. In our approach, the exponential cooling scheme is used, which is given as

$$T_{k+1} = \lambda T_k, \quad k \geq 0, \tag{9}$$

where λ is set as 0.2. The number of steps of SA for the $(k+1)$-th temperature is calculated as $N_{k+1} = N_k * 2$, which also means the length of the Markov chain increases with the temperature approaching 0. The number of different temperature states can be calculated as

$$p = \left\lceil \frac{\ln(T_e/T_0)}{\ln \lambda} \right\rceil + 1. \tag{10}$$

With the total computational cost as C, N_0 can be calculated as

$$N_0 = \left\lfloor \frac{C}{2^p - 1} \right\rfloor. \tag{11}$$

Local Move

2-opt is one of the most used local search algorithms for the TSP, not only for its efficiency but also for the simplicity of implementation. We apply the local move used in 2-opt in our IDGA. The local move considers removing two edges from the current tour and adding two new edges to form a new tour. If the new tour is acceptable according to the transition probability of SA, the current tour will be updated by the new tour.

IDGA

With all the above features, the detailed algorithm of IDGA for the TSP is given in the appendix. Within this algorithm, *crossover* is the function to generate the new solution by performing crossover on S_f and S_m; *2-opt-move* is the function to generate a new tour S'_j through applying a 2-opt move on the old tour S_j; *rand* is the function to generate a real number randomly in $(0, 1)$ according to uniform distribution; N_d is used to record the length of the current Markov chain. When the population decreases to one solution, the additional cost $C - \overline{C}$ will be added to the depth for the last drilling.

5 Computational Study

We implemented IDGA with Visual C++ 6.0 on the Windows XP platform. All experiments are conducted on a laptop PC with a PIV 1.6 GHz Intel dual-core processor and 1 GB memory.

As mentioned above, only the number of steps of the SA is counted as computational cost. In all experiments, the cost limit is given as 12499968 (set α, β, K as 2, 10, 8), where the cost limit can be set as any other value instead and the results are consistent.

We randomly selected 11 TSP instances from the set of 67 EUC2D TSP instances posted in TSPLIB [23], where the number of cities ranged from 52 to 2392. We will first present the results on different ID-pyramids discussed in Sect. 3.2, followed by detailed experimental results of comparing IDGA with GA, SA, and GASA.

5.1 Selecting the ID-Pyramid

The shapes of different ID-pyramids are given in Fig. 2, where some insights on the ID-pyramids are discussed in Sect. 3.2. The average deviation from the optimal solutions and average running time for different ID-pyramids on the 11 TSP instances selected are recorded in Table 1. Note that for ID-pyramid $(2, 2)$ the initial population size is calculated as 524288, which will be computationally prohibitive. Hence, for all cases, we set the limit for the initial population size as 5000.

Table 1. Computational results for different ID-pyramids

ID-pyramids	(1.2,1.2)	(1.5,1.5)	(2,2)	(2,4)	(4,2)	(4,4)	(2,10)	(2,20)
Avg. dev. (%)	5.15	5.03	4.65	3.83	5.73	4.97	3.36	4.56
Avg. time	81.08	54.13	45.29	25.28	39.88	39.92	13.98	12.71

In Table 1, we observe that even with the same number of steps of SA, more computational time is required for smaller values of α and β (such as ID-pyramids $(1.2, 1.2)$, $(1.5, 1.5)$ and $(2, 2)$). This is because the aggregate population sizes for those ID-pyramids are large, and hence more computational cost of GA

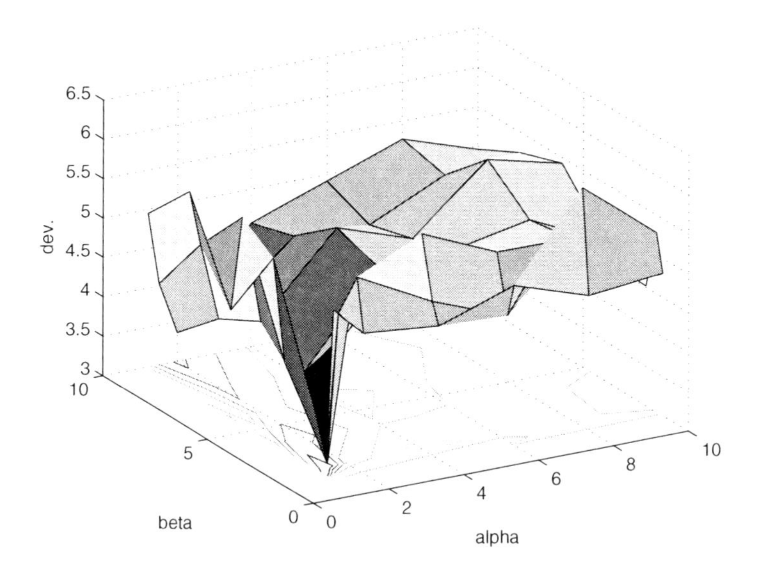

Fig. 3. Results of different ID-pyramids

operators will be incurred. We note that the main computational cost of ODM is the drilling process. Therefore, in IDGA, we only count the steps of SA as the computational cost. We also discover that the top 3 ID-pyramids are $(2, 10)$, $(2, 4)$, and $(2, 20)$, while ID-pyramid $(4, 2)$ returns the worst result. The reason is that ID-pyramid $(4, 2)$ focuses on diversification so that the last level of ID-Pyramid $(4, 2)$ is much lower than the other ID-pyramids, which is also indicated in Fig. 2. We also observe that too much strength on intensification is also not good, as the results of ID-pyramid $(2, 20)$ is worse than ID-pyramid $(2, 10)$. It is quite clear from Table 1, that ID-pyramid $(2, 10)$ should be selected for IDGA to achieve balanced strength of intensification and diversification within a reasonable computational time.

To further investigate whether there are multiple balanced points for the strength of intensification and diversification, we have sampled 49 different ID-pyramids, where the values of α and β are taken from the set $\{1.2, 1.5, 2, 4, 6, 8, 10\}$. The deviations from the optimal solutions of those ID-pyramids are illustrated in Fig. 3. We observe there is a valley phenomenon in Fig. 3, which suggests that while there are multiple balanced points (best ID-pyramids), the best ones are those with an α value around 2. We also observe that the worst results are achieved by those with $\alpha > \beta$, which shows that for the TSP intensification should have a stronger strength than diversification. The best result, which is 3.04% from the optimal solutions, is achieved by setting α and β to be 1.5 and 2.0 respectively. Note however that the average computational time for this case is 53.20 seconds. On the other hand, the ID-pyramid $(2, 10)$ is a good choice if the computational budget is limited to be within, say 20 seconds.

5.2 Experimental Results

Detailed results for comparing IDGA with SA on the 11 TSP instances are presented in Table 2. We observe that IDGA outperforms SA on 9 cases, with 1 equal result and only 1 worse result. On average, the solutions of IDGA are 1.39% better than the solutions of SA. We also observe that IDGA outperforms SA significantly on the two largest instances, where IDGA yields 9% better than SA for the TSP instance pr2392. Note that SA can be viewed as the ID-pyramid $(1, C)$ which mainly focuses on intensification, while IDGA has been designed to achieve balanced strength in diversification and intensification.

Table 2. Comparing IDGA with SA on the TSP instances

		SA			IDGA			
TSP instance	opt.	results	dev. opt. (%)	time (sec)	results	dev. opt. (%)	time (sec)	dev. SA (%)
pr2392	378032	457374	20.99	14.20	416042	10.05	18.77	-9.04
pr1002	259045	290080	11.98	13.52	281675	8.74	14.70	-2.90
pr107	44303	45177	1.97	14.34	44580	0.63	14.77	-1.32
rd100	7910	8141	2.92	13.98	8083	2.19	14.39	-0.71
berlin52	7542	7542	0.00	13.97	7542	0.00	13.95	0.00
pr76	108159	109841	1.56	13.91	108638	0.44	13.89	-1.10
a280	2579	2720	5.47	11.83	2710	5.08	11.92	-0.37
kroa100	21282	21567	1.34	14.05	21369	0.41	14.03	-0.92
ch150	6528	6791	4.03	12.52	6695	2.56	12.50	-1.41
ei151	426	426	0.00	10.45	426	0.00	10.36	0.00
pcb442	50778	52924	4.23	14.06	54256	6.85	14.55	2.52
Average	80598.55	91143.91	4.95	13.35	86546.91	3.36	13.98	-1.39

IDGA uses the same operators as SA, GA, and GASA. However, the population of solutions in IDGA decreases and the steps of SA in IDGA increase for each iteration. For IDGA, the total SA step limit is set to 12499968 (set α, β, K as $2, 10, 8$). For pure SA, the total SA step limit is also set to 12499968. For pure GA, the population size is set to 150 and the number of generations is set to 100. For GASA, the population is set to 60, the number of generations is set to 30, and the number of SA steps is set to 6944 (12499968/60/30).

The results of GA and GASA are given in Table 3. Here, GA can be viewed as the ID-pyramid $(1, 0)$, where we try to provide same computation time to GA by assigning the number of population as 150 and the number of generations as 100, while GASA can be viewed as ID-pyramid $(1, 1)$. The population is set to 60, the number of generations is set to 30 and the number of SA steps for each GA sequence is set to 6944 ($12499968/(60 \times 30)$). We observe in Table 3 that, on average, the results for IDGA are 2.10% better than the results of GASA, and 14.24% better than the results of GA. It is noteworthy that the TSP on planar graphs exhibits the well-known big valley property [4], and hence GA typically returns worse results than SA. The comparison between IDGA and GA, SA, and

Table 3. Comparing IDGA with GA and GASA on selected TSP instances

	GA			GASA			IDGA	
TSP instance	results	dev. opt. (%)	time (sec)	results	dev. opt. (%)	time (sec)	dev. GA (%)	dev. GASA(%)
pr2392	461170	21.99	35.72	461170	21.99	17.75	-9.79	-9.79
pr1002	331103	27.82	21.75	331103	27.82	14.55	-14.93	-14.93
pr107	46242	4.38	14.24	44595	0.66	14.74	-3.59	-0.03
rd100	9887	24.99	13.83	7910	0.00	15.19	-18.25	2.19
berlin52	8736	15.83	13.63	7542	0.00	15.17	-13.67	0.00
pr76	138484	27.61	13.70	108792	0.59	15.00	-21.55	-0.14
a280	3149	22.10	14.66	2697	4.58	11.69	-13.94	0.48
kroa100	27132	27.49	13.91	21344	0.29	15.17	-21.24	0.12
ch150	8191	25.47	14.16	6571	0.66	12.84	-18.26	1.89
ei151	468	9.86	13.63	427	0.23	10.20	-8.97	-0.23
pcb442	61979	22.06	16.34	55760	9.81	14.45	-12.46	-2.70
Average	99685.55	20.87	16.87	95264.64	6.06	14.25	-14.24	-2.10

Table 4. Average results of 67 instances from TSPLIB

Avg. results	IDGA	SA	GASA	GA
Length	77416.97	78927.42	82854.39	88553.10
Dev. opt (%)	5.99	6.92	8.88	23.01
Time	12.66	11.73	10.72	17.33
IDGA improvements (%)	0.00	-0.81	-2.21	-13.66

Table 5. Comparing IDGA with memetic algorithms on TSP instances, where γ indicates average fitness per generation at each run, averaged over 100 runs, and η indicates best fitness values achieved during these runs

	SSMA_HC		TGMA_HC		IDGA	
TSP instance	γ	η	γ	η	γ	η
C20	149.799	62.575	153.128	62.575	116.367	62.575
C30	186.314	62.716	198.448	62.716	120.025	62.716
C40	243.077	62.768	309.523	62.768	125.721	62.768
S21	129.523	60.000	149.920	60.000	121.439	60.000
F32	157.291	89.288	172.591	84.180	124.640	84.180
F41	205.571	68.168	239.901	68.168	125.461	68.168

GASA has illustrated the superior performance of ODM and IDGA, and more importantly, demonstrated that balancing diversification and intensification can certainly improve the performance over pure and standard hybrid algorithms.

The average results for all the 67 instances from TSPLIB are presented in Table 4. We observe that IDGA achieved best average results overall, with 0.81% improvement over SA, 2.21% over GASA, and 13.66% over GA.

We also compare IDGA with state-of-art memetic algorithms for the TSP [21], including a steady state memetic algorithm with hill climbing (SSMA_HC) and a trans-generational memetic algorithm with hill climbing (TGMA_HC). The TSP instances C20, C30, C40, S21, F32, and F41 proposed by Ender Ozcan and Murat Erenturk [21] are used here. The detailed results can be found in Table 5.

As shown in Table 5, IDGA has found optimal solutions for all the test instances. IDGA also achieved better average fitness value than the other two approaches, which implies that IDGA converges faster than other methods.

6 Conclusion

An oil drilling model (ODM) has been proposed in this chapter for solving combinatorial optimization problems. Following ODM, we designed a new iterative deepening genetic annealing (IDGA) algorithm to solve the TSP. We also proposed the notion of ID-pyramids to illustrate different intensification-diversification schemes, where the best ID-pyramid is selected according to experimental results. Detailed experimental results on the TSP also show that IDGA outperforms pure GA, SA, and a standard hybrid of GA and SA, which illustrates the effectiveness of ODM and IDGA and demonstrates that it is possible to combine different heuristics to achieve a balanced point of diversification and intensification.

To further investigate on IDGA and ODM, more experiments on different kinds of combinatorial optimization problems should be conducted in the future. Moreover, different sieving and local search methods other than GA and SA can also be used. The ID-pyramids may be used as a tool to illustrate the strength of intensification and diversification for other hybrid heuristics. Since different problems may have different balance points of diversification and intensification, we think that an interesting future work is to design an intelligent method that is capable of performing automated selection of the best ID-pyramid for a specified problem.

References

1. Applegate, D., Bixby, R., Chvátal, V., Cook, W.: On the solution of traveling salesman problems. In: Documenta Mathematica, Extra Volume Proceedings ICM III, pp. 645–656 (1998)
2. Applegate, D., Bixby, R., Chvátal, V., Cook, W.: The Traveling Salesman Problem: A Computational Study. Princeton University Press, Princeton (2007)
3. Blum, C., Roli, A.: Metaheuristics in combinatorial optimization: Overview and conceptual comparison. ACM Computing Surveys 35(3), 268–308 (2003)
4. Boese, K.D.: Models for Iterative Global Optimization. PhD Thesis, UCLA Computer Science Department (1996)
5. Chen, D.J., Lee, C.Y., Park, C.H.: Hybrid genetic algorithm and simulated annealing (HGASA) in global function optimization. In: 17th IEEE International Conference on Tools with Artificial Intelligence, pp. 126–133 (2005)

6. Dorigo, M., Maniezzo, V., Colorni, A.: The Ant System: Optimization by a colony of cooperating agents. IEEE Transactions on Systems, Man, and Cybernetics – Part B 26(1), 29–41 (1996)
7. Glover, F.: Tabu search – Part I. ORSA Journal on Computing 1(3), 190–206 (1989)
8. Glover, F.: Tabu search – Part II. ORSA Journal on Computing 2(1), 4–32 (1990)
9. Gomory, R.: Solving linear programs in integers. In: Bellman, R.E., Hall Jr., M. (eds.) Combinatorial Analysis, Proceedings of Symposia in Applied Mathematics X, pp. 211–216. American Mathematical Society, Providence (1960)
10. Helsgaun, K.: An effective implementation of the Lin-Kernighan traveling salesman heuristic. European Journal of Operational Research 126(1), 106–130 (2000)
11. Holland, H.J.: Adaption in Natural and Artificial Systems. University of Michigan Press, Ann Arbor (1975)
12. Kirkpatrick, S., Gelatt Jr., C.D., Vecchi, M.P.: Optimization by simulated annealing. *Science* 220, 671–680 (1983)
13. Korf, R.E.: Depth-first iterative-deepening: An optimal admissible tree search. Artificial Intelligence 27, 97–109 (1985)
14. Krasnogor, N., Smith, J.E.: A tutorial for competent memetic algorithms: Model, taxonomy and design issues. IEEE Transactions on Evolutionary Computation 9(5), 474–488 (2005)
15. Lin, S., Kernighan, B.W.: An effective heuristic algorithm for the traveling-salesman problem. Operations Research 21, 498–516 (1973)
16. Liu, Z., Kang, L.: A hybrid algorithm of n-OPT and GA to Solve Dynamic TSP. In: Li, M., Sun, X.-H., Deng, Q., Ni, J. (eds.) GCC 2003. LNCS, vol. 3032, pp. 1030–1033. Springer, Berlin (2004)
17. Martin, O., Otto, S.W.: Combining simulated annealing with local search heuristics. Annals of Operations Research 63, 57–75 (1996)
18. Merz, P., Freisleben, B.: Genetic local search for the TSP: New results. In: Proceedings of the 1997 IEEE International Conference on Evolutionary Computation, pp. 159–164 (1997)
19. Moscato, P.A.: On evolution, search, optimization, genetic algorithms and martial arts: Towards memetic algorithms. Technical Report Caltech Concurrent Computation Program, Report 826. Caltech (1989)
20. Nguyen, Q.H., Ong, Y.S., Lim, M.H., Krasnogor, N.: A study on the design issues of Memetic Algorithm. In: Proceedings of the 2007 IEEE International Conference on Evolutionary Computation, pp. 2390–2397 (2007)
21. Ozcan, E., Erenturk, M.: A brief review of memetic algorithms for solving Euclidean 2D traveling salesrep problem. In: Proceedings of the 13th Turkish Symposium on Artificial Intelligence and Neural Networks, pp. 99–108 (2004)
22. Skiscim, C.C., Golden, B.L.: Optimization by simulated annealing: A preliminary computational study for the TSP. In: Proceedings of the 15th Conference on Winter Simulation, vol. 2, pp. 523–535 (1983)
23. Library of Sample Instances for the TSP (1997), `http://elib.zib.de/pub/mp-testdata/tsp/tsplib/tsplib.html`
24. Zachariasen, M., Dam, M.: Tabu search on the geometric traveling salesman problem. In: Osman, I.H., Kelly, J.P. (eds.) Metaheuristics: Theory and Applications. Proceedings Metaheuristics International Conference 1995, Colorado, pp. 571–587 (1995)

Appendix

Algorithm 2. Iterative Deepening Genetic Annealing Algorithm (IDGA)

1: Generate candidate set for each vertex
2: Set initial value for T_0 and T_e
3: Calculate the number of generations K with binary search
4: Calculate POP_0 and $STEP_0$
5: **for** $i \leftarrow 1$ to POP_0 **do**
6: Generate Solution S_i randomly according to candidate sets
7: **end for**
8: Calculate N_0 according to T_0, T_e, and C
9: $i \leftarrow 0$
10: $c \leftarrow 0$, $d \leftarrow 0$
11: **while** $i < K$ **do**
12: **for** $j \leftarrow 1$ to $4 * POP_i$ **do**
13: Select S_f and S_m
14: $S_{j+POP_i} \leftarrow crossover(S_f, S_m)$
15: **end for**
16: Select best POP_i solutions
17: **for** $j \leftarrow 1$ to POP_i **do**
18: **for** $u \leftarrow 1$ to $STEP_i$ **do**
19: $S'_j \leftarrow$ 2-opt-move(S_j)
20: $p \leftarrow rand(0, 1)$
21: **if** $p < e^{(S_j - S'_j)/T_k}$ **then**
22: $S_j \leftarrow S'_j$
23: **end if**
24: $c \leftarrow c + 1$
25: **if** $c = N_d$ **then**
26: $d \leftarrow d + 1$
27: $N_d \leftarrow 2 * N_{d-1}$
28: $c \leftarrow 0$
29: **end if**
30: **end for**
31: **end for**
32: $POP_{i+1} \leftarrow POP_i / \alpha$
33: $STEP_{i+1} \leftarrow STEP_i * \beta$
34: **if** $POP_{i+1} = 1$ **then**
35: $STEP_{i+1} \leftarrow STEP_{i+1} + C - \overline{C}$
36: **end if**
37: $i \leftarrow i + 1$
38: **end while**

Online Transportation and Logistics Using Computationally Intelligent Anticipation

Peter A.N. Bosman and Han La Poutré

Centre for Mathematics and Computer Science, P.O. Box 94079, 1090 GB Amsterdam, The Netherlands
{Peter.Bosman,Han.La.Poutre}@cwi.nl

Summary. With advances in technology in communication and navigation, the ability to make decisions online (i.e. as time goes by) becomes increasingly important in transportation and logistics. In this chapter, we focus on online decision making in these areas. First, we point out the importance of anticipation when optimizing decision processes online. Anticipation is the possibility to take into account future events and the influence of decisions taken now on those future events. Second, we discuss how computational intelligence (CI) can be used to design approaches that perform anticipation. We illustrate this particular use of CI techniques in two different applications: dynamic vehicle routing (transportation) and inventory management (logistics). In both cases the use of anticipation is found to lead to substantial improvements. This demonstrates our main conclusion that the ability to perform anticipation in online transportation and logistics is very important.

Keywords: Online optimization, Estimation-of-distribution algorithm, Evolutionary algorithm, Vehicle routing, Inventory management.

1 Introduction

Transportation and logistics play an important role in many companies. Optimizing the processes that are involved directly influences a company's efficiency and hence can lead to better revenues. With the advances in technology in communication and navigation, companies can exert an increasing amount of direct control on their transportation and logistics processes. For instance, using global positioning systems the whereabouts of trucks or goods in general can be tracked 24 hours a day. Using global communication, new instructions for moving the goods can be issued at any time. Such technological advances hold a promise to service customers faster because new servicing orders can be given out immediately.

Exploiting these new abilities in the best possible way is therefore an important issue. Traditionally, transportation and logistics problems are optimized using a static model, i.e. plans are made ahead and then executed. If needed, new plans are made for a new period. The main point is that the plans are not made to be adjusted *online*, i.e. while the plans are being executed. Typically,

A. Fink and F. Rothlauf (Eds.): Advances in Computational Intelligence, SCI 144, pp. 185–208, 2008.
springerlink.com

this results in plans that are not very flexible and hence cannot easily accommodate for changes that may be required. This is a myopic, i.e. "near-sighted", approach. The quality of plans is taken only to be how good they are for the current situation. Optimizing plans while keeping in mind that they might need to be adjusted during execution, or optimizing the plans completely online, is however not trivial and is an important field on its own [10, 17].

The optimization problems under consideration are thus dynamic, meaning that they change with time. Moreover, they have to be solved in an online fashion, i.e. as time goes by. Typical examples in transportation and logistics are vehicle routing [21, 27] and inventory management [23, 26]. This type of problem is often hard, even for a single point in time. Also, there is typically not much time between two subsequent decision moments. For these reasons, restarting optimization from scratch is often undesirable. Instead, the tracking of (near-) optima, once they have been found, is often desired. To be able to do this, the optimization algorithm needs to have a proper degree of adaptivity. Evolutionary algorithms (EAs) [16], an important optimization methodology in computational intelligence, are good candidates to this end. The reason for this is that EAs employ a set of solutions rather than just a single solution. Adaptivity is then a virtue of issues such as maintaining diversity around (sub)optima and continuously searching for new regions of interest that may appear over time [10].

Tracking optima alone is not enough however. Tracking optima corresponds to building (optimal) plans only for the current situation and repeating this continuously or whenever something changes. Decisions taken now however have consequences in the future. For instance, a consequence of executing a particular plan may be it does not leave enough flexibility for future requests to be served (e.g. new loads to be picked up). Another example is that refusing service or not being able to service a customer (e.g. sell goods) may lead to decreasing repeated servicing requests in the future. Because of such future consequences of current decisions, a myopic approach can perform poorly in the long run. Anticipation of future situations is needed to be able to make well-informed decisions. Consider the problem of online dynamic vehicle routing. Intuitively, one can construct a more efficient routing, resulting in the delivery of more loads, if one would know beforehand when the loads will become ready for transport. It would then be possible to send a vehicle to a customer that currently has no load available, but it is known that a load will be available at that customer upon arrival of the vehicle. A myopic approach would only consider currently available loads. Alternatively stated, this information allows us to see that a current decision to send a vehicle to pick up a currently available load may in the end not be the best thing to do. However, this optimal information about future introduction of new loads is not readily available. The only option left is to learn to predict it.

In this chapter we discuss how computational intelligence can be used to design approaches that perform anticipation in online optimization. Specifically we combine EAs with another key area of computational intelligence: statistical/machine learning (ML/SL). The ML/SL techniques are used to explicitly predict for future times the observed values of problem-specific variables and/or the quality of decisions. The EA optimizes decisions not only with respect to the current situation but also future decisions with respect to future, predicted, situations (i.e. it builds a plan). We illustrate this approach for two applications: vehicle routing and inventory management. In both cases the use of anticipation is found to lead to substantial improvements. This demonstrates our main conclusion that the ability to perform anticipation in practical online transportation and logistics is very important.

The remainder of this chapter is organized as follows. In Sect. 2 we sketch the approach of performing anticipation with computational intelligence techniques in some more detail and discuss related literature. We apply the methodology to vehicle routing in Sect. 3 and to inventory management in Sect. 4. We conclude this chapter in Sect. 5.

2 Intelligent Online Optimization by Anticipation

In dynamic optimization, the goal is to optimize some function $\mathfrak{F}$ over a period of time. The variables to optimize over within this time span, represent the decisions to be made (e.g. which truck to use for picking up a certain load). Function $\mathfrak{F}$ can be seen as the *real world*. Function $\mathfrak{F}$ cannot be evaluated beyond the current time t^{now}. The reason why we require anticipation is the existence of time-dependence. Time-dependence means that $\mathfrak{F}$ may depend on previous decisions. If only the current situation is taken into account, the decision that immediately leads to the highest reward is optimal. But because function $\mathfrak{F}$ may change as a result of a decision, it is in general suboptimal to only consider the current situation. Future changes in $\mathfrak{F}$ as a result of decisions taken now must thus be taken into account also. This means that we must optimize $\mathfrak{F}$ not only for the current situation, but also for future situations. Only then does optimization reveal the true value of a certain decision for the current situation. Anticipation is thereby automatically performed. However, this implies that we evaluate $\mathfrak{F}$ also for the decisions pertaining to future situations, which is not possible. The only way in which we can still take into account the future is to *predict* the value of $\mathfrak{F}$. Summarizing, the approach is to build and maintain (i.e. learn online) an approximation of $\mathfrak{F}$ (i.e. $\hat{\mathfrak{F}}$) and to optimize decisions for the present and for the approximated future simultaneously using the approximation. Approximation function $\hat{\mathfrak{F}}$ can be seen as a *simulation* of the real world.

The nice thing about this approach is that under the condition of perfect prediction, optimal decisions can be taken. For a typical application it is not required that the entire function needs to be learned. Instead, only specific parameters need to be estimated. In logistics settings, typical parameters to be learned include the rate of customer demand or parameters describing the distribution of customer demand.

Some parallels can be drawn here with other research found in literature. Optimizing future, simulated situations is not the only approach to performing anticipation. Anticipation can also be performed by heuristically changing $\mathfrak{F}$

into $\tilde{\mathfrak{F}}$ so that $\tilde{\mathfrak{F}}$ measures not only the quality of the current plan, but also takes into account additional information such as the flexibility, robustness and sensitivity of a solution. Although this approach typically does not have the property of being able to reach optimality of the original problem, such heuristic adaptations of $\mathfrak{F}$ can still be very effective. Examples of this approach include scheduling [12], where an emphasis is placed on schedule flexibility by scheduling new jobs as quickly as possible to ensure free machines for new, yet unknown, jobs, and vehicle routing [13], where an emphasis is placed on flexibility by allowing vehicles to wait before moving on to the next location.

The idea and importance of optimizing for future, simulated situations when solving dynamic optimization problems, is in itself not a new idea. In most problems related to real-world applications, the dynamism is caused by stochasticity (e.g. the appearance of new customers is often a Poisson process). In the expectation method [14], multiple future scenarios are sampled. For each possible decision d that can be made for the current situation, an optimal path of future decisions is computed for each of the sampled scenarios that starts with decision d. The decision $d^{\star}$ for the current situation that leads to the highest expected value of the profit (i.e. average) is then actually taken. The expectation method bears much resemblance to the sample average approximation method for non-dynamic stochastic programming where the quality of a solution consists of a deterministic part a and a stochastic part. The deterministic part gives the (immediate) quality of a plan. The stochastic part gives a penalty for changing the plan to accommodate as best possible the scenario that has actually become reality. To find the expected best solution, one must average over many scenarios. The sample average approximation method can be applied for instance to stochastic vehicle routing [31]. It has been shown recently that the expectation method can provide high-quality solutions in the expected-value sense [22]. It is important though that the expected difference between the optimal choice for any scenario and the optimal expected-value choice does not become too large. It has already been argued that this assumption is satisfied in most real-world problems. It has also been proved to be the case for a real-world problem (packet scheduling) [22]. Because the expectation method requires optimization for many combinations of decisions and scenarios, methods that approximate the expectation method have also been developed [4, 5]. These methods are typically faster, but result in solutions of a lesser quality.

In most of the approaches in the literature, the stochasticity is assumed to be known so that scenarios can be sampled. Hence, there is no learning. Also, the application of optimizing future scenarios so far has been limited to sampling scenarios beforehand and then optimizing the future decisions in these scenarios. Time-dependence is thus explicitly not considered in full. The type of time-dependence that is tackled is the direct influence of one decision on another decision. For example, a decision to drive to customer c_i puts the truck at that location instead of the location where it would otherwise have stayed. What is not taken into account is the possible influence of a decision on the future response of the system, e.g. the stochastics. For example, deciding to drive to customer c_i may lead to a higher frequency of new orders from customer c_i in the near future. With the existence of time-dependence of this kind, sampling events in a scenario beforehand is unacceptable because the events may change as a result of decisions made.

The general approach requires a minor modification to allow for time-dependence in any of its forms to be tackled. Optimization with scenarios can be done by randomly choosing $N^{\text{scenarios}}$ random seeds and then optimizing the simulation for each random seed. This essentially makes the approximation (i.e. simulation) deterministic, allowing it to be re-evaluated under the exact same circumstances during optimization. The computational implications of this change are however not minor. Finding a solution (i.e. future trajectory) for each and every scenario requires solving an interactive problem (i.e. an online dynamic optimization problem without stochastics). To be able to do this efficiently complex algorithmic design is required. Alternatively, exhaustive search can be used, but this is very inefficient. For this reason the problem instances solved using global (enumerative) optimization methods are relatively small. Also, a problem exists with most methods if the decision to be made concerns *continuous* (e.g. real-valued) variables. It is then not possible to optimize decision trajectories for all possible values for the current decision. It is not even possible to optimize the entire trajectory and then choose for the current situation the decision that has maximum average profit. For continuous decision variables it is namely not likely that optimal values will be the same in different scenarios. Discretization can be a solution, but doing this properly is typically very hard.

Clearly, there is a need for computationally intelligent algorithms that are able to come up with good solutions at any time and are not restricted to solving discrete optimization problems with only a few alternative decisions to choose from at any point in time. Evolutionary algorithms offer a way of doing just that as they are a means of performing optimization by continuously adapting a set of solutions. Thus EAs always have a solution available. Moreover, EAs are known to be able to successfully optimize many different types of solution. In the following sections we shall give two examples of using EAs to solve dynamic optimization problems online in the area of transportation and logistics. In both applications, the EA is continually run to evolve a plan. A plan can be either a list of decisions (yet) to be executed or a strategy on the basis of which decisions are made (i.e. what to do under which circumstances). Whenever a plan is evaluated, it is not only evaluated for the current situation, but also for the predicted future.

3 An Application to Transportation: Vehicle Routing

In this section we focus on the dynamic vehicle routing problem. In this problem, routes have to be planned for a fleet of vehicles to pick up loads at customers. The problem is dynamic because the loads to be transported are announced while the vehicles are already on-route [15].

A few studies currently exist in which information is used about future loads [6, 7, 13, 18, 24, 25, 30]. Most approaches employ a waiting strategy. For each vehicle, upon its arrival at a customer, a waiting window is defined within which a new load is expected to arrive at that customer or at a nearby customer. During that waiting period, the vehicle does not move because it anticipates on having to move only a little in the near future to pick up a new load. In this section, similar to [30], we opt for an approach in which the vehicles keep driving, unless they are at a centrally located depot. The rationale behind this idea is the principled notion that as long as there are loads to be transported, we do not want to have any vehicles waiting around. To move the vehicles as efficiently as possible, we propose to learn the distribution of load announcements at the customers. We use this information to predict the number of future expected loads at a certain customer. By directly integrating this expected value into the fitness of solutions, i.e. vehicle routes, the EA that we use here is able to make informed decisions about anticipated moves (i.e. moves to customers that currently do not have a load ready).

3.1 Problem Definition

The definition of the dynamic vehicle routing problem that we use here is the same as the one used by Van Hemert and La Poutré [30]. Here we shall restrict ourselves to an intuitive, yet concise, description of the problem at hand. Exact mathematical details can be found in [30].

A set of customers is predefined. Each customer has a certain location defined by 2D coordinates. The distance between two locations is the Euclidean distance. The goal in solving the problem is to deliver as many loads as possible. Each load has to be picked up at a certain customer and must be delivered to the central depot. A load has a certain announcement time (i.e. the time from which it is available for pickup). Each load must be delivered to the depot within a certain delivery window, starting from the moment of announcement. Beyond this delivery window the load is no longer accepted by the depot. The size of the delivery window is fixed and is denoted Δ.

To transport the loads, a fleet of vehicles is available. All vehicles have the same capacity. All loads have the same size. Both the size of the loads and the capacity of the vehicles is integer. Initially, all vehicles are at the depot.

At any time t^{now}, the solver must be able to return a list of actions to be performed; one separate action for each vehicle in the fleet. Actions are either to go and pick up a load at a customer, to go to a certain customer without a pickup assignment (i.e. an anticipated move) or to go back to the depot to drop off all loads that are currently being carried.

To ensure that loads are only picked up if they can be delivered on time and to furthermore ensure that loads that have actually been picked up are indeed delivered on time, constraints exist to ensure that such solutions are infeasible. The optimization approach must now only return feasible solutions.

3.2 Optimization Approach

The Dynamic Solver

The dynamic solver updates the optimization problem whenever a change occurs, i.e. when a new load becomes available for pick up. In addition, the dynamic solver controls the EA. It runs the EA and announces changes to the EA so that these changes may be accounted for in the solutions that the EA is working with. It also requests the currently best available solution from the EA whenever changes occur and presents that solution to the real world as the plan to be executed. In our case, the problem changes whenever a new load is announced, whenever a load is picked up or whenever loads are delivered. In addition, the currently executed solution changes whenever a vehicle arrives at a customer, regardless of whether a load is to be picked up there.

The EA is run between changes (also called events). In practice, the time that is available for running equals the time between events. Because computing fitness evaluations takes up most of the time, in our simulated experiments we ensured that the number of evaluations that the EA was allowed to perform between two subsequent events is linearly proportional to the time between events in the simulation. For each time unit of the simulation the EA may perform one generation. Since the population size will be fixed, the EA will thus perform a fixed number of evaluations in one simulation run.

The whole simulation operates by alternatively running the EA and the simulated routing problem. The routing simulator calculates when the next event will occur, e.g., a vehicle will pick up or deliver a load, or, a load is announced for pickup. Then, the EA may run up until this event occurs. This way we simulate an interrupt of the EA when it needs to adapt to changes in the real world. The best individual from the last generation before the interrupt is used to update the assignments of the vehicles in the routing simulation. Then, the routing problem is advanced up until the next event. Afterward, the individuals of the EA are updated by removing assignments that are no longer applicable (i.e. delivered loads or loads that have passed their delivery window) and by adding assignments to pick up loads that have recently been made available.

Base EA: Routing Currently Available Loads

With the exception of the selection method, the base EA that we use is the same as the one used by Van Hemert and La Poutré [30].

Representation: The representation is a set of action lists, one separate list for each vehicle in the fleet. An action list describes all actions that the vehicle will perform in that order. In the case of the base EA, this action list contains only pickup actions. The first action in the list for a specific vehicle is currently being executed by that vehicle. Properties such as the number of loads that a vehicle already carries is stored in the simulation and is not subject to search.

New loads: Whenever new loads are announced, these loads are injected randomly into the action list of a single vehicle in each member of the population.

Variation: Only mutation is considered. Two vehicles are chosen randomly. These two vehicles may be the same vehicle. From these two vehicles, two actions from their lists are chosen randomly. These actions are swapped. This operator allows visits to customers to be exchanged between vehicles or to be re-ordered in the route of a single vehicle directly.

Selection: The selection scheme that we employ ensures elitism. We use truncation selection to select half of the population. The other half is discarded. Using variation, the discarded half is replaced with new individuals. Hence, elitism of the best half of the population is employed. Although this selection operator is rather strict, enough diversity is introduced as a result of mutation and the random introduction of new loads. As a result, selecting more strictly allows the EA to weed out bad solutions more efficiently.

Decoding: It is important to note that as a result of load introduction and of variation, action lists may come to represent an infeasible solution. For example the action list may cause the vehicle to be on-route too long for the loads that it is carrying to be returned to the depot within the delivery window. For this reason a decoding mechanism is used that decodes the representation into a valid solution, i.e., a solution where none of the constraints are violated. The representation itself is not altered. Assignments that violate one or more time constraints are ignored upon decoding. When a vehicle reaches its capacity or when adding more assignments will violate a time constraint, the decoder inserts a visit to the depot into the action list. Afterward, this vehicle may be deployed to service customers again. This procedure is the same as used in [20]. The fitness of the individual will be based on the decoded solution. Although the decoding process may have a large impact on the fitness landscape, it is necessary as in a dynamic environment we must be able to produce valid solutions on demand.

Fitness: The fitness of an individual corresponds to the number of loads that is returned to the depot, i.e. the number of loads picked up when executing the current decoded action lists for all vehicles. It should be noted that this representation already provides, in part, a way to oversee future consequences of current decisions. To see this, note that the only decision required to be taken at each point in time from the problem's perspective is what to let each vehicle in the fleet do next. By having a list of actions to perform after the first next move, planning ahead is made possible, which consequently allows for more efficient routing. However, this anticipation at time t^{now} covers only the non-stochastic information about the problem that is available at time t^{now}. It might be possible to improve the benefits of anticipation further by considering some of the stochastic information about the problem at time t^{now}. This is the goal of introducing anticipated moves.

Enhanced EA I: Implicit Anticipated Moves

In this approach, there is no explicit link between making an anticipated move and the expected reward to be gained from that move [30]. Instead, the mechanism behind anticipated moves is implicit and focuses on ensuring that anticipated moves do not cause the resulting routing to violate any constraints. Ultimately, this results in slight deviations from routes that are built based upon currently available loads, otherwise the loads will not be returned to the depot in time. It is only over multiple generations and over time that in retrospect an anticipated move can be found to have been advantageous.

Variation: To guide the introduction of anticipated moves, an anticipated-move-rate α is used. Upon mutation, this parameter represents the probability of introducing an anticipated move into the route of a single vehicle. Similar to mutation in evolution strategies [2], this parameter is subject to self-mutation.

Fitness: To prevent selection of individuals with a large α that cause many constraint-violating anticipated moves to be introduced, the fitness function is extended with a penalty term. The penalty term grows linearly with the number of constraint-violating anticipated moves in a solution.

Enhanced EA II: Explicit Anticipated Moves

The results reported in [30] already indicate an improvement over the base EA for certain values of the delivery window Δ. However, there is no directly apparent reason that the anticipated moves will actually result in the collection of more loads. A bigger improvement is to be expected if a proper, direct and explicit reason for making anticipated moves is introduced. To this end, we opt for an explicit means of anticipation. We still use the basic strategy of introducing anticipated moves randomly, i.e. we use the same variation technique. To bias the search toward feasible anticipated moves with a positive number of expected future loads, we alter the fitness function.

Fitness: First, assume that we have an oracle that can tell us for each customer exactly when future loads will become available at that customer. In that case the fitness of a decoded action list can be computed by not only counting the number of loads that are picked up along the route as a result of premeditated pickup actions, but also the number of loads that are available at customers upon arrival there. Care must be taken that the capacity of the vehicle is not exceeded in computing the number of additional loads. Moreover, only the loads that can still be brought back to the depot on time should be counted. Also, each load should only be counted once to avoid overrating the goodness of anticipated moves when two or more vehicles have planned a visit to the same customer. As we now know exactly how fruitful certain anticipated moves are, a much more efficient search of anticipated moves becomes possible.

In practice we do not have such perfect information. For each customer we therefore propose to estimate the distribution of the time between two subsequent loads becoming available for transport. To estimate this distribution, we use the normal distribution. How to compute maximum-likelihood estimates for the normal distribution is well known from the literature [1, 29]. The expected number of loads that will become available at a certain customer c_i between the current time t^{now} and the time t^{arrive} of arrival of a vehicle at c_i is just $(t^{\text{arrive}} - t^{\text{now}})/\mu^{c_i}$ where μ^{c_i} is the mean of the distribution of the time between two subsequent loads at customer c_i. Similar to the case of perfect information we must only count the expected loads that can still be brought back to the depot in time. Also the capacity of the vehicle and the possibility of multiple vehicles planning an anticipated trip need to be taken into account. This can be done in the same way as in the case of perfect information.

3.3 Experiments

Experimental Setup

In practice, customers are often clustered into regions as opposed to scattered around uniformly [28]. We therefore use a particular arrangement of the customers by clusters, similar to the arrangement used in [30]. First a set of points called the set of cluster centers C is created by randomly selecting points (x, y) in the 2-dimensional space such that these points are uniformly distributed in that space. Then for each cluster center $(x, y) \in C$ a set of locations $R(x, y)$ is created such that these locations are scattered around the cluster center by using a Gaussian random distribution with an average distance of τ to choose the diversion from the center. This way we get clusters with a circular shape. The set of customer nodes N is defined as $N = \{n | n \in R(x, y) \wedge (x, y) \in C\}$. The set of locations form the nodes of the graph $G = (N, E)$. This graph is a full graph and its edges E are weighted with the costs to traverse them. For each $(n_1, n_2) \in E$, this cost is the Euclidean distance between n_1 and n_2.

A set of loads is randomly generated, which represents the work that needs to be routed. Every load starts at a customer node and needs to be carried to a central depot, which is located in the center of the map. Each customer generates loads where the time between two subsequent loads is normally distributed with a mean of μ^{Loads} and a standard deviation of σ^{Loads}. Typically customers are manufacturers. Therefore, the internal process of preparing loads is often quite regular. The larger σ^{Loads}, the less regular the process is assumed to be and the more randomly generated the loads will appear to be.

We have randomly generated 25 problem instances and have run the EA without anticipation, the EA with implicit anticipation, the EA with optimal-information anticipation and the EA with learned-information anticipation for $1 \cdot 10^5$ time units. We have varied the standard deviation of the time between subsequent loads, the delivery window and the capacity of the vehicles. An overview of all parameters used in our experimental setup is given in Table 1.

Table 1. Parameter settings used in our experiments

Parameter	Value
Maximum width and height of the map	200×200
Number of locations	$\|N\| = 50$
Number of clusters	$\|C\| = 5$
Spread of locations in a cluster	$\tau = 10$
Number of vehicles	$\|V\| = 10$
Capacity constraint	$q \in \{1, 5\}$
Delivery time constraint	$\Delta \in \{20, 40, \ldots, 400\}$
Average time spread of loads	$\mu^{\text{Loads}} = 400$
Standard dev. time spread of loads	$\sigma^{\text{Loads}} \in \{20, 40, \ldots, 200\}$

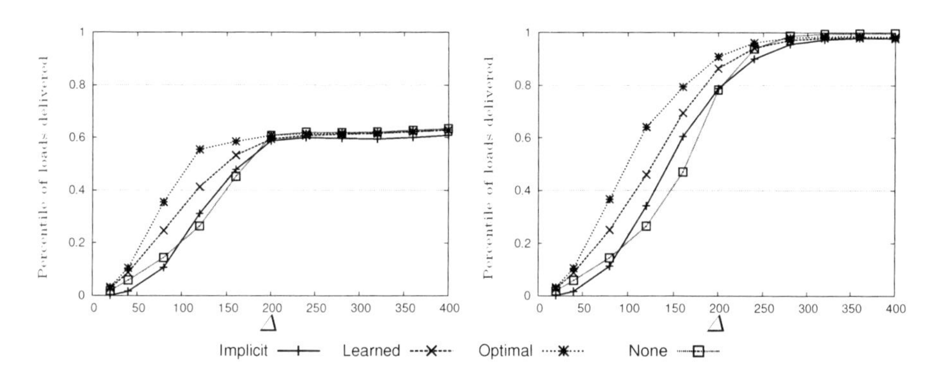

Fig. 1. Routing efficiency in percentage of loads delivered as a function of the delivery window for all EAs and a standard deviation of the time spread of the loads of $\sigma^{\text{Loads}} = 40$. Vehicle capacity is 1 in the left graph and 5 in the right graph.

Results

Figure 1 shows the efficiency of the various EAs with respect to the problems in our test suite for a standard deviation of the time spread of the loads of 40. There is a clear shift in problem difficulty when varying the length of the delivery time window. If this time window is very small, anticipatory routing only pays off if one is certain that there will be loads that can be picked up upon arrival at a certain customer. The number of loads that can be picked up and delivered on time is so small that uninformed anticipatory moves directly cause a drop in the number of loads that could have been delivered otherwise. Indeed, if the learned information or the perfect information is used, an improvement can be found over not using anticipatory moves, where the perfect information of course leads to the biggest improvements. For an average Δ there is much room for improvement. Indeed all anticipatory approaches are capable of obtaining better results than the non-anticipatory approach. However, the use of explicit learning and predicting the announcement of new loads is able to obtain far

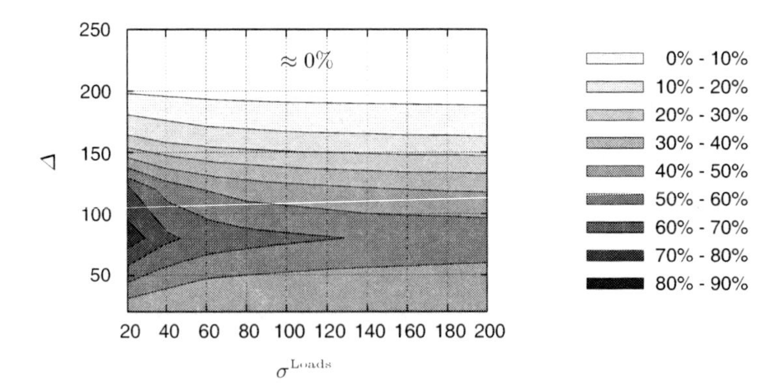

Fig. 2. Relative performance increase of our proposed anticipation-by-explicit-learning EA over the non-anticipatory EA for various values of the delivery window and the standard deviation of the time spread of the loads

better results than when the implicit means of anticipation is used. If the delivery window becomes too large, there is ample time to fully use the capacity of the fleet to the maximum and hence there is no longer any difference between using anticipation and using no anticipation. The problem thus becomes easier.

Figure 2 shows a contour graph of the relative performance increase that can be obtained when using our explicit prediction approach to anticipation as compared to the EA that does not use anticipatory moves. This height-map shows clearly the phase-shift with respect to the delivery time window. There are clear bounds within which an improvement can be obtained. This graph also shows the influence of the randomness of the problem. The larger the standard deviation of the time between subsequent loads, the smaller the performance increase becomes. Clearly, the largest performance increase can be obtained if the variance goes down to 0, which corresponds to the case of using the optimal information. Although the best results only comprise a small area of the graph and thus correspond only to a specific type of problem settings, the range of problem settings for which an improvement can be obtained is large and rather robust with respect to an increase in randomness. Hence we can conclude that our explicit anticipatory approach provides a robust means of improving the quality of online dynamic vehicle routing that is generally speaking preferable compared to an implicit means of anticipatory routing.

4 An Application to Logistics: Inventory Management

General Description

We employ a commonly used definition of inventory management (IM) problems [26]. In Fig. 3 a schematic overview is given. Buyers, also called customers and denoted C_i, buy goods from a vendor. The number of buyers is denoted n_c. The number of goods and the frequency of buying are called the demand and is

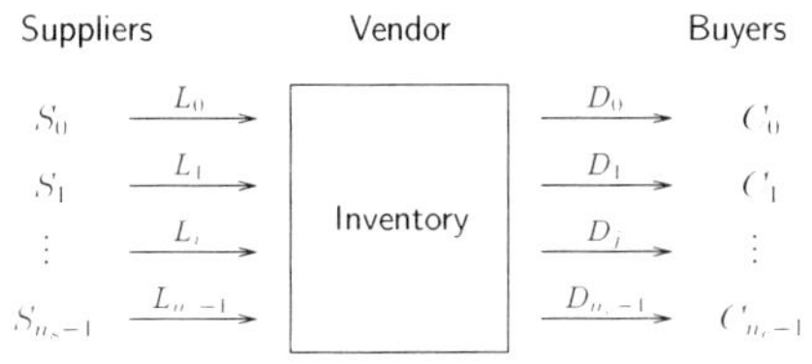

Fig. 3. Schematic overview of inventory management

denoted D_i for customer C_i. To prevent going out of stock, the store keeps an inventory. Inventory must be replenished from time to time. Because delivery of new stock from the store's suppliers also takes time (called the lead time, denoted L_j for supplier S_j), the replenishment order must be placed before going out of stock. The number of suppliers is denoted n_s.

Base Definition

Although IM poses problems that are continuous through time, when solving these problems we typically discretize them. In this chapter, we will use a discretization of time into units of a minute. Let $M(t)$ be the money that the vendor has at time t. The decisions to be taken are how much to order and which suppliers to order from at a any given point in time. Hence, a general formulation of IM is

$$\max_{\boldsymbol{x}(t)} M(0) + \left\{ \sum_{t=0}^{t^{\text{end}}} \Delta M(t) \right\} \tag{1}$$

where $\Delta M(t)$ is the change in the money of the vendor at time t, $M(t)$ depends on the decisions $\boldsymbol{x}(t)$ and t^{end} is the end of the planning horizon. The expenses of the vendor are the costs of holding the inventory and the ordering of new supplies. The income of the vendor is based on sales made:

$$\Delta M(t) = \mathit{Sales}(t) - \mathit{HoldingCost}(t) - \mathit{OrderingCost}(t) \tag{2}$$

The holding cost can be computed per time unit and depends on the size of the inventory $I(t)$. Typically, holding costs are a fixed price p^H per unit of inventory per time unit:

$$\mathit{HoldingCost}(t) = p^H I(t) \tag{3}$$

Typically, the cost of an order that is placed at some supplier is paid for when the order is delivered. The ordering cost at time t therefore depends on earlier decisions $\boldsymbol{x}(t')$, $t' < t$. The cost of a replenishment order at supplier S_i typically consists of two parts: a fixed part c_i^O and a part that increases with the size of the order. Typically, a fixed price p_i^O per unit of ordered goods is charged. Let $x_i^{\text{OrderQuantity}}(t)$ be the quantity of goods ordered from supplier i at time t and let

$L_i(x_i^{\text{OrderQuantity}}(t'))$ be the time it takes supplier S_i to deliver that order, then we have:

$$OrderingCost(t) = \sum_{t'<t} \sum_{i=0}^{n_s-1} o(i,t',t) \tag{4}$$

where

$$o(i,t',t) = \begin{cases} c_i^O + p_i^O x_i^{\text{OrderQuantity}}(t') & \text{if } x_i^{\text{OrderQuantity}}(t') > 0 \\ & \text{and } t' + L_i(x_i^{\text{OrderQuantity}}(t')) = t \\ 0 \text{ otherwise} & \end{cases}$$

Note that in an implementation, it is not efficient to use Equation 4 directly. Instead, an event queue can be used to check whether a new supplier order has arrived at time t and hence whether payment is due.

The income of sales at time t depends on whether the demand of a buyer at time t can be met. Only if there are enough goods in inventory, a sale is made. Partial sales are typically excluded [26]. The goods are sold at a fixed unit price p^S. Let $D(t)$ be the demand of the buyers at time t, then we have:

$$Sales(t) = \begin{cases} p^S D(t) & \text{if } D(t) \leq I(t) \\ 0 & \text{otherwise} \end{cases} \tag{5}$$

We note that there are many other ways in which the above costs and gains can be computed, but most are just minor variations of the equations presented above. Also, other costs and gains besides the ones mentioned above may be taken into account such as the directly computable cost of having more demand than inventory (called excess demand cost) and the subsequent possible loss of sales (called lost sales cost). Such functions are similar to the ones above. Although they indeed make the model more involved, it does not increase the level of problem difficulty, especially for a general problem solving approach that we shall use for this type of problem (see Sect. 4.1).

Although now $\Delta M(t)$ has been defined, there are still functions in the above definitions that are left undefined, namely $I(t)$, L_i and $D(t)$. The inventory (or stock) level $I(t)$ usually is assumed to have the property $I(0) = 0$, but this is not a requirement. In addition, the inventory level rises with $x_i^{\text{OrderQuantity}}(t')$ at time $t' + L_i(x_i^{\text{OrderQuantity}}(t'))$ (i.e. when the supplier order arrives). The inventory level falls with $D(t)$ when a sale is made at time t. The formulations for the rise and fall of $I(t)$ are very similar to those of $\Delta M(t)$. Function L_i for the delivery time for an order placed at supplier i typically follows a certain probability distribution. The same holds for the demand function $D(t)$. The demand function has two sources of stochasticity however: both the quantity Dq of the demand and the frequency of the demand Df. In this chapter we shall assume all these functions to be normally distributed at any time t with means of respectively $\mu^L(t)$, $\mu^{Dq}(t)$ and $\mu^{Df}(t)$ and variances of respectively $\sigma^{2,L}(t)$, $\sigma^{2,Dq}(t)$ and $\sigma^{2,Df}(t)$ where $\mu^{Df}(t)$ and $\sigma^{2,Df}(t)$ describe the distribution of the time between two subsequent customer visits. Note that all these distributions are restricted to $\mathbb{R}^+$.

Time-dependence plays an important role even in the base definition of IM. The decision of whether or not to place a replenishment order at a certain point in time has great future consequences because it determines future inventory levels. Also, a decision to place an order at a supplier leads to a future event (delivery of goods) that is a response to placing the order. Still, although time-dependence is already important here, it is of a rather trivial nature because the only function that is affected by decisions made earlier is the level of the inventory. Given a fixed demand, a single supplier, and a fixed lead time for that supplier, the best strategy for placing replenishment orders is straightforward to compute [26]. Although this is already no longer possible if there is more than one supplier [23], an extension of the model that defines a second, but also very practically relevant, level of time-dependence, makes the problem even harder.

Extended Definition: Customer Satisfaction

Customer satisfaction is important in doing business with customers. A higher level of customer satisfaction will most likely result in a growing frequency of customer transactions, either from the same customer or new customers as satisfied customers will spread the word. In our model this means that whether or not a customer is satisfied when requesting goods from the vendor influences the stochastic model that underlies the customer demand behavior. We can integrate this in the above model by changing the parameters that describe the distribution of the time between two subsequent customer visits, $\mu^{Df}(t)$ and $\sigma^{2,Df}(t)$. If a sale can be made (there is enough inventory, see Equation 5), the customer frequency increases and thus the time between two subsequent customer visits decreases. If the customer cannot be satisfied (no sale is made), the frequency decreases:

$$\begin{aligned} \mu^{Df}(t+1) &= \max\left\{1, C(t)\mu^{Df}(t)\right\} \\ \sigma^{2,Df}(t+1) &= \max\left\{1, C(t)\sigma^{2,Df}(t)\right\} \end{aligned} \tag{6}$$

$$C(t) = \begin{cases} \frac{1}{2} \text{ if } D(t) \leq I(t) \text{ and } D(t) > 0 \\ 2 \text{ if } D(t) > I(t) \text{ and } D(t) > 0 \\ 1 \text{ otherwise } (D(t) = 0) \end{cases} \tag{7}$$

The influence of customer satisfaction now is one of true time-dependence, which can be seen as follows. A decision whether or not to order new goods from a supplier has a future effect on the size of the inventory. Depending on the level of the inventory, a sale can either be in the future made or not. Although this is already a form of time-dependence, customer satisfaction brings a secondary level of time-dependence. Whether or not a future sale can be made influences further future sales indirectly because of an altered customer frequency. This secondary effect is very important however, because it determines the rate of change in future profits. If a higher frequency of customer visits can be met as a result of proper inventory management, profits can be made faster. However, this type of time-dependent influence is often overlooked in literature [3]. Without overseeing the future effects on the frequency of customer visits, the strategy optimizer will see no need to change the strategy beyond one that meets with the current expected rate of customer frequency. Although this does not mean that profits become losses, the strategy can clearly not be optimal.

4.1 EA Design

Many models exist for simple to hard problems in IM. For the simplest problems, exact optimal strategies are known. For practical problems however, there are typically multiple store suppliers to choose from with different delivery times, order magnitudes and costs. Also demands and lead times tend to be stochastic rather than deterministic. Although these aspects increase the benefit of a non-myopic view, they also make the problem harder to solve [23]. Consequently, no exact optimal strategies exist for the harder and more general cases. For specific cases, specific heuristics exist. There is no general flexible approach however that is applicable to a variety of IM problems. Here we again follow the principled idea of optimizing the future to oversee the consequences of current decisions.

We again use EAs as the base optimization technique. We take a slightly different approach to finding the decisions than we did in the previous section for vehicle routing however. It is common practice in IM to find a strategy instead of individual decisions. The strategy that we employ is a common one in IM. The strategy is a so-called (s, Q) strategy [26]; s is called the re-order point and Q the order-up-to size. One such strategy is used for each supplier. Hence, the genotype contains $2n_s$ real values to be optimized, where n_s is the number of suppliers. If the stock drops below the re-order point s_i of supplier i, and no order is currently outstanding for supplier i, a new order is placed at supplier i of size $Q - \mathit{stocklevel}$. Thus, in the case of two suppliers, if an order from the cheaper supplier is running late, the stock level will drop further and the rule for the more expensive, emergency supplier becomes active. It is not known whether this strategy can be optimal for the case of two suppliers, but it is an often used, sensible choice.

The EA thus finds a strategy. In addition to a population, a current best strategy is maintained. This allows the EA to be run continuously. Whenever a decision needs to be made, the current best strategy can be applied. Evaluation of a strategy is done by running that strategy in the simulation multiple times using different random seeds (i.e. in different scenarios). Note that in the previous section, we used the expected values of the distributions for prediction instead of averaging over multiple scenarios. It was recently shown that the optimization of strategies using EAs works better when averaging over multiple sampled future scenarios is used instead of the expected value [9]. The quality of a strategy is measured by its average evaluation value, averaged over all sampled future scenarios. The variance is also stored. The variance is required to compare the best strategy in the population with the current best strategy. Because multiple random seeds are used, corresponding to multiple drawings from the probability distribution that underlies the problem, statistical hypothesis tests are required to be certain that an improvement has been obtained. The statistical hypothesis test that we used in our experiments is the Aspin-Welch-Satterthwaite (AWS) T-test at a significance level of $\alpha = 0.05$. The AWS T-test is a statistical hypothesis test for the equality of means in which the equality of variances is not assumed [19].

The EA must optimize the parameters of the (s, Q) strategies. Since these are real values, we use a real-valued EA. Specifically, we use a recent Estimation-of-Distribution Algorithm (EDA) called SDR-AVS-IDEA [8]. EDAs estimate the distribution of the selected solutions and sample new solutions from this estimated distribution in an attempt to perform variation more effectively. In SDR-AVS-IDEA normal distributions are used. For this problem we did not learn any covariances. This means that SDR-AVS-IDEA computes for each parameter the mean and variance in the selected set of solutions and resamples new solutions with these parameters using a separate one-dimensional normal distribution for each parameter to be optimized. It should be noted that other EAs can be used as well. Particularly of interest are EAs that have been designed to tackle dynamically changing environments without anticipation such as the self-organizing scouts [11]. The focus of this chapter however is on the extended design of EAs to facilitate anticipation. For the problems at hand, the relatively simple EA that we employ suffices.

4.2 Experiments

Problems

We use four IM problems. For each problem, inventory is to be managed for 129600 minutes, i.e. 90 days. Orders can be placed any minute of the day.

Problem I represents problems of the type for which an optimal strategy can be computed beforehand. There is one supplier and one product. Product quantities are integer. The product is sold to the buyers at a price of 50 and bought from the supplier at a price of 20. A fixed setup cost for each order placed at a supplier is charged at a price of 50. Inventory holding costs are 1 per day per unit. The lead time of the supplier is fixed to 3 days. The demand is fixed to an order of 1 item every hour.

Problem II represents problems for which there is not a known optimal strategy. There are two suppliers. One supplier is cheaper than the other. The more expensive supplier can supply immediately, but costs twice as much. This type of setting is popular in IM research. It is typically known as IM with emergency replenishments and is known to be a hard problem [23]. The second supplier is used only if the stock has become really low and stock outs are imminent. To add to the difficulty of the problem, we have made the lead time of the cheapest supplier both stochastic and periodically changing. The lead time of the slower supplier is normally distributed with mean (in minutes) of $4320\left(\cos((2\pi t)/43200) + 1\right)/2$, i.e. it varies between 0 and 3 days and the period-length of the cosine is 30 days. The variance is $1440^2\left(\cos((2\pi t)/43200) + 1\right)/2$, i.e. it varies between 0 and 1440 days, corresponding to a maximum standard deviation of 38 days with the same period-length as the mean. The periodically changing lead time causes the optimal strategy to change with time as well. The maximum size of the standard deviation is not very likely to occur in practical situations. The experiments in this chapter are however meant to illustrate the working principles of the framework. The current setup will show whether changes can be anticipated. The demand is now also stochastic. The time between two subsequent orders is normally distributed with a mean of one hour and a variance of 60 hours. The amount ordered is also normally distributed, with a mean of 3 products and a variance of 9 products. For this setting, there are not any known heuristics.

Problem III equals Problem I but has customer satisfaction (Equation 6).

Problem IV equals Problem II but has customer satisfaction (Equation 6).

Algorithmic Setup

We used two different EA settings, a "small" setting and a "big" setting. The small setting corresponds to a situation in which there is only very little time to do optimization and thus the EA resources are small. The big setting corresponds to a situation in which there is more time and thus the EA resources are larger. In the small settings, the population size is 50, scenario-evaluation simulates 10 days into the future, 5 generations of the EA can be done per day, and 10 scenarios are used. In the big settings, all settings are three times bigger. The population size is 150, scenario-evaluation simulates 30 days into the future, 5 generations of the EA can be done every eight hours and 30 scenarios are used. To facilitate the simulation of the future, the EA learns the distribution behind the stochasticity of the buyer using maximum likelihood estimates. The stochasticity of the supplier is assumed to be known. All results were averaged over 100 independent runs.

Results

Problem I: In Fig. 4 the average profit obtained is shown for both EA settings. The approach can be seen to be a scalable technique (also on Problem II) in the sense that allowing more resources results in better solutions. Investing in computing power thus results in a better policy for a vendor. Moreover, even the small settings for the EA lead to very good profits. The maximum profit that can be obtained on problem I is 58888. This profit corresponds to a setting of the strategy ($(s, Q) = (143, 143)$) that is far outside the range in which we initialized the EA ($(s, Q) \in [0, 25] \times [0, 50]$). Out of all settings in the initialization range, the maximum profit is only 25705. The EA is thus also capable of finding much better solutions when initialization is suboptimal. The big EA settings lead to near-optimal results.

Figure 4 also shows the strategies obtained with the big EA settings for both problems in a typical run of the EA. The lack of stochasticity in problem I

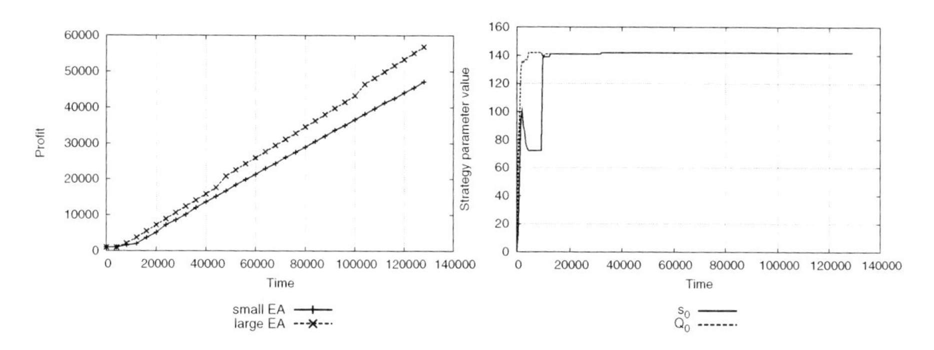

Fig. 4. *Left:* Results on inventory management problem I; *Right:* Strategies evolved in a typical run of the EA on problem I

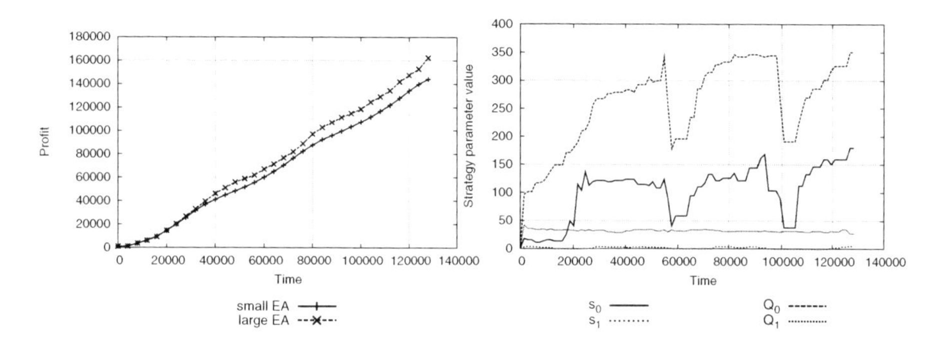

Fig. 5. *Left:* Results on inventory-management problem II; *Right:* Strategies evolved in a typical run of the EA on problem II

translates into finding a stable strategy by the EA very quickly and maintaining that strategy throughout the run.

Problem II: In Fig. 5 the average profit obtained is shown for both EA settings. The profits on problem II are higher than on problem I. The average demand in problem II is three times higher than in problem I. Indeed, the EA is able to obtain a profit of about 3 times higher than on problem I even though problem II is far more difficult.

Figure 5 shows that the adaptive capacity of the EA allows the algorithm to continuously adapt the strategies and find better solutions for the situation at hand as the lead time of the cheapest supplier changes with time. The periodic change of the lead time of the cheapest supplier (S_0) is clearly translated into a periodic change in strategy. When the average and variance of the lead time of the cheapest supplier are small, less products need to be ordered and the threshold can be lower. The threshold for emergency replenishments can even become 0. When the lead time is the largest, emergency replenishments may become

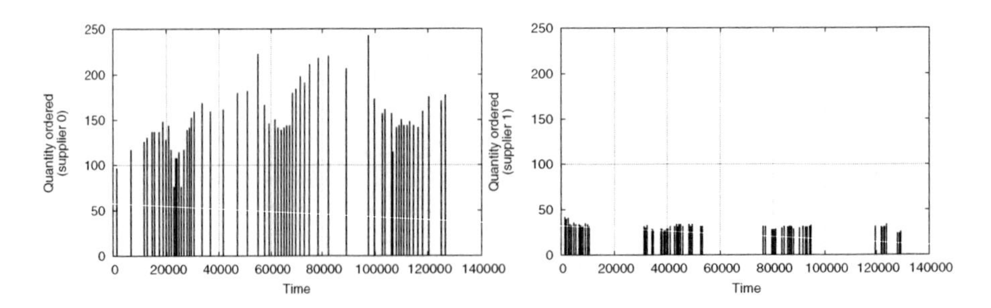

Fig. 6. Quantities ordered in a typical run of the EA on problem II

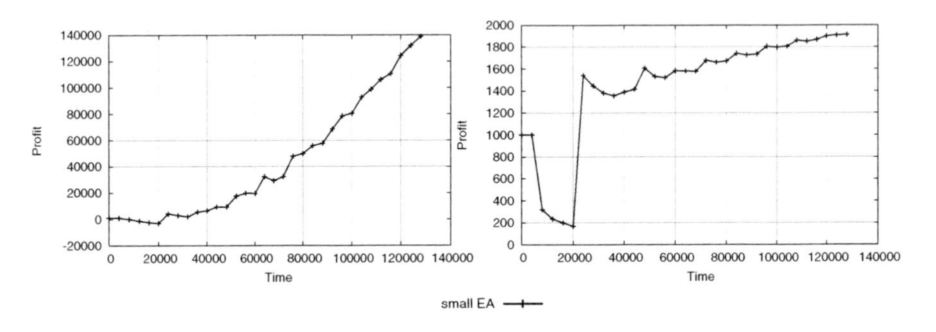

Fig. 7. Results on inventory management problem III with (*Left*) and without (*Right*) anticipation of customer satisfaction

necessary and concordantly, the EA proposes a strategy in which emergency replenishments are made before the stock runs out completely. Also, in this case the re-order point for the cheapest supplier is much higher as is the number of products ordered from that supplier. It can be seen in Fig. 6 that emergency replenishments are indeed made during the periods when the cheapest supplier is less reliable. Furthermore, note that the strategies are not exactly the same in each period. Note that while the EA is optimizing strategies, it is also still learning distributions. Learning converges to the true distribution over time. Finally, the periodic change in the lead time of the cheapest supplier can also be seen in the obtained profits in Fig. 5. When the lead time of the cheapest supplier is the smallest, the EA finds a strategy that uses this supplier more and therefore obtains more profit, resulting in a steeper slope of the profit-versus-time graph at these moments.

Problem III: From the results on problems I and II it is now clear that giving more resources to the EA improves the results. For this reason we now only continue our experiments with the small EA settings. Figure 7 shows the average profit that was obtained if the EA is run with anticipation and without anticipation. The difference between the results is very large. Without anticipation,

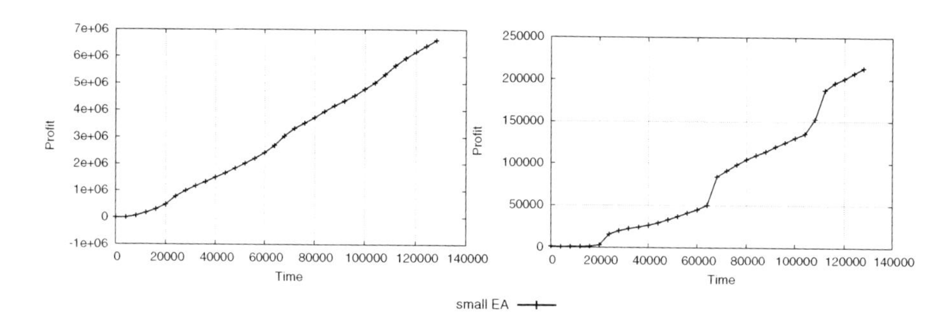

Fig. 8. Results on inventory management problem IV with (*Left*) and without (*Right*) anticipation of customer satisfaction

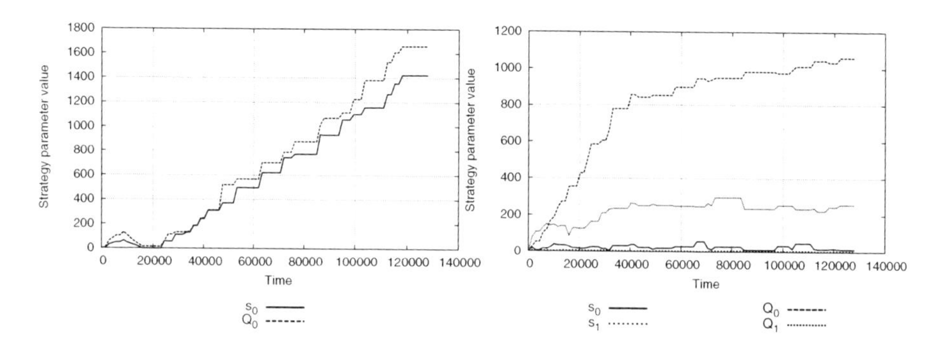

Fig. 9. Strategies evolved in a typical run of the EA with anticipation on problem III (*Left*) and problem IV (*Right*)

the EA is still able to obtain a profit. With anticipation however, the EA notices that having a larger inventory eventually leads to a higher frequency of sales and the result is a much faster growing profit. The growth is exponential until the customer frequency (one customer per minute) can no longer increase.

Inventory growth as a result of ordering more supplies (i.e. a change in strategy) can clearly be seen in Fig. 9. The strategy moves far away from its initialization range and continues to raise the number of goods to order as time goes by until customer satisfaction can no longer be increased (which happens approximately at the end of the run).

Problem IV: A difference that is similar in magnitude can be seen in Fig. 8 between the EA that uses anticipation and the EA that does not use anticipation on problem IV. The main difference with the results on problem III (besides the effect due to the the periodic change in the supply time of the main supplier) is that the increase in the rate of profit is almost immediate. On problem III it takes some time before the strategy matches the maximum customer satisfaction level. The reason for this difference is the emergency supplier. The emergency supplier can supply goods immediately. Ordering more from the emergency supplier therefore has a much faster effect on the customer satisfaction level. In Fig. 9 it can indeed be seen that the strategy for ordering from the emergency supplier is changed very quickly: an increase in the number of goods to be ordered can be observed. This increase allows to immediately meet with the quickly growing demand of the customers. Meanwhile, the strategy for ordering from the main supplier also increases the number of goods to order, making more profit in the end by meeting the maximum level of customer satisfaction mostly through the supplies bought from the main (and cheaper) supplier rather than the emergency supplier.

5 Discussion and Conclusions

In this chapter we have focused on designing intelligent algorithms for solving dynamic optimization problems in the transportation and logistics domain in an online fashion. We have indicated that it is not enough to only track optima as they shift with time. The optimization algorithm is also required to perform anticipation, i.e. take into account consequences of decisions taken earlier. To this end, we investigated a principled approach in which optimization is performed not only for the current decision but also, simultaneously, for future decisions in future, predicted situations.

We have applied this approach to key problems in transportation and logistics, specifically vehicle routing and inventory management, and found significant improvements over traditional approaches. By analyzing carefully what it is that should be predicted, learning this information from past experience and explicitly incorporating it in anticipating the consequences of current decisions, better results can be obtained than when using no prediction or when using an implicit means of prediction.

In addition to the positive results on the problems in this chapter, one of the most beneficial aspects of the approach that we used is that when the problems are changed the overall approach remains the same and is able to obtain positive results. No in-depth redesigning needs to be done when something changes in the definition of the problem. Often, well-designed heuristics need to be completely redesigned when transferring a problem from theory into practice because of discrepancies between the theoretical case and the practical case and the fact that the heuristics are very problem-specific.

Although there are many important questions still to be answered about anticipation in general such as whether the length of the time-interval that is required to use future predictions over can also be detected online, we can conclude that online dynamic optimization and the use of anticipation represent an important avenue of research.

References

1. Anderson, T.W.: An Introduction to Multivariate Statistical Analysis. Wiley, New York (1958)
2. Bäck, T.: Evolutionary Algorithms in Theory and Practice. Oxford University Press, Oxford (1996)
3. Beamon, B.: Supply chain design and analysis: Models and methods. International Journal of Production Economics 55, 281–294 (1998)
4. Bent, R., Van Hentenryck, P.: Online stochastic and robust optimization. In: Maher, M.J. (ed.) ASIAN 2004. LNCS, vol. 3321, pp. 286–300. Springer, Heidelberg (2004)
5. Bent, R., Van Hentenryck, P.: Regrets only! Online stochastic optimization under time constraints. In: McGuinness, D.L., Ferguson, G. (eds.) Proceedings of the National Conference on Artificial Intelligence – AAAI 2004, pp. 501–506. AAAI Press, Menlo Park (2004)
6. Bent, R., Van Hentenryck, P.: Scenario-based planning for partially dynamic vehicle routing with stochastic customers. Operations Research 52(6), 977–987 (2004)
7. Bent, R., Van Hentenryck, P.: Waiting and relocation strategies in online stochastic vehicle routing. In: Veloso, M.M. (ed.) Proceedings of the International Joint Conference on Artificial Intelligence – IJCAI 2007, Hyderabad, pp. 1816–1821 (2007)
8. Bosman, P.A.N., Grahl, J., Rothlauf, F.: SDR: A better trigger for adaptive variance scaling in normal EDAs. In: Thierens, D., et al. (eds.) Proceedings of the Genetic and Evolutionary Computation Conference – GECCO 2007. ACM Press, New York (2007)
9. Bosman, P.A.N., La Poutré, H.: Learning and anticipation in online dynamic optimization with evolutionary algorithms: The stochastic case. In: Thierens, D., et al. (eds.) Proceedings of the Genetic and Evolutionary Computation Conference – GECCO 2007, pp. 1165–1172. ACM Press, New York (2007)
10. Branke, J.: Evolutionary Optimization in Dynamic Environments. Kluwer, Norwell (2001)
11. Branke, J., Kaußler, T., Schmidt, C., Schmeck, H.: A multi–population approach to dynamic optimization problems. In: Parmee, I.C. (ed.) Adaptive Computing in Design and Manufacture – ACDM 2000, pp. 299–308. Springer, Berlin (1999)
12. Branke, J., Mattfeld, D.: Anticipation and flexibility in dynamic scheduling. International Journal of Production Research 43(15), 3103–3129 (2005)
13. Branke, J., Middendorf, M., Noeth, G., Dessouky, M.: Waiting strategies for dynamic vehicle routing. Transportation Science 39(3), 298–312 (2005)
14. Chang, H., Givan, R., Chong, E.: Online scheduling via sampling. In: Chien, S., et al. (eds.) Proceedings of the Fifth International Conference on Artificial Intelligence Planning Systems – AIPS 2000, pp. 62–71. AAAI Press, Menlo Park (2000)
15. Ghiani, G., Guerriero, F., Laporte, G., Musmanno, R.: Real-time vehicle routing: Solution concepts, algorithms and parallel computing strategies. European Journal of Operational Research 151(1), 1–11 (2004)
16. Goldberg, D.E.: Genetic Algorithms in Search, Optimization and Machine Learing. Addison Wesley, Reading (1989)
17. Grötschel, M., Krumke, S.O., Rambau, J. (eds.): Online optimization of large scale systems. Springer, Berlin (2001)
18. Ichoua, S., Gendreau, M., Potvin, J.-Y.: Exploiting knowledge about future demands for real-time vehicle dispatching. Transportation Science 40, 211–225 (2006)

19. Kendall, M.G., Stuart, A.: The Advanced Theory Of Statistics, Inference And Relationship, vol. 2. Griffin, London (1967)
20. Laporte, G., Louveaux, F., Mercure, H.: The vehicle routing problem with stochastic travel times. Transportation Science 26, 161–170 (1992)
21. Larsen, A.: The Dynamic Vehicle Routing Problem. PhD thesis, Technical University of Denmark, Denmark (2000)
22. Mercier, L., Van Hentenryck, P.: Performance analysis of online anticipatory algorithms for large multistage stochastic programs. In: Veloso, M.M. (ed.) Proceedings of the International Joint Conference on Artificial Intelligence – IJCAI 2007, Hyderabad, pp. 1979–1984 (2007)
23. Minner, S.: Multiple-supplier inventory models in supply chain management: A review. International Journal of Production Economics 81(82), 265–279 (2003)
24. Mitrovic-Minic, S., Krishnamurti, R., Laporte, G.: Double-horizon based heuristics for the dynamic pickup and delivery problem with time windows. Transportation Science B 38, 669–685 (2004)
25. Mitrovic-Minic, S., Laporte, G.: Waiting strategies for the dynamic pickup and delivery problem with time windows. Transportation Science B 38, 635–655 (2004)
26. Nahmias, S.: Production and Operations Analysis. Irwin, Homewood (1997)
27. Savelsbergh, M.: DRIVE: Dynamic routing of independent vehicles. Operations Research 46(4), 474–490 (1998)
28. Solomon, M.: The vehicle routing problem and scheduling problems with time window constraints. Operations Research 35, 254–265 (1987)
29. Tatsuoka, M.M.: Multivariate Analysis: Techniques for Educational and Psychological Research. Wiley, New York (1971)
30. van Hemert, J.I., La Poutré, J.A.: Dynamic routing problems with fruitful regions: Models and evolutionary computation. In: Yao, X., Burke, E.K., Lozano, J.A., Smith, J., Merelo-Guervós, J.J., Bullinaria, J.A., Rowe, J.E., Tiňo, P., Kabán, A., Schwefel, H.-P. (eds.) PPSN 2004. LNCS, vol. 3242, pp. 690–699. Springer, Heidelberg (2004)
31. Verweij, B., Ahmed, S., Kleywegt, A.J., Nemhauser, G., Shapiro, A.: The sample average approximation method applied to stochastic routing problems: A computational study. Computational Optimization and Applications 24(2-3), 289–333 (2003)

Part III

Supply Chain Management

Supply Chain Inventory Optimisation with Multiple Objectives: An Industrial Case Study

Lionel Amodeo, Haoxun Chen, and Aboubacar El Hadji

ICD – LOSI (FRE CNRS 2732), University of Technology of Troyes, 12 rue Marie Curie, 10012 Troyes, France
lionel.amodeo@utt.fr, haoxun.chen@utt.fr

Summary. Effective inventory management across a supply chain is very important for reducing inventory costs while improving services to customers. One problem for the management is to determine an optimal inventory policy for each stock in the supply chain. The problem is difficult to solve not only because a supply chain is a multi-echelon inventory system with multiple interrelated stocks but also because it involves conflicting objectives. Finding a set of pareto-optimal solutions for the problem requires a robust and efficient method that can efficiently search the entire solution space of the problem. Genetic algorithms (GAs) seem to be suited for this task because they process multiple solutions in parallel, possibly exploiting the similarities of the solutions by recombining them. In this chapter, supply chain inventory polices are optimised using a multi-objective optimisation approach that combines a genetic algorithm with a Petri net-based simulation tool for performance evaluation. The supply chain considered is first modeled as a batch deterministic and stochastic Petri net, and a simulation-based optimisation method is developed for parameter optimisation of inventory policies of the supply chain with a multi-objective optimisation approach as its search engine. In this method, the performance of a supply chain is evaluated by simulating its Petri net model, and a Non dominated Sorting Genetic Algorithm (NSGA2) is used to guide the optimisation search process toward high-quality solutions. An application to a real-life supply chain demonstrates that our approach can obtain inventory policies better than ones currently used in practice in terms of inventory cost and service level.

Keywords: Supply chain management, Petri nets, Simulation, Multi-objective optimisation, NSGA-II.

1 Introduction

A supply chain is a network of material suppliers, manufacturers, distribution centres/warehouses, and retailers through which products or services are produced and delivered to customers. One important issue in supply chain management is inventory management. The goal of this study is to develop a practical tool that can help companies to reduce their inventory costs and improve their customer services by optimising the inventory policies of their supply chains. The supply chain inventory optimisation problem is difficult to solve because supply chains are multi-echelon inventory systems with multiple interrelated stocks and

A. Fink and F. Rothlauf (Eds.): Advances in Computational Intelligence, SCI 144, pp. 211–230, 2008.
springerlink.com

conflicting objectives. For example, a stock of a manufacturer depends on the stock of its supplier from which the downstream stock is replenished, and the reduction of inventory cost and the improvement of service level are two conflicting objectives for any stock.

Petri nets have been proved a very powerful tool for modelling and analysis of discrete event systems such as manufacturing systems. The tool is also applicable to supply chains since they are discrete event systems as well from a high level of abstraction. In the literature, supply chains are usually described as multi-echelon inventory systems [19] but most existing models can only describe a restricted class of supply chains with simplifications.

In our previous work [3, 14], we have developed a new Petri net model, called Batch Deterministic and Stochastic Petri Nets (BDSPN), for supply chain modelling. With this model, material, information and financial flows of a supply chain can be described in a graphical, concise and integrated way, where operational policies of the chain form one part of the flows. A simulation tool has been developed for performance evaluation of the model. This study aims at developing an effective approach for supply chain inventory optimisation based on the model and the tool.

As we know, for large scale complex systems such as supply chains, it is very difficult to develop analytical methods for performance evaluation and optimisation. As an alternative, many researchers have been seeking for simulation-based optimisation methods, which combine optimisation techniques with performance evaluation by simulation [4, 10]. The optimisation techniques used in these methods include metaheuristics such as genetic algorithms, tabu search, simulated annealing and other stochastic optimisation methods. By appropriately combining these metaheuristics with simulation, near optimal solutions of a stochastic optimisation problem can be found in a reasonable computation time.

In this chapter, we present a simulation-based optimisation approach for parameter optimisation of inventory policies of supply chains. The approach combines the Petri net-based simulation tool for performance evaluation with a genetic algorithm for optimisation. Two objectives are considered in the inventory policy optimisation: inventory cost and service level. Since the problem considered is a multi-objective optimisation problem, a multi-objective optimisation approach based on Non dominated Sorting Genetic Algorithms (NSGA-II) is developed. In the approach, a supply chain is modeled as a batch deterministic and stochastic Petri net, and the performance of the supply chain is evaluated by simulating its Petri net model. The NSGA-II is used to guide the optimisation search process toward global optima. Our approach is tested by using a real-life instance – an industrial supply chain for producing and distributing electrical connectors. Numerical results show that the approach can obtain inventory policies significantly better than ones currently used in practice with a reduced cost and an improved service level.

The remainder of this chapter is organized as follows: Batch deterministic and stochastic Petri nets are briefly introduced in Sect. 2. The multi-objective genetic algorithm is presented in Sect. 3. A real-life supply chain is described in Sect. 4.

Our simulation-based optimisation approach is applied to the optimisation of the inventory policies of the supply chain in Sect. 5 with numerical results and analysis. Several tests are made to evaluate our approach. Concluding remarks are given in Sect. 6.

2 Batch Deterministic and Stochastic Petri Nets

In our previous work [3, 14], we have proposed a new class of stochastic Petri nets – Batch Deterministic and Stochastic Petri Nets (BDSPN) – to meet the modelling needs of supply chains, where inventory replenishment and distribution operations are usually performed in a batch mode triggered by purchase orders and customer orders, respectively. Batch deterministic and stochastic Petri nets extend Deterministic and Stochastic Petri Nets (DSPN) [16] by introducing batch places and batch tokens. The motivation for such an extension is that in systems like supply chains, purchasing, production, and distribution are usually performed in a batch way where customer orders play an important role. In BDSPN, there are two types of places, discrete places and batch places. Tokens in a discrete place are viewed indifferently as those in standard Petri nets, while tokens in a batch place, which have sizes, are viewed as different entities and are represented by Arabic numbers. The tokens of the second type are called batch tokens. We use different ways to represent the marking of a discrete place and the marking of a batch place. The former is represented by a nonnegative integer as in standard Petri nets, while the latter is represented by a set of nonnegative integers. The set may have identical elements and each integer in the set represents a batch token with a given size. We use a vector (or mapping) μ to represent the marking of a BDSPN with $\mu(p)$ denoting the marking of place p, where $\mu(p)$ is either a nonnegative integer or a set of nonnegative integers, depending on the type of the place. Moreover, another type of marking, called M-marking, is also introduced for the definition of BDSPNs. For each discrete place, its M-marking is the same as its μ-marking, while for each batch place its M-marking is defined as the total size of the batch tokens in the place.

Definition: A batch deterministic and stochastic Petri net is a nine tuple:

$$N = (P, T, I, O, V, W, P, D, \mu_0)$$

where $P = P_d \cup P_b$ is a set of places consisting of the discrete places in set P_d and the batch places in set P_b; $T = T_i \cup T_d \cup T_s$ is a set of transitions consisting of the immediate transitions in set T_i, the deterministic timed transitions in set T_d, and the stochastic timed transitions in set T_s; $I \subseteq (P \times T)$ and $O \subseteq (T \times P)$ define the input arcs and the output arcs of the transitions, respectively, as in standard Petri nets; $V \subseteq (P \times T_i)$ defines the inhibitor arcs for immediate transitions with $V \cap I = \emptyset$; W defines the weights for all ordinary arcs and inhibitor arcs. For each arc $a \in I \cup O \cup V$, its weight $W(a)$ is a linear function over the M-marking of the net with integer coefficients, i.e., $W(a) = \alpha + \sum_{p \in P} \beta_p M(p)$, where $\alpha, \beta_p (p \in P)$ are integers, and $W(a) \geq 0$. $W(a)$ is assumed to be constant for each arc a

associated with a timed transition of $T_d \cup T_s$; $\Pi : T \longrightarrow \aleph$ is a priority function assigning a priority to each transition, where $\aleph$ is the set of nonnegative integers. Timed transitions are assumed to have the lowest priority, i.e., $\Pi(t) = 0$ for all $t \in T_d \cup T_s$, and $\pi(t) \geq 1$ for all $t \in T_i$. $D : T \longrightarrow \{0\} \cup \Re^+ \cup \Omega$ defines the firing times of all transitions, where $\Re^+$ is the set of positive real numbers, Ω is the set of random variables with a given distribution, $D(t) = 0$ for all $t \in T_i$, $D(t) \in \Re^+$ for all $t \in T_d$, and $D(t) \in \Omega$ for all $t \in T_s$; $\mu_0 : P \longrightarrow \aleph \cup 2^{\aleph}$ is the initial μ-marking, where $2^{\aleph}$ is a superset consisting of all subsets of $\aleph$, $\mu_0(p) \in \aleph$ if $p \in P_d$, and $\mu_0(p) \in 2^{\aleph}$ if $p \in P_b$.

In graphical representation, discrete places and batch places are represented by single circles and squares with an embedded circle, respectively. Immediate, deterministic, and stochastic transitions are represented by thin bars, filled rectangles, and empty rectangles, respectively. Inhibitor arcs are represented by arrows ending with a small circle. Ordinary tokens are represented by dots, while batch tokens are represented by Arabic numbers that indicate the sizes of the tokens.

The state of the net is represented by its μ-marking $\mu : P \longrightarrow \aleph \cup 2^{\aleph}$. The corresponding M-marking is denoted by $M(\mu) : P \longrightarrow \aleph$, or simply by M, where $M(p) = \mu(p)$ for $p \in P_d$, and $M(p) = \sum_{b \in \mu(p)} b$ for $p \in P_b$. In the following, a place that is connected with a transition by an arc is referred to as an input, output, and inhibitor place, depending on the type of the arc. The set of input places, the set of output places, and the set of inhibitor places of transition t are denoted by ${}^\bullet t$, $t^\bullet$, and ${}^\circ t$, respectively, i.e., ${}^\bullet t = \{p \mid (p,t) \in I\}$, $t^\bullet = \{p \mid (t,p) \in O\}$, and ${}^\circ t = \{p \mid (p,t) \in V\}$.

To well define the semantics of the net, it is assumed that: 1) Inhibitor arcs are only associated with immediate transitions; 2) The weights of all arcs associated with timed transitions (i.e., transitions of $T_d \cup T_e$) are constants; 3) No place which is both inhibitor and input for the same transition exists (i.e., ${}^\bullet t \cap {}^\circ t = \varnothing$).

2.1 Transition Enabling and Firing

The state or the μ-marking of the net is changed with two types of transition firing called *batch firing* and *discrete firing*. They depend on whether a transition has no batch input place.

Discrete Enabling and Firing Rules

If a transition has no batch input place, its enabling condition is the same as that for a transition of DSPNs. That is, a transition $t \in T$ with ${}^\bullet t \cap P_b = \varnothing$ is enabled at marking μ (its corresponding M-marking) if and only if:

$$\forall p \in {}^\bullet t, \quad M(p) \geq W(p,t) \tag{1}$$

$$\forall p \in {}^\circ t, \quad M(p) < W(p,t) \tag{2}$$

Note that in this case, the M-marking of each input place of transition t is the same as the μ-marking of the place. An enabled transition can be fired. If transition t is fired, $W(p,t)$ tokens will be "frozen" in each input place p (i.e., the tokens cannot be used to enable and fire other transitions). After a time delay given by $D(t)$, the firing will finish and tokens will be removed from its input places and created in its output places. The number of ordinary tokens removed from an input discrete place or created in an output discrete place is determined by weight of the arc connecting the place with the transition. For each output batch place, a batch token with the size equal to the weight will be created. That is, the state of the net after the firing of transition t will become:

$$\forall p \in {}^\bullet t, \quad \mu'(p) = \mu(p) - W(p,t) \tag{3}$$
$$\forall p \in t^\bullet \cap P_d, \quad \mu'(p) = \mu(p) + W(t,p) \tag{4}$$
$$\forall p \in t^\bullet \cap P_b, \quad \mu'(p) = \mu(p) + \{W(t,p)\} \tag{5}$$

Batch Enabling and Firing Rules

A transition with at least one input batch place is called a batch transition. A batch transition t is said to be enabled at μ-marking μ if and only if there is a batch firing index (positive integer) $q \in \aleph$ $(q > 0)$ such that:

$$\forall p \in {}^\bullet t \cap P_b, \quad \exists b \in \mu(p), \quad q = b/W(p,t) \tag{6}$$
$$\forall p \in {}^\bullet t \cap P_d, \quad M(p) \geq q \times W(p,t) \tag{7}$$
$$\forall p \in {}^\circ t, \quad M(p) < W(p,t) \tag{8}$$

The firing of t leads to a new μ-marking μ':

$$\forall p \in {}^\bullet t \cap P_d, \quad \mu'(p) = \mu(p) - q \times W(p,t) \tag{9}$$
$$\forall p \in {}^\bullet t \cap P_b, \quad \mu'(p) = \mu(p) - \{q \times W(p,t)\} \tag{10}$$
$$\forall p \in t^\bullet \cap P_d, \quad \mu'(p) = \mu(p) + q \times W(t,p) \tag{11}$$
$$\forall p \in t^\bullet \cap P_b, \quad \mu'(p) = \mu(p) + \{q \times W(t,p)\} \tag{12}$$

A batch transition t is said to be enabled if : (a) Each batch input place p of the transition has a batch token with size b such that all these batch tokens have the same batch firing index q defined as $b/w(p,t)$ for the transition, (b) Each discrete input place of the transition has enough tokens to simultaneously fire the transition for a number of times given by the index, (c) The number of tokens in each inhibitor place of the transition is less than the weight of the inhibitor arc connecting the place to the transition.

For any batch output place, the firing of an enabled batch transition generates a batch token with the size given by the multiplication of the batch firing index and the weight of the arc connecting the transition to the batch place. For any discrete output place, the firing of the transition generates a number of discrete tokens with the number given by the multiplication of the batch firing index and the weight of the arc connecting the transition to the discrete place.

2.2 Inventory System Modelling

Figure 1 shows the BDSPN model of an inventory system with continue review (R, Q) policy, where R is the reorder point and Q is the fixed order quantity. In the model, discrete place p_1 represents the on-hand inventory of the stock and place p_3 represents outstanding orders. Discrete place p_2 represents the on-hand inventory of the stock plus its outstanding orders (the orders that are placed by stock p_1 but not filled yet) that is, $M(p_2) = M(p_1) + M(p_3)$. The inventory position of the stock equals to $M(p_1) + M(p_3) - M(p_4) = M(p_2) - M(p_4)$. Inventory replenishment decisions are based on the position. Batch place p_4 represents backorders of the stock (total unfilled customer demand). The operations of the system such as generation of replenishment orders (t_4), inventory replenishment (t_2), and order delivery (t_1) are performed in a batch way because of the batch nature of customer orders recorded in batch place p_4 and the batch nature of the outstanding orders recorded in batch place p_3. The fulfillment of a customer order will decrease on-hand inventory of the stock as well as its inventory level. This is described by the arcs from places p_1, p_4 and p_2 to transition t_1. If a customer order cannot be filled because of stock-out, it will become a backorder (a batch token in place p_4). The continuous inspection of inventory position, i.e., $M(p_2) - M(p_4)$, is represented by immediate transition t_4 and its associated inhibitor arc. When this position is below the reorder point R, i.e., $M(p_2) - M(p_4) < R$, or equivalently $M(p_2) < R + M(p_4)$, an order with size Q (batch order) will be placed to the supplier (a batch token with the size Q will be created in batch place p_3). If there is a batch token (order) in place p_3, transition t_2 will be fired after its associated delay, which will replenish stock p_1 by delivering the order to it.

Immediate transitions fire in constant zero time whereas timed transitions fire after either an exponentially distributed or a deterministic firing delay. When some timed transitions are enabled in a marking, the transition with the minimum firing delay will cause the marking change. This is called a race condition [1]. Further, as in DSPN, it is assumed that after a marking change each timed

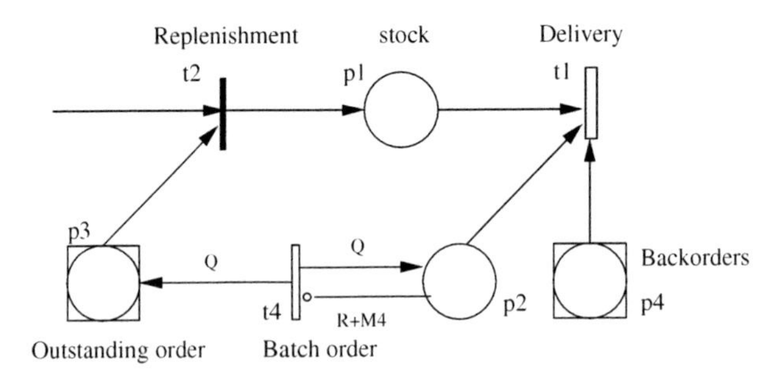

Fig. 1. BDSPN model of an (R, Q) inventory control policy

transition newly enabled samples a remaining firing time from its firing delay distribution. Each timed transition which has already been enabled in the previous marking and is still enabled in the current marking keeps its remaining firing time. This stochastic behavior corresponds to the race policy with enabling memory of stochastic Petri nets [1].

For a more detailed explanation of the BDSPN model (i.e., firing rules, conflicts, temporal policies and behavior) please refer to our previous paper [3].

3 Multi-objective Genetic Algorithm

Multi-objective optimisation is a research topic attracting much attention because many optimisation problems involve multiple and conflicting objectives and a compromise may have to be made among these objectives. The most important concept in multi-objective optimisation is Pareto optimality. A solution is Pareto optimal if it is not dominated by any other solution in terms of all objectives considered. Since a number of solutions may be Pareto optimal, the task of multi-objective optimisation is to find as many as possible such non dominated solutions, and this task is quite complex. In this section, a simulation-based multi-objective optimisation method that combines simulation evaluation of performance and meta-heuristic search with multi-objective genetic algorithms called NSGA-II is adapted for the optimisation of inventory policies of supply chains. We use a multi-objective optimisation method because there are more than one objective in supply chain inventory optimisation especially the inventory cost and the service level.

GA is a search technique based on the mechanism of natural selection and reproduction introduced by Holland and is used to search large, non-linear solution spaces where expert knowledge is lacking or difficult to encode and where traditional optimisation techniques fall short. It starts with an initial set of randomly generated solutions called population and each individual in the population is called chromosome representing a solution. Each chromosome is composed of a set of elements called genes. At each iteration (generation) of GA, all newly generated chromosomes are evaluated using a fitness function to determine their qualities. High quality chromosomes are selected to produce offspring chromosomes through genetic operators, namely, crossover and mutation. After a number of generations, the GA converges to a chromosome which is very likely to be an optimal solution or a solution close to the optimum.

Although standard GAs cannot efficiently solve multi-objective optimisation problems, multi-objective GAs such as NSGA-II can. The choice of NSGA-II is motivated by its potential as a stochastic search method and the facility of encoding inventory policy parameters of supply chains. One of its major advantages for multi-objective optimisation, like in this study, is their ability to find multiple Pareto optimal solutions in one single simulation run. Also, we obtain at the end of the search process a diversified population of good solutions. This diversity is a useful element because it makes possible to have more varied possibilities. Therefore, NSGA-II is very well adapted to treat multi-objective optimisation problems. The first Multi-Objective method based on Genetic Algorithm is VEGA (Vector Evaluated Genetic algorithm) [18]. After VEGA, many multi-objective optimisation methods based on GA are proposed: MOGA (Multi-Objective Genetic Algorithm) [9] in which each individual of the population is arranged according to the number of individuals who dominate it, Horn et al. [12] presented NPGA (Niched Pareto Algorithm). This method uses a tournament based on the concept of Pareto dominance. Another method, NSGA (Non-dominated Sorting Genetic Algorithm) [8] is proposed by Deb and Srivinas.

However these approaches are not elitist, the generated Pareto front is not diversified, and convergence is slow. In order to solve these problems, some new elitist techniques are developed [6]. They keep the best individuals of old generations after genetic operators crossover and mutation. This is realized by copying the best individuals in the population of each generation to the next generation. Among those methods, one can cite RPSGAe (Reduced Pareto Set Genetic Algorithm) [11], SPEA (Strengh Pareto Evolutionary Algorithm) [20], PESA (Pareto Envelope based Selection Algorithm) [13], NSGA-II (Non-dominated Sorting Genetic Algorithm-2) [7], where Deb tries to solve all criticisms of NSGA [8] approach: complexity, lack of elitism and need for specifying sharing parameters. We choose NSGA-II for our supply chain inventory optimisation with multiple objectives because of following reasons : (i) it is an elitist and fast method because it uses non dominated sorting procedure, (ii) its modular and flexible structure, (iii) it can be applied to a large variety of problems.

Moreover, any multi-objective evolutionary algorithm (MOEA) has a formal proof of convergence, with a large diversity, to real Pareto optimal solutions [15]. Indeed, in recent years, extremely robust methods are proposed [2, 5] to solve multi-objective optimisation problems using genetic algorithms. According to the studies which have evaluated several multi-objectives algorithms, the NSGA-II is classified among the best and the most popular algorithm in term of convergence and diversity of solutions.

The following two subsections introduce the principle of NSGA-II and its adaptation to bi-objective supply chain inventory optimisation.

3.1 NSGA II Principle

NSGA-II computes successive generations of a population of solutions belonging to non-dominated fronts. The non-dominated set is identified and constitutes the non-dominated front of level 1 or front 1. In order to find the individuals in the next non-dominated front, the solutions of front 1 are discounted temporarily and the above procedure is repeated. This process continues until all fronts are identified. In order to maintain diversity in the population, the crowding-distance is used. The overall structure of the NSGAII is specified by Algorithm 1.

Algorithm 1: NSGAII overall structure

```
Create the initial population P of size n
Evaluate the n solutions using simulation
Sort P by non domination
Compute the crowding distance of each solution
repeat
  Create and add n children to P (using genetic operators: selection, crossover and
  mutation of two parents)
  Sort P by non domination
  Compute the crowding distance of each solution
  newP ← ∅
  i ← 1
  while |newP| + |front(i)| ≤ n do
    Add front(i) to newP
    i ← i + 1
  end while
  missing ← n − |newP|
  if missing ≠ 0 then
    Sort the solutions by descending order of the crowding distance
    for j ← 1 to missing do
      Add the jth solution of front(i) to newP
    end for
    P ← newP
  end if
until Stopping Criterion
```

3.2 GA Components for Bi-objective Supply Chain Inventory Optimisation

In this study, we consider a supply chain composed of s interrelated stocks. Each stock uses a batch ordering policy (R, Q) for its inventory replenishment. The supply chain has two objectives, namely total inventory cost and service level, to optimise. Basic operations that characterize our adapted NSGA-II are explained as follows.

Encoding

For batch ordering policies, only two parameters are involved: the reorder point R and the order quantity Q. For a supply chain with s stocks controlled by batch ordering policies, each chromosome is encoded as $[R_1, Q_1, \ldots, R_i, Q_i, \ldots, R_s, Q_s]$, where R_i and Q_i are the reorder point and the order quantity of the batch ordering policy of the i_{th} stock, respectively, $i = 1, 2, \ldots, s$. For a supply chain with other types of inventory policy or a mixture of multiple types of policies, the chromosome can be similarly encoded. The length of the chromosome is the total number of the parameters of all inventory policies of the supply chain. To make genetic operations easier and effective, R_i and Q_i are encoded in binary.

Initial Population

The NSGA-II starts the search by generating a population of candidate solutions. In our implementation, this population is randomly generated according to uniform distributions. That is, the parameters (gene values) R_i and Q_i are randomly generated according to uniform distributions $U[R_{imin}, R_{imin}]$ and $U[Q_{imin}, Q_{imin}]$, respectively, where R_{imin}, Q_{imin} and R_{imax}, Q_{imax} are respectively the minimum and the maximum possible values of R_i and Q_i.

Chromosomes Evaluation and Selection

In our study, each chromosome is evaluated through simulation. The simulation is controlled by two control parameters: the number of simulations per chromosome N and the length of each simulation run T. For each chromosome i, the fitness value f_i is evaluated on its average value over the N simulations.

Selection is a process in which chromosomes are chosen according to their fitness function value or their rank value. In this study, the tournament parent selection is used. Tournament selection is one of selection methods in genetic algorithms which runs a *tournament* among a few individuals chosen at random from the population and selects the winner (the one with the best fitness) for crossover. The tournament size used in our computational experiments is 2.

Crossover and Mutation

The crossover produces new offspring chromosomes from parent individuals. Two new chromosomes are created by exchanging some genes of two parent chromosomes. We use in our implementation the single-point crossover. This kind of crossover creates a pair of offspring by exchanging parts of the parents after a randomly chosen crossover point.

The mutation introduces some extra variability into the current population. The function of mutation is to maintain the diversity of the population in order to prevent too fast convergence of the algorithm. The probability of mutation is taken as $1/l$ where l is the string length of our binary-coded variables.

Crowding Distance Calculation

As the overall population size of P is $2n$, we cannot accommodate all fronts in the new parent population ($newP$) of size n. The non accommodated fronts are simply deleted. When the last allowed front is considered, there may exist more solutions in the last front than the remaining slots in the new population.

In order to avoid arbitrarily choosing individuals, we choose the individuals that can assure diversity between the considered ones; that is what we call a niching strategy. For this reason, we calculate the crowding distance measuring the Euclidean distance between the neighboring individuals in each front (Fig. 2). The Algorithm 2 describes the crowding distance procedure.

Algorithm 2: Crowding distance calculation

Let ζ the number of solutions
for each solution i **do**
  Initialize the distance to 0
  for each objective m **do**
    Sort the solutions according to the objective m
    $d_1 \leftarrow 0$, $d_\zeta \leftarrow \infty$
    for $i \leftarrow 2$ to $(\zeta - 1)$ **do**
      $d_i \leftarrow d_i + (f_m^{i+1} - f_m^{i-1})/(f_m^{MAX} - f_m^{MIN})$
    end for
  end for
end for

where f_m^i is the value of the objective function m of the ith solution. Thereafter, those strings with largest crowding distance values are chosen to become the new parent population. Once the non-dominated sorting is over, the new parent population, $newP$, is created by choosing solutions of different non-dominated fronts.

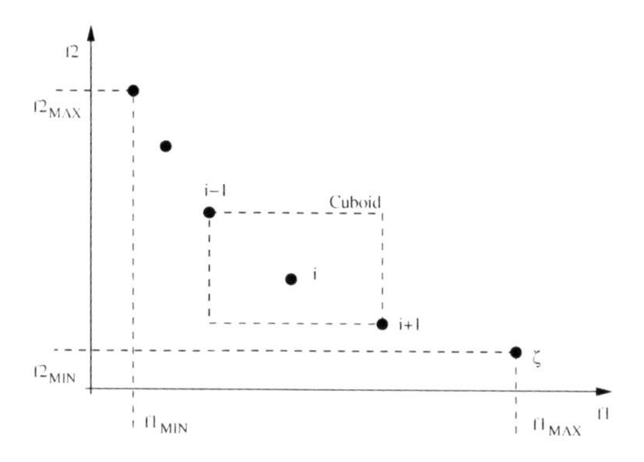

Fig. 2. Crowding distance

Stopping Conditions

There are no universal stopping conditions accepted for multi-objective genetic algorithms. In this study, we simply stop our algorithm after a given number of iterations (Ng).

4 Industrial Case

In this section, a real life supply chain is presented. For confidential reason, the name of the company concerned is not mentioned. The supply chain is composed of three suppliers, three upstream transporters with one for each supplier, a manufacturer, a downstream transporter for the manufacturer and a set of customers (see Fig. 3).

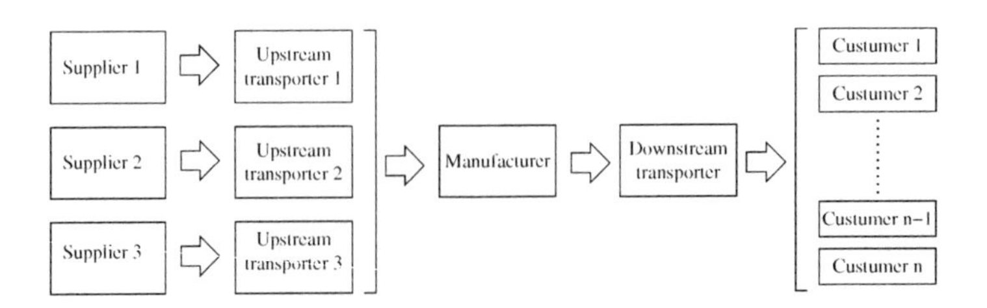

Fig. 3. Structure of an industrial supply chain

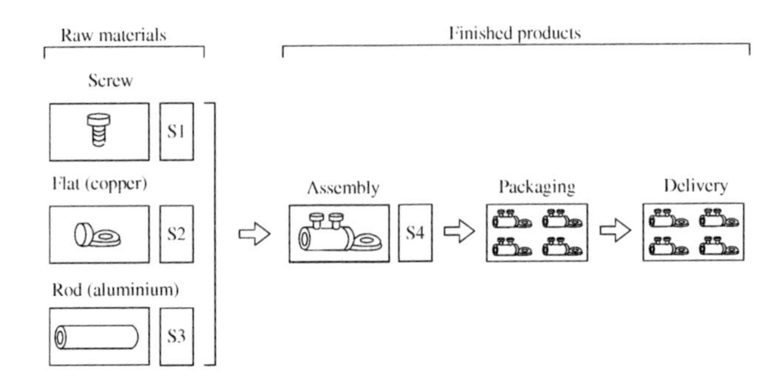

Fig. 4. Manufacturing process of an electrical connector

The manufacturer produces an electrical connector for high voltage lines. There are three types of flow in the supply chain: material flows, information flows and financial flows. For material flows (see Fig. 4), the manufacturer needs three raw materials for the production of the connector: flats, rods, and screws. They are purchased from suppliers 1, 2 and 3, respectively. These materials are delivered to the manufacturer by the corresponding upstream transporters. At the manufacturer site, rods of aluminum are cut into shafts with a pre-specified length, flats are bored and ground, and the finished product is then produced by assembling a shaft, a flat, and two screws. The product will be further packaged and delivered to customers by the downstream transporter.

For information flows, the manufacturer receives customer demand in the form of orders. When the inventory position of the finished product (Stock 4) is below a pre-specified reorder level, an assembly order with a given batch size will be released to the assembly line of the manufacturer. For the raw material stocks (S1, S2 and S3), when their inventory positions are below their reorder levels, a purchase order with a given quantity will be placed to their corresponding suppliers. Each purchase or assembly order contains the information such as order release time and order quantity, which are determined by an inventory policy involved. The inventory policies of all the stocks are periodic review batch ordering policy. For financial flows, customers pay the manufacturer within a given time period after receiving their ordered finished products. The manufacturer pays its

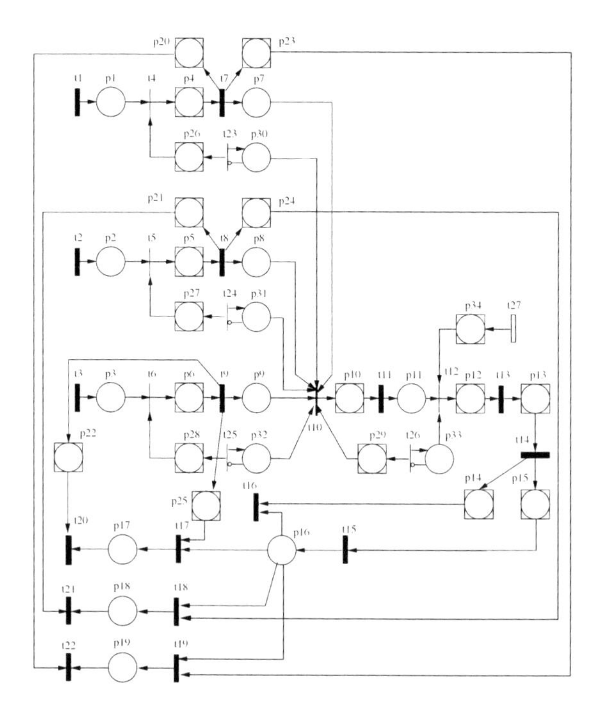

Fig. 5. BDSPN model of the industrial supply chain

suppliers similarly. The suppliers and the manufacturer pay their transporters within a given time period after the delivery of raw materials or finished products.

Figure 5 shows the BDSPN model of the industrial supply chain. In the model, the material flows are represented by timed transitions t_1, t_2 and t_3 (inventory replenishments of the suppliers), t_7, t_8 and t_9 (deliveries of raw materials from the suppliers to the manufacturer), t_{11} (assembly operation), t_{13} (delivery preparation of the manufacturer), t_{14} (delivery of finished products from the manufacturer to customers), and their associated places and arcs. The information flows are represented by immediate transitions t_{23} to t_{26}, their associated discrete places p_{30} to p_{33} and batch places p_{26} to p_{29}, the arcs connecting the places with the transitions, and the weights of the arcs. The financial flows are represented by transitions t_{15} to t_{22}, discrete and batch places p_{14} to p_{25}, and their associated arcs, where each transition represents a payment operation from one company to another while the markings of the discrete and batch input places represent the money available in a company and the order delivery information received from a company.

5 Computational Experiments

In this study, we have programmed our adapted NSGA-II algorithm on a LINUX station using C++ under KDevelop environment. The algorithm is tested on real

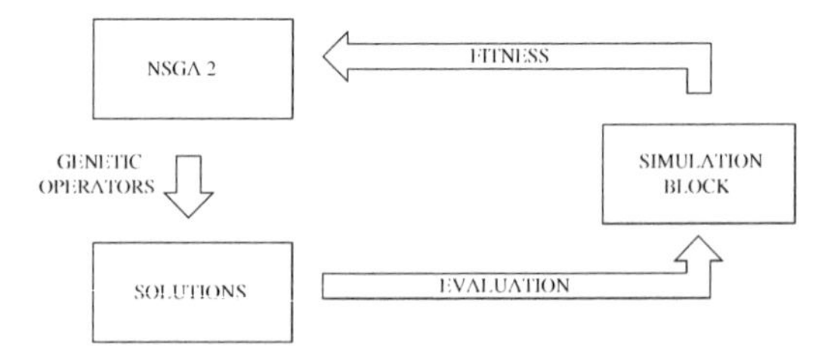

Fig. 6. Structure of the simulation based optimisation tool

data of an industrial case. Figure 6 shows the coupling between the BDSPN simulator and the NSGA2 optimisation method.

With the BDSPN model, the performance of the current real-life supply chain is evaluated by simulation using a BDSPN simulator (a C++ program) developed by us. All data in our simulation are taken from the company concerned. The mean replenishment lead times for stocks S1, S2, S3, and S4 are 40, 20, 70, and 20 days, respectively. The mean transportation lead times for all transporters are 2 days. The standard deviations of these lead times are negligible. Customer orders arrive randomly with the inter-arrival times subject to an exponential distribution with mean value 0.0355. Actually, the reorder points of stocks S1, S2, S3, S4 are 2300, 590, 2000, 400, respectively, while the order quantities of these stocks are 5000, 3000, 9300, 2000, respectively. The annual holding costs of raw materials and final products in stocks S1, S2, S3, S4 are 12% of their prices which are 0.3€, 0.6€, 0.16€, and 3€, respectively. The performance criteria of the supply chain considered include average inventory level and service level of each stock, where the service level is defined as the probability that customer orders can be filled on time. The first criterion is easy to obtain since it corresponds to the average number of tokens in the discrete place representing the on-hand inventory of a stock. For the evaluation of the service level, it needs to obtain the total time that the discrete place has no token while the corresponding batch place that represents customer orders is not empty in each simulation run. This can be done by observing the markings of the two places during the simulation.

5.1 Performance Evaluation

Because of the stochastic nature of the model, the simulation has to be replicated for many times with a long time horizon (simulation length) to get reliable estimates of the performance indices. Each index is evaluated as its average value over all simulation replications. The accuracy of the evaluation depends on the number of simulations performed and the simulation length adopted.

Thus, several instances were tested while varying the simulation time and the number of replications. The simulation time T and the number of replications N take respectively the following values $T \in [1000, 2000, 3000, 4000, 5000]$ and $N \in [5, 10, 25, 50, 75, 100, 150, 200]$. To choose a relevant value of T and N, 5

Table 1. Simulation results with $T = 2000$

N	CT	σ_{CT}	SL	σ_{SL}	Time (s)
5	516.05	2.98	0.8260	0.0082	0.203
10	516.35	3.27	0.8268	0.0082	0.343
25	**517.39**	3.46	**0.8286**	0.0077	**0.781**
50	517.35	3.30	0.8283	0.0074	1.593
75	517.54	3.35	0.8286	0.0072	2.296
100	517.36	3.31	0.8283	0.0070	3.031
150	517.26	3.02	0.8280	0.0065	4.500
200	517.37	3.06	0.8283	0.0066	6.031

Table 2. Performance of the industrial supply chain with $T = 2000$ days and $N = 25$ replications

Stock	Average inventory level	Average backorder size	Service level	Average inventory cost
S1	3355.33 (11.28)	0.00 (0.00)	1.0000 (0.0000)	120.79 (0.41)
S2	1189.43 (5.22)	178 (0.18)	0.6845 (0.0015)	85.64 (0.37)
S3	5071.65 (14.76)	0.00 (0.00)	0.9232 (0.0231)	97.37 (0.28)
S4	593.30 (9.39)	20.81 (1.95)	0.8286 (0.0077)	213.59 (3.38)
Supply chain performance			0.8286 (0.0077)	517.39 (3.46)

performance indices are taken into account: the total inventory cost TC, the variance of the total inventory cost σ_{TC}, the customer service level SL, the variance of the customer service level σ_{SL}, and the computation time (*time*). The choice of the simulation parameters of T and N is a compromise between the computation time and the quality of the performances. Table 1 gives the performances obtained with $T = 2000$ days and $N \in [5, 10, 25, 50, 75, 100, 150, 200]$ replications. The computation time is reasonable for a coupling with an optimisation module for $N = 25$. Beyond $N = 25$ replications, the performances (TC, σ_{TC}, SL, σ_{SL}) do not vary significantly any more.

According to these simulation results, the simulation horizon for the optimisation procedure is set as 6 years ($T = 2000$) with a warm up period of 6 months (10% of the simulation horizon) in order to reach a steady state, and the number of replications is set to $N = 25$.

The complete performance of the supply chain is shown in Table 2, where each entry without bracket enclosed is the estimated mean value of a performance criterion, while the entry enclosed with a bracket below the mean value is the estimated standard deviation of the criterion. From Table 2, we can find that the standard deviation is quite small, usually within 1% of its corresponding mean value. This implies that the obtained results are trustful. The performances of the supply chain are **517.39** for the mean total inventory cost per day, which is the sum of the mean inventory costs of the four stocks, and the mean service level to the final customers are **82.86 %**.

The current customer service level is relatively low. This is the main motivation driving the manufacturer to improve its supply chain. The goal of the manufacturer is to maximise the service level to the final customers and to minimize the total inventory cost. To reach this goal, the simulation-based optimisation approach is applied to optimise the inventory policies of the supply chain.

5.2 Multiobjective Optimisation

Our algorithm was run with different values of the generation number (from 25 to 1000), with an initial population size of 100 and with the following parameters referring to the NSGA-II [7]: the probability of crossover is 0.9 and the probability of mutation is $1/l$, where l is the length of the binary coded decision variables. Also the genetic operations such as the Pareto dominance ranking procedure and the elitist selection are used. The elitist approach is very important. It considers both parent and child population for selecting better candidates for mating pool. This is supported by the fact that subsequent generations do not allow fitter chromosomes to get lost in the crowd due to the randomness involved in the selection operation of NSGA-II.

Evaluation Criteria

In the resolution of a multi-objective optimisation problem based on genetic algorithms, the central issue is to compare two fronts obtained by resolving this problem with different parameters of the algorithm. Two evaluation criteria are used:

- Zitzler Measure ($C(F_1, F_2)$): This measure has been proposed by Zitzler [20]. Let F_1, F_2 be two Pareto fronts. The function C maps the ordered pair $C(F_1, F_2)$ to the interval $[0, 1]$:

$$C(F_1, F_2) := \frac{|\{x_2 \in F_2; \exists x_1 \in F_1 : x_1 \geqslant x_2\}|}{|F2|} \tag{13}$$

  $C(F_1, F_2)$=1 means that all solutions in F_2 are dominated by or equal to solutions in F_1. The opposite, $C(F_1, F_2)$=0 represents the case where none of the solutions in F_2 are dominated by the solutions of F_1. Since this measure is not symmetrical, it is necessary to calculate $C(F_2, F_1)$. Therefore, F_1 is better than F_2 if $C(F_1, F_2) < C(F_2, F_1)$. This measure can be used to show one Pareto front dominates another; however, it does not tell how much better this is. The second measure proposed by Riise [17] overcomes this drawback. It calculates numerically the sum distance between two Pareto fronts, which is given by:
- μ Distance:

$$\mu = \sum_{i=1}^{N} d_i \tag{14}$$

  where d_i is the distance between a solution $i \in F_1$ and its orthogonal projection on F_2. The μ value is negative if F_1 is below F_2, and positive otherwise.

Table 3. Optimisation results with $T = 2000$ and $N = 25$

	Ng	Time (s)	$\|F\|$	$C(F_i; F_1)$	$C(F_1; F_i)$	μ	$\bar{\mu}$
F_1	25	1416	78	0	0	0	0
F_2	50	2758	100	2	73	3.863	0.0495
F_3	75	4190	100	0	78	7.318	0.0938
F_4	100	5543	100	0	78	8.366	0.1072
F_5	250	13580	100	0	78	9.636	0.1235
F_6	500	27132	100	0	78	9.875	0.1266
F_7	750	41107	100	0	78	9.627	0.1234
F_8	1000	55762	100	0	95	9.433	0.1209
IS	-	1	1	100	0	-	-18.659

Since μ depends on the number of solutions in $|F_1| = \zeta$, a normalized measure is generally taken: $\bar{\mu} = \frac{\mu}{\zeta}$

Results

To test our optimisation method, several sets of parameters are used and the corresponding results are presented in Table 3. Ng is the number of generations. F_i is the optimal Pareto front and IS is the industrial solution described before with the performances (517.39€, 82.86%). The number of solutions $|F|$ in all Pareto optimal fronts F_i is equal to 78 or 100. Each optimal Pareto front F_i and the solution IS are compared with the reference front F_1 on two evaluation criteria, the Zitzler measures $C(F_i; F_1)$ and $C(F_1; F_i)$ and the μ distance. Results show that with all parameters settings, the algorithm can obtain solutions better than the industrial solution IS but with a longer computation time (more than 18 hours for F_8). The longer computation time is acceptable since the optimisation of the inventory policies can be done off time.

The results in Table 3 show that after 100 generations, there are no more noticeable improvements of the front. Thus, for the robustness analysis of the algorithm hereafter, the number of generations is fixed to 250.

Graphically, Fig. 7 gives an example of the optimal Pareto distribution of the front F_8. Each point represents a specific solution that is Pareto-optimal. We can also discover that the points are evenly distributed along the front. The IS point is below the optimal Pareto front. Figure 8 compares two optimal Pareto fronts F_1 and F_8.

5.3 Robustness Analysis

The robustness of our method was tested by performing multiple simulation runs with different standard deviations of demand Cv_d and lead time Cv_{lt}, in order to see the variability on the quality of non dominated solutions for cost minimization and service level maximization. The different Pareto fronts obtained using different standard deviations (10%, 20%, 30%) of demand and (0%, 10%, 20%) of lead time are presented in Table 4. Each optimal Pareto front F_i are

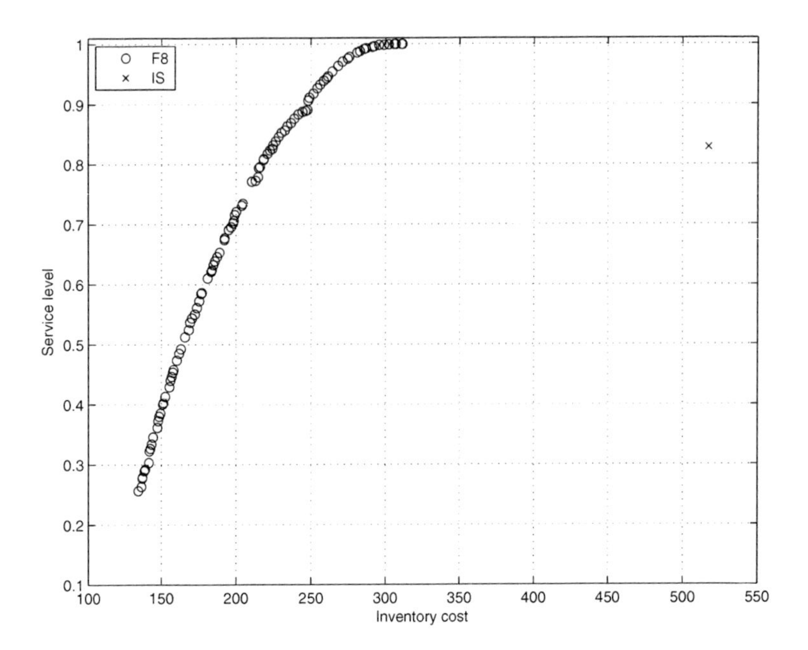

Fig. 7. Distribution of F_8 and IS point

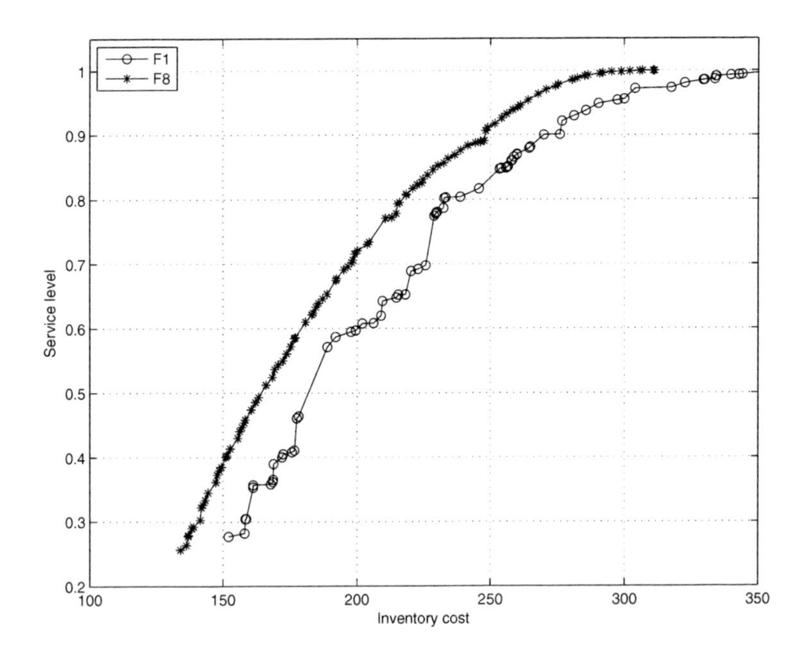

Fig. 8. Comparison between F_1 and F_8

Table 4. Robustness results with $T = 2000$, $N = 25$ and $Ng = 250$

	Cv_d	Cv_{lt}	$\|F\|$	$C(F_i; F_{10})$	$C(F_{10}; F_i)$	μ	$\bar{\mu}$
F_{10}	10%	0%	100	0	0	0	0
F_{11}	10%	10%	100	99	0	-0.596	-0.00596
F_{12}	10%	20%	100	100	0	0.364	0.00364
F_{13}	20%	0%	100	62	7	-0.033	-0.00033
F_{14}	20%	10%	100	91	1	0.104	0.00104
F_{15}	20%	20%	100	99	0	0.436	-0.00436
F_{16}	30%	0%	100	85	1	0.031	0.00031
F_{17}	30%	10%	100	92	1	0.467	0.00467
F_{18}	30%	20%	100	100	0	0.576	0.00576

compared with the reference front F_{10} on two evaluation criteria, the Zitzler measures $C(F_i; F_1)$ and $C(F_1; F_i)$ and the μ distance. It can be observed that there is no significant difference between all Pareto fronts. This indicates that all Pareto fronts resulting of demand variability have little variation, thus this shows the robustness of our method.

6 Conclusion

In this chapter, we have developed a simulation-based method for the optimisation of the inventory policies of supply chains with multiple objectives. This method combines simulation evaluation of performance with a genetic algorithm that guides an optimisation search process toward global optima. The simulation evaluation is based on a batch deterministic and stochastic Petri net modelling tool we developed, while the genetic algorithm is developed based on NSGA2. Our method is tested by an industrial case study that optimises inventory policies for a real-life supply chain. The test is performed with different parameter settings of the algorithm. The different solutions obtained by the algorithm and represented by Pareto fronts are compared with two evaluation criteria: μ distance and Zitzler measure. Numerical results show that our method can obtain inventory policies much better than the ones currently used in practice with a reduced inventory cost and an improved service level. The application of the method to other problems in supply chains such as the inventory routing problem that minimizes the total cost of inventory and transportation is a topic for future research.

References

1. Ajmone Marsan, M., Balbo, G., Conte, G., Donatelli, S., Franceschinis, G.: Modelling with Generalized Stochastic Petri Nets. Wiley, Chichester (1995)
2. Burke, E.K., Landa Silva, J.D.: The influence of the fitness evaluation method on the performance of multiobjective search algorithms. European Journal of Operational Research 169(3), 875–897 (2006)

3. Chen, H., Amodeo, L., Chu, F., Labadi, K.: Modelling and performance evaluation of supply chains using batch deterministic and stochastic petri nets. IEEE Transactions on Automation Science and Engineering 2(2), 132–144 (2005)
4. Coello Coello, C.A.: An updated survey of GA-based multiobjective optimization techniques. ACM Computing Surveys 32(2), 109–143 (2000)
5. Daniel, J.S.R., Rajendran, C.: Heuristic approaches to determine base-stock levels in a serial supply chain with a single objective and with multiple objectives. European Journal of Operational Research 175(1), 566–592 (2005)
6. Deb, K.: Multi-Objective Optimization Using Evolutionary Algorithms. Wiley, Chichester (2001)
7. Deb, K., Pratap, A., Agarwal, S., Meyarivan, T.: A fast and elitist multi-objective genetic algorithm: NSGA-II. IEEE Transactions on Evolutionary Computation 6(2), 182–197 (2002)
8. Deb, K., Srinivas, N.: Multiobjective optimization using non-dominated sorting in genetic algorithms. Evolutionary Computation 2(3), 221–248 (1994)
9. Fonseca, C., Fleming, M.: Genetic algorithm for multiobjective optimization: formulation, discussion and generalization. In: Proceedings of the Fifth International Conference on Genetic Algorithms, San Mateo (1993)
10. Fu, M.C.: Simulation optimization. In: Proceeding of the 2001 Winter Simulation Conference, pp. 53–61 (2001)
11. Gaspar-Cunha, A., Covas, J.A.: RPSGAe-reduced pareto set genetic algorithm: Application to polymer extrusion. In: Gandibleux, X., Sevaux, M., Sorensen, K., T'kindt, V. (eds.) Metaheuristics for Multiobjective Optimisation, pp. 221–249. Springer, Berlin (2004)
12. Horn, J., Nafpliotis, N., Goldberg, D.: A niched pareto genetic algorithm for multiobjective optimisation. In: Proceedings of the 1st IEEE Conference on Evolutionary Computation (1994)
13. Knowles, J.D., Corne, D.W., Oates, M.J.: The pareto-envelope based selection algorithm for multiobjective optimization. In: Deb, K., Rudolph, G., Lutton, E., Merelo, J.J., Schoenauer, M., Schwefel, H.-P., Yao, X. (eds.) PPSN 2000. LNCS, vol. 1917, pp. 839–848. Springer, Berlin (2000)
14. Labadi, K., Chen, H., Amodeo, L.: Modeling and performance evaluation of inventory systems using batch deterministic and stochastic petri nets. IEEE Transactions on Systems, Man, and Cybernetics, Part C: Applications and Reviews 37(6), 1287–1302 (2007)
15. Laumanns, M., Thiele, L., Deb, K., Zitzler, E.: Combining convergence and diversity in evolutionary multiobjective optimization. Evolutionary Computation 10(3), 263–282 (2002)
16. Lindemann, C.: Performance Modeling with Deterministic and Stochastic Petri Nets. Wiley, Chichester (1998)
17. Riise, A.: Comparing genetic algorithms and tabu search for multiobjective optimization. In: Abstract Conference Proceedings of IFORS 2002, Edinburgh, p. 29 (2002)
18. Schaffer, J.: Multi-objective optimisation with vector evaluated genetic algorithms. In: Proceedings of the First International Conference on Genetic Algorithms and Their Applications, pp. 93–100 (1985)
19. Tayur, S., Ganeshan, R., Magazine, M.: Quantitative Models for Supply Chain Management. Kluwer Academic Publishers, Boston (1998)
20. Zitzler, E., Thiele, L.: Multiobjective evolutionary algorithms: A comparative case study and the strength pareto approach. IEEE Transactions on Evolutionary Computation 3(4), 257–271 (1999)

Decomposition of Dynamic Single-Product and Multi-product Lotsizing Problems and Scalability of EDAs

Jörn Grahl[1], Stefan Minner[1], and Franz Rothlauf[2]

[1] Department of Logistics, University of Mannheim, Schloss, 68131 Mannheim, Germany
{joern.grahl,minner}@bwl.uni-mannheim.de

[2] Department of Information Systems and Business Administration, University of Mainz, Jakob-Welder-Weg 9, 55128 Mainz, Germany
rothlauf@uni-mainz.de

Summary. In existing theoretical and experimental work, Estimation of Distribution Algorithms (EDAs) are primarily applied to decomposable test problems. State-of-the-art EDAs like the Hierarchical Bayesian Optimization Algorithm (hBOA), the Learning Factorized Distribution Algorithm (LFDA) or Estimation of Bayesian Networks Algorithm (EBNA) solve these problems in polynomial time. Regarding this success, it is tempting to apply EDAs to real-world problems. But up to now, it has rarely been analyzed which real-world problems are decomposable. The main contribution of this chapter is twofold: (1) It shows that uncapacitated single-product and multi-product lotsizing problems are decomposable. (2) A state-of-the-art EDA is used to solve both problems. The problems are fundamental in inventory management and their fitness functions can be expressed as additively decomposable functions. It is analyzed how a search distribution of Boltzmann-type factorizes for these functions and it is shown that the factorizations of the Boltzmann distribution are polynomially bounded. Consequently, experimental scalability analysis is conducted for both problems with a state-of-the-art EDA. The total number of fitness evaluations required to reliably solve the problems to optimality is found to grow with a low-order polynomial depending on the problem size. The results confirm existing scalability theory for decomposable problems.

Keywords: Estimation of Distribution Algorithms, Decomposition, Lotsizing.

1 Introduction

Genetic Algorithms (GA) [16, 9] and Estimation of Distribution Algorithms (EDA) [29] show high performance when solving decomposable problems. A problem is decomposable if its dimensionality can be reduced by splitting it into possibly interacting subproblems that are easier to solve and from whose the overall solution can be constructed. In recent years, much progress has been made in solving decomposable problems. Artificial test problems that are intractable for simple GAs can reliably be solved to optimality by EDA such as LFDA [27],

A. Fink and F. Rothlauf (Eds.): Advances in Computational Intelligence, SCI 144, pp. 231–251, 2008.
springerlink.com

hBOA [32] or EBNA [21] in a scalable manner. Although there exist numerous theoretical and experimental results that indicate the efficiency of EDA in solving decomposable problems, successful applications of EDA to problems of industrial interest are still rare.

The main contribution of this chapter is twofold. *First*, the practical relevance of decomposability assumptions is demonstrated. Two lotsizing problems are studied that are fundamental in inventory management and of practical relevance. These are the single-product lotsizing problem [44] and the dynamic joint replenishment problem (JRP, [1]). The single-product lotsizing problem considers the placement of replenishment orders for a single product over time and is polynomial-time solvable, see [43]. The dynamic joint replenishment problem extends the single-product problem to a multi-product case where ordering costs are linked between the products. The JRP is NP-complete [2]. It is shown that both problems are decomposable and that their fitness functions can be formulated as additively decomposable functions. It is assessed how Boltzmann-type search distributions factorize into marginal distributions for these decompositions and it is shown that the factorizations of the Boltzmann distribution are polynomially bounded for both problems.

Second, the problems are solved with a state-of-the-art EDA in an experimental scalability analysis. The total number of evaluations required to reliably solve the problems to optimality grows with a low-order polynomial depending on the problem size. The results confirm existing scalability theory for decomposable problems.

This chapter is structured as follows. In Sect. 2.1, a brief introduction to lotsizing problems is given. The single-product lotsizing problem is presented in Sect. 2.2, the dynamic joint replenishment problem in Sect. 2.3. In Sect. 3 we review existing results on the influence of problem decomposition on the scalability of Evolutionary Algorithms, especially GAs and EDAs. In Sect. 4, we propose decompositions for the single- and the multi-product lotsizing problem. In Sect. 5, experimental scalability analyses for both problems are conducted. The paper ends with concluding remarks.

2 Lotsizing

2.1 Introduction

Almost any organization has to cope with inventories to exploit economies of scale, guarantee a certain service level or to decouple processes to name just a few reasons. A central problem of managing inventories is matching replenishment ordering decisions with customer demand over time. Lotsizing problems are solved to balance cost trade-offs that arise from placing replenishment orders in an inventory system at different points in time.

Assume that customer demand for a product is known over time, e.g., for the next 12 weeks. A company can place replenishment orders for the product at various discrete time points, e.g., every Monday. Received but not used goods can be stored as inventories. Customer demand has to be satisfied completely from the available sources.

The total costs that are caused by placing replenishment orders consist of (1) fixed ordering costs and (2) inventory holding costs. Fixed ordering costs arise for each order that is placed (e.g. shipping costs). They are independent from the amount of goods that is actually ordered. Inventory holding costs arise, if goods are put on stock and are carried as inventory. The sum of the fixed ordering costs *grows* with the number of orders that are placed. The sum of the inventory holding costs *decreases* with the number of orders that are placed. This trade-off has to be balanced in order to minimize the total costs that is caused by placing replenishment orders.

2.2 Single-Product Lotsizing

The standard single-product dynamic, discrete time lotsizing problem (also known as the Wagner-Whitin problem) was introduced in [44]. This fundamental problem addresses placement of replenishment orders for a single product over time such that the sum of fixed ordering costs and inventory holding costs is minimized. The single-product lotsizing problem is the starting point for several extensions into various directions that address issues arising in practice.

In the single-product lotsizing problem, time is discretized into T time-points. The t-th period denotes the time interval between time-point t and time-point $t+1$. Orders of amount $q_t \geqslant 0$ are placed at each time point $t \in \{1, 2, \ldots, T\}$. This means that orders are placed at the beginning of period t. An order of $q_t > 0$ causes a fixed costs of c that is independent of the amount that is actually ordered. The ordered goods are immediately available. They can be stored in a warehouse with unlimited capacity. The inventory level of the warehouse at the end of period t is denoted by y_t. We assume that initial inventory y_0 is zero. Holding a single item of inventory from one time-point to the next causes costs of h. After an order q_t has been placed, demand occurs of size z_t. Without loss of generality we assume throughout the paper that $z_1 > 0$ in the single-product lotsizing problem because initial periods with zero demand can be neglected. Demand z_t has to be satisfied completely from the order q_t or from stock (thereby reducing y_t).

The objective is to place the orders q_t such that the sum of fixed ordering costs and inventory holding costs over the planning horizon is minimized. Therefore, the following two questions have to be answered:

1. For which time-points $t \in \{1, 2, \ldots, T\}$ should $q_t > 0$?
2. If $q_t > 0$, how large should q_t be?

Using a mixed integer linear program, the single-product lotsizing problem can be formulated as follows.

$$\min \quad f = \sum_{t=1}^{T} (c \cdot x_t + h \cdot y_t) \tag{1}$$

$$y_t = y_{t-1} + q_t - z_t \qquad \forall t = 1, \ldots, T \tag{2}$$

$$y_0 = 0 \tag{3}$$

$$q_t \leqslant x_t \cdot \sum_{i=t}^{T} z_i \qquad \forall t = 1, \ldots, T \tag{4}$$

$$x_t \in \{0, 1\} \qquad \forall t = 1, \ldots, T$$

$$y_t, q_t \geqslant 0 \qquad \forall t = 1, \ldots, T$$

In this formulation, binary variables x_t are introduced to indicate whether an order is placed at time-point t. If $x_t = 1$, then an amount of $q_t > 0$ is ordered. If $x_t = 0$, then no order is given at time-point t. In that case, inequality (4) restricts the order amount q_t to zero. The inventory balance equation (2) states that the inventory level at the end of period t equals the inventory level at the end of period $t-1$ plus items that are ordered in period t (q_t) minus demand that occurs in period t (z_t). Equality (3) states the initial condition that starting inventory is zero.

The single-product lotsizing problem can be solved in polynomial time. Wagner and Whitin [44] proposed a dynamic programming formulation that has a complexity of $O(T^2)$. More efficient algorithms with $O(T \log T)$ complexity have been proposed by [8] and [43]. Several heuristics for the problem have been proposed, overviews are available, e.g., in [46] and [5].

2.3 Dynamic Joint Replenishment Problem

Standard lotsizing problems become more complex in a multi-product context with dependencies between different products. Examples in practice are products that are manufactured on the same machines (multi-item lotsizing and scheduling problems) or products that share a common warehouse. In the dynamic joint replenishment problem, the ordering costs depend on the mix of multiple products that are replenished jointly.

The JRP considers $K > 1$ products. For each product $k = 1, \ldots, K$, a single product lotsizing problem (see Sect. 2.2) has to be solved. Fixed ordering costs c^k (also referred to as minor setup costs), inventory holding costs h^k, and customer demand z_t^k are product-specific. We assume for the JRP throughout the paper that initial periods where all products have zero demand are neglected. It follows that at least a single product has positive demand in the first period. The products are linked as follows. If at least one product is replenished at time-point t, then an additional order cost of c^o arises (also referred to as major setup costs), independent of the number of products that are actually replenished at time-point t. If no product is replenished at time point t, c^o does not arise at time point t. Note that if several products are ordered at time point t, product specific fixed ordering costs c^k arise for each of these products but c^o arises once.

One example in practice is that several products are jointly shipped in a single truck, causing transportation costs of c^o.

If $c^o = 0$, the JRP decomposes into K single-product lotsizing problems that can be solved independently. If $c^k = 0 \; \forall \; k = 1, \ldots, K$, the JRP can be transferred into one single-product lotsizing problem.

The dynamic joint replenishment problem can be formulated as a mixed-integer linear program. The notation used follows the notation that has been introduced in Sect. 2.2. Decision variables and problem parameters are assigned a product index k. The fitness function includes all ordering and inventory holding costs.

$$\min \quad f = \sum_{k=1}^{K} \sum_{t=1}^{T} \left(c^k \cdot x_t^k + h^k \cdot y_t^k \right) + \sum_{t=1}^{T} c^o \cdot w_t \tag{5}$$

$$\begin{aligned} & y_t^k = y_{t-1}^k + q_t^k - z_t^k && \forall t = 1, \ldots, T; k = 1, \ldots, K \\ & y_0^k = 0 && \forall k = 1, \ldots, K \\ & q_t^k \leqslant x_t^k \cdot \sum_{i=t}^{T} z_i^k && \forall t = 1, \ldots, T; k = 1, \ldots, K \\ & x_t^k \leqslant w_t && \forall t = 1, \ldots, T; k = 1, \ldots, K \end{aligned} \tag{6}$$

This formulation is a direct extension of the single-product lotsizing mixed-integer linear program (see Sect. 2.2) for K products. If at least one product is replenished in period t, an indicator variable w_t is forced to 1 and ordering costs c^o arise in the fitness function (5). This coupling of the ordering costs is modeled in inequality (6).

The computational complexity of the JRP was analyzed in [2]. It is NP-complete. Algorithms that determine an optimal solution using dynamic programming approaches were developed by [39] and [19]. A branch-and-bound method was proposed by [7]. A dual-based method was developed by [38], heuristics were proposed in [17]. Efficient integer programming formulations are available in [3] and [30].

3 Problem Decomposition and Scalability of Evolutionary Algorithms

Genetic Algorithms (GA) [16, 9] are population-based stochastic search strategies that mimic evolutionary concepts like gene recombination, mutation, and selection to solve optimization problems. It is commonly assumed that solutions are represented as binary strings of fixed length l. Without any further assumptions the size of the search space grows exponentially with l. In a significant amount of work, GAs are applied to additively decomposable test problems [37, 31, 23, 45]. Decomposing a problem aims at reducing the dimensionality of the search space. In inventory management (see, e.g., [8]) the term decomposition is used as well, however in the non-related context of planning horizon theorems. According to [25] the fitness function $f(x)$ is additively decomposable if it can be formulated as

$$f(x) = \sum_{i=1}^{m} f_i(x_{s_i}). \quad (7)$$

$f(x)$ is additively defined over m subset of the alleles. The $s_1, s_2, \ldots, s_m$ are index sets, $s_i \subseteq \{1, 2, \ldots, l\}$. The f_i are sub-problems that are only defined on the alleles x_j with $j \in s_i$. The subproblems can be non-linear. The x_{s_i} are subsets of all alleles. These subsets can overlap.

Equation (7) exhibits a modular structure. It consists of m components that can, but may not, be coupled. If the s_i are disjoint, $s_i \cap s_j = \emptyset \; \forall \; i \neq j$, the functions do not overlap and the overall problem is called *separable*. Separable problems can be solved by solving the m subproblems f_i and summing up the results. Depending on the size of the subproblems, separation reduces the dimensionality of the search space significantly. Assuming that $|s_i| = k \; \forall \; i$ the dimensionality is reduced from l to k and the size of the solution space is reduced from 2^l to $m2^k$. Problem (7) is called *decomposable* if some sets s_i, s_j exist for which $s_i \cap s_j \neq \emptyset$. In this case, a strict separation of the sub-functions is no longer possible because a single decision variable influences more than one sub-function.

What makes decomposable problems hard to solve? This is a non-trivial question and several answers can be found in the literature. Most obviously the hardness of the sub-functions directly contributes to the overall complexity of the problem. Deceptive problems (see [6]) are hard to solve for GA and EDA and are often assumed as sub-functions for testing purposes. Deceptive functions are typically harder to solve for GA and EDA than non-deceptive functions. Further, subproblems can contribute to the overall fitness on a similar scale, or the scaling of the sub-functions can differ greatly. In the first case, all sub-functions of equal importance and convergence towards the partial solutions will happen simultaneously. If the sub-functions are exponentially scaled however, the most salient of them will converge first. The other sub-functions may converge later and some instantiations might already be lost at that time. Additively decomposable functions with exponentially scaled sub-functions are harder to solve for GA and EDA – they require a higher population size, see [40]. [18] discuss whether the size $|s_i|$ of the sets influences the hardness of a problem. This can be the case, if for solving $f_i(s_i)$ all associated variables must be regarded simultaneously. It may not be the case however, if interactions are not very strong and only some of the dependencies are important. The size of the sets can thus be a source for the hardness of a problem but the degree of connectivity and importance of the dependencies appears to be a more important source for the GA- or EDA-complexity of a function.

The decomposition of a problem defines which bits in a solution depend on each other and which are independent from each other. This information is also called linkage information [15]. We refer to high-quality configurations of alleles that belong to the sets s_i as building blocks (BBs, see [16, 10]). Building blocks can be characterized by their size $k_i = |x_{s_i}|$ and their defining length

$\delta_i = \max s_i - \min s_i$. The defining length of a BB is the distance between its two outermost bits.

A different approach to assess the problem complexity for GAs was introduced using the notion of schemata. It was found in [16] that GA using uniform crossover promote BBs with small k_i and perform worse on problems with large k_i. GA using one-point or multi-point crossover promote BBs with small δ_i and perform worse on problems with large δ_i. In both cases, low performance is due to crossover operators that disrupt BBs instead of recombining them correctly.

Crossover operators that do not disrupt BBs are called linkage-friendly. The use of linkage-friendly crossover operators significantly enhances GA performance [41]. However, to design linkage-friendly crossover operators, the decomposition of the problem has to be known. This is seldom the case, and often, linkage-information has to be learned by the algorithm [15]. This observation has triggered the development of techniques that learn linkage information by adaptively changing the positioning of the bits [12, 11, 20] or adapting the crossover operators.

One of the latter approaches that has received considerable attention uses density estimation and sampling as the major source of variation. These algorithms are commonly referred to as Estimation of Distribution Algorithms [29]. In EDA, crossover and mutation are replaced by the following two steps:

1. Estimate the joint density of high quality solutions.
2. Generate new solutions by sampling from the estimated density.

In the estimation step, linkage information is learned from the selected individuals [13, 35] and encoded into a density estimate. New solutions are sampled from this search distribution which respect to the captured linkage information. EDAs perform an iterative procedure of estimation, sampling, and selection.

For EDA success it is crucial that solutions can efficiently be generated by sampling. In the following paragraphs we refer to work that has been developed in the theory of decomposable graphs and probability theory and has been adapted to EDA theory, e.g, in [25] to illustrate for which decompositions this is possible. Assume that a fitness function of type (7) is given and one tries to solve the l-dimensional optimization problem $\boldsymbol{x}^{opt} = \text{argmax } f(x)$ by sampling the $\boldsymbol{x}$ from a search distribution. A good candidate for the search distribution is the Boltzmann distribution, which is given as follows [24]:

$$p_\beta(\boldsymbol{x}) = \frac{e^{\beta f(\boldsymbol{x})}}{\sum_{\boldsymbol{y}} e^{\beta f(\boldsymbol{y})}}.$$

The Boltzmann distribution with so-called temperature $1/\beta$ has the appealing property that for increasing β, it focuses only on the global optima of $f(\boldsymbol{x})$. For $\beta \to \infty$, only global optima have positive probabilities. Unfortunately, sampling from the Boltzmann distribution needs exponential effort and is no tractable search strategy.

If the fitness function is additively decomposable, the sampling effort can be reduced by sampling from a factorization of the Boltzmann distribution. A factorization is a decomposition of a multi-dimensional density into some marginal densities of smaller sizes. If it can be shown that for a fitness function $f(x)$ the Boltzmann distribution can be decomposed into smaller marginal distributions, generating solutions for $f(x)$ from a factorized Boltzmann distribution can potentially be an efficient search strategy.

To analyze whether this is the case, we define the sets d_i, b_i and c_i for the index sets $s_1, s_2, \dots, s_m$ for $i = 1, 2, \dots, m$ as follows:

$$d_i = \bigcup_{j=1}^{i} s_j \qquad b_i = s_i \setminus d_{i-1} \qquad c_i = s_i \cap d_{i-1}.$$

If the following Factorization Theorem [25, 28, 26] holds for a given decomposable function, the Boltzmann distribution can be factorized into some of its marginal distributions.

Factorization Theorem: Let the fitness function $f(\boldsymbol{x}) = \sum_{i=1}^{m} f_i(\boldsymbol{x}_{s_i})$ be an additive decomposition. If

$$b_i \neq 0 \quad \forall\, i = 1, \dots, m \tag{8}$$

and

$$\forall\, i \geq 2\, \exists\, j < i \text{ such that } c_i \subseteq s_j, \tag{9}$$

then

$$q_\beta(x) = \prod_{i=1}^{m} p_\beta(x_{b_i} | x_{c_i}) = p_\beta(x). \tag{10}$$

Condition (9) is called the running intersection property (RIP). If conditions (9) and (10) hold, the Boltzmann distribution can be obtained from an exact factorization into some of its marginals. But, it is only reasonable to sample new solutions from (10) in order to solve (7), if sampling new solutions from (10) is computationally easier than solving (7) directly. This is not the case if the marginal distributions are of arbitrary dimensionality, because the sampling effort grows exponentially with their size. However, it is indeed the case, if the size of the sets b_i and c_i is bounded by a constant independent of l. Then, the factorization is called polynomially bounded.

A major result of EDA theory is that if the factorization of the Boltzmann distribution for a given problem is polynomially bounded, new solutions can efficiently be generated and an EDA can theoretically solve the decomposable problem with a polynomial number of fitness evaluations [26]. During the last years, this potential has been turned into scalable optimizers that outperform standard GA on a wide range of problems. For overviews on EDA instances and historical developments, the reader is referred to [22], [36] and [4].

4 Decomposition of Lotsizing Problems

4.1 Single-Product Case

In this section, we reformulate the fitness function (1) as an additively decomposable function of type (7). Then, we show that the Factorization Theorem holds for the decomposition and that the size of the marginal distributions is bounded with the consequence that a state-of-the-art EDA should be able to solve the problem in polynomial time.

The proposed decomposition is constructed by exploiting the zero inventory property (see [44]). This property states that in an optimal solution for the single-product lotsizing problem, replenishment orders $q_t > 0$ are placed at time t if and only if $y_{t-1} = 0$. From this it follows that order quantities q_t in optimal solutions are batches of aggregate consecutive future demands. In the previous formulation of the problem (see Sect. 2.2), a company has to decide when replenishment orders should be placed *and* how much should be ordered. If we exploit the zero inventory property, it is *only* necessary to decide when the orders have to be placed and not how much should be ordered. The order amounts are completely derived from the times when orders are placed. If an order q_t is placed at time t and the next order u is placed at time $t < u < T$, then $q_t = \sum_{i=t}^{u-1} z_i$. Consequently the last order amount is $q_u = \sum_{i=u}^{T} z_i$.

Exploiting the zero-inventory property allows us to represent a solution on a binary string; see Fig. 1. The ordering decision variables x_t are binary (either an order is placed at time t or not) and their values are aligned on the string. The ordering decision x_t is represented by the t-th bit on a string of length T.

Note that for lotsizing problems considered in this paper an order has to be placed in period one as initial inventory is zero and $z_1 > 0$. This means that solutions where $x_1 = 1$ are feasible and solutions where $x_1 = 0$ are infeasible. For all following reformulations and analysis we assume feasibility of the solutions.

x_1	x_2	x_3	...	x_t	...	x_{T-1}	x_T	Bitstring
1	2	3		t		T-1	T	Periods

Fig. 1. Coding for the single-product lotsizing problem

The total costs caused by m replenishment orders over T periods of time can be calculated by adding up m costs values that cover the costs that occurs in m lots over disjoint and aligned sets of periods. Thus, the single-product lotsizing problem is decomposed along batches of aggregate consecutive future demands.

We reformulate (1) to an additively decomposable function of type (7) as follows:

$$\min f = \sum_{i=1}^{m} f_i(x_{s_i}) \tag{11}$$

with

$$p = \{p_1, p_2, \dots, p_m\}$$
$$p_i \in \{1, 2, \dots, T\}$$
$$p_1 = 1$$
$$s_i = \bigcup_{j=p_i}^{p_{(i+1)}-1} j.$$
$$f(x_{s_i}) = c + h \cdot \left[\sum_{j=0}^{|x_{s_i}|-1} j \cdot z_{(p_i+j)} \right]$$

Fitness function (11) is additively defined over m sets of the ordering decision variables x_{s_i}. m denotes the number of $1 \leqslant m \leqslant T$ orders that are placed. The x_{s_i} are disjoint subsets of all ordering decision variables and x_{s_i} consist of all variables that are associated with the order point p_i. Order points $p_i, i = 1, 2, \dots, m$, are time points t where $x_t = 1$. The set p contains all order points.

Example 1. Let the number of periods T be 12. Orders are placed at the beginning of periods 1,5,6, and 10. In this case, $m = 4$, $p = \{1, 5, 6, 10\}$, $p_1 = 1$, $p_2 = 5$, $p_3 = 6$ and $p_4 = 10$. The sets s_i have the following structure: $s_1 = \{1, 2, 3, 4\}$, $s_2 = \{5\}$, $s_3 = \{6, 7, 8, 9\}$, and $s_4 = \{10, 11, 12\}$. The decision variables x are grouped into sets of $x_{s_1} = \{x_1, x_2, x_3, x_4\}$, $x_{s_2} = \{x_5\}$, $x_{s_3} = \{x_6, x_7, x_8, x_9\}$, and $x_{s_4} = \{x_{10}, x_{11}, x_{12}\}$. These groups can be denoted as building blocks of the problem instance.

The example illustrates that the building blocks of single-product lotsizing problem instances have a well-defined structure. They consist of a leading 1 that denotes the setup decision plus subsequent 0s, if any. An illustration of the BB structure is given in Fig. 2. The BBs of a problem instance need not be equally sized. But the average size of the BBs will increase with the setup costs c and decrease with h. The more expensive an order is, the more is ordered in a

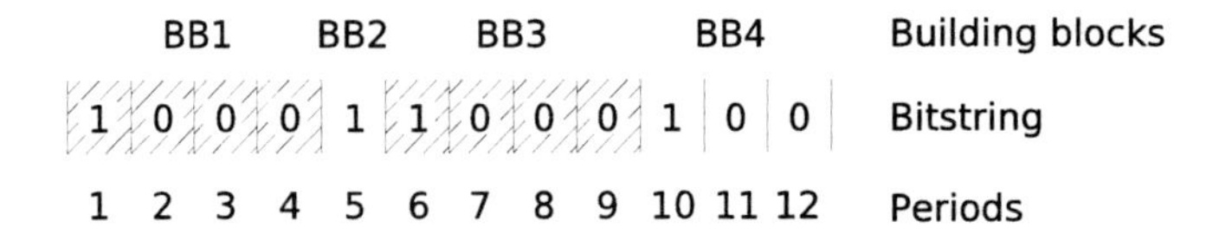

Fig. 2. Building blocks of a single-product lotsizing instance

single batch to avoid frequent ordering and the more demand of future periods is covered by a single order.

We now show that the Factorization Theorem holds for this decomposition and that the factorization of the Boltzmann distribution is polynomially bounded (see 3).

First, we show that $b_i \neq \emptyset$ for all $i = 1, 2, \ldots, m$.

Obviously, all $s_i \neq \emptyset$, because the s_i always include at least one index p_i. Since $s_i \cap s_j = \emptyset$ for all $i \neq j$, it follows that

$$c_i = s_i \cap d_{i-1} = s_i \cap \bigcup_{j=1}^{i-1} s_j = \emptyset.$$

Thus, $b_i = s_i \setminus d_{i-1} = s_i$. Since $s_i \neq \emptyset$, it follows that $b_i \neq \emptyset \; \forall i = 1, 2, \ldots, m$. The RIP is fulfilled because the subproblems do not overlap, the problem is separable.

Additionally, it is desired that the factorization is polynomially bounded. This means that the size of the marginal distributions is bounded by a constant independent of T. The size of a marginal distribution relates directly to the number of periods $|s_i|$ whose demand a replenishment order placed in p_i covers. It is reasonable to assume that the ordering costs c and the inventory holding costs h are bounded and positive. Assume now that $T \to \infty$. Under these assumptions, the number of periods that any replenishment order covers is bounded. Any values chosen for c and h will make it beneficial to order more than once if $T \to \infty$. The size of any s_i is bounded. b_i is bounded since $b_i = s_i$ and c_i is bounded because $c_i = \emptyset$. The factorization of the Boltzmann distribution is polynomially bounded. The planning horizon theorem from [44] yields similar results by stating independence of partial solutions spanning periods t to $t + H$, if the optimal solution for periods t to $t + H + 1$ includes an additional order in period $t + H + 1$ and the optimal order amount in period t remains unchanged.

4.2 Joint Replenishment Case

In this section, we will reformulate (5) as an additively decomposable function of type (7). Then we show that the Factorization Theorem holds for this decomposition and that the factorization of the Boltzmann distribution is polynomially bounded.

The zero-inventory property also holds for the dynamic joint replenishment problem [42]. This allows us to extend the additively decomposable reformulation of the single-product lotsizing problem (see Sect. 4.1) to the multi-product case.

The reformulation introduces an artificial product with index $k = 0$ that is used to model the coupling of ordering costs. Inventory holding costs for the artificial product are zero. If at least one real product is replenished at time point t, then the artificial product is replenished as well at time point t. This causes fixed ordering costs of c^o that are independent of the number of products that are actually replenished in period t. If no real product is replenished at time-point t, then the artificial product is not replenished at time-point t.

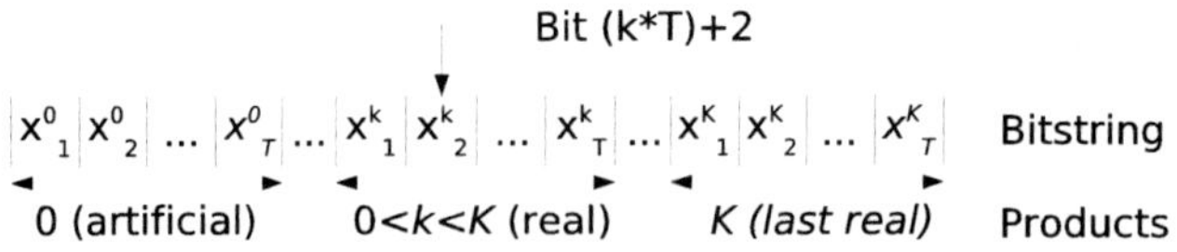

Fig. 3. Coding for the dynamic joint replenishment problem

Solutions of the JRP are represented as a binary string as follows. Like in the single-product case we do not need to encode the amounts that are ordered. Due to the zero-inventory property, we only need to encode binary ordering timing decisions. A single product $k = 0, 1, \ldots, K$ is exactly encoded as described in 4.1. The setup decisions for all products are aligned on the binary string. This means, that if 3 products are given and $T = 12$, then $4 \cdot 12 = 48$ bits are needed to encode a single solution. The first T bits are used to represent the order decision of the artificial product. Bits $T+1$ to bit $2 \cdot T$ represent ordering decisions of the first product, second, and so forth. In general, the ordering decision for period t of product k is represented by bit $T \cdot k + t$. The coding is illustrated in Fig. 3.

We reformulate function (5) as an additively decomposable function of type (7) as follows:

$$\min f(x) = \sum_{k=0}^{K} \sum_{i=1}^{m^k} f_i^k \left(x_{s_i^k} \right) \tag{12}$$

with

$$\begin{aligned}
c^0 &= c^o; \quad h^0 = 0 \\
p^k &= \{p_1^k, p_2^k, \ldots, p_{m^k}^k\} && \forall k = 0, 1, \ldots, K \\
p_i^k &\in \{1, 2, \ldots, T\} && \forall k = 0, 1, \ldots, K; i = 1, 2, \ldots, m^k \\
p_1^k &= 1 && \forall k = 1, 2, \ldots, K \\
p^k &\subseteq p^0 && \forall k = 1, \ldots, K \qquad (13) \\
s_i^k &= \bigcup_{j=p_i^k}^{p_{(i+1)}^k - 1} [\, j + (k \cdot T)] && \forall k = 0, 1, \ldots, K; i = 1, 2, \ldots, m^k.
\end{aligned}$$

$$f_i^k(x_{s_i^k}) = c^k + h^k \cdot \left[\sum_{j=0}^{|x_{s_i^k}|-1} j \cdot z_{(p_i^k + j)}^k \right]$$

Fitness function (12) is additively defined over all products $k = 0, 1, \ldots, K$ and all associated sets of decision variables $x_{s_i^k}$ for each of the products. (13) links the ordering decisions of the real products with the ordering decisions of the artificial product.

Example 2. Let the number of periods T be 12 and the number of real products K be 3. The first product is replenished in periods 1,5, and 6. The second product is replenished in periods 1, 6, and 11 and the third product is replenished in periods 1 and 3. The artificial product is therefore replenished in periods 1,3,5,6, and 11. For this setting, $p^1 = \{1,5,6\}$, $p^2 = \{1,6,11\}$, $p^3 = \{1,3\}$ and $p^0 = \{1,3,5,6,11\}$. The set $s_1^0 = \{1,2\}$, $s_2^0 = \{3,4\}$, $s_3^0 = \{5\}$, $s_4^0 = \{6,7,8,9,10\}$, $s_5^0 = \{11,12\}$, $s_1^1 = \{13,14,15,16\}$, $s_2^1 = \{17\}$, $s_3^1 = \{18,19,20,21,22,23,24\}$, $s_1^2 = \{25,26,27,28,29\}$, $s_2^2 = \{30,31,32,33,34\}$, $s_3^2 = \{35,36\}$, $s_1^3 = \{37,38\}$, $s_2^3 = \{39,40,41,42,43,44,45,46,47,48\}$.

For the following analysis, we assume only feasible solutions.

We have shown in Sect. 4.1 that the Factorization Theorem holds for a single product. We transfer this result to the dynamic joint replenishment problem. The above reformulation of the JRP separates the products from each other. The index sets s_i^k do not overlap between the products. This means that for each product that is modelled in the JRP, the single-product results apply. Because the products of the JRP are separated, the Factorization Theorem also holds for the JRP as a whole.

We have shown in Sect. 4.1 that the Boltzmann distribution is polynomially bounded for a single product. In our JRP formulation we separate the products from each other and no additional complexity is introduced for all feasible solutions. Thus, this result is valid individually for every "real" product $k > 0$. Note, that the artificial product is special because it has zero-inventory holding costs. The size of the inventory holding costs does not affect the analysis of the decomposition. Therefore, the results for a single real product apply for the artificial product as well.

An instance of the JRP is made up by a combination of a bounded number of real products and one single artificial product. If we let $T \to \infty$, the size of no set s_i^k can tend to infinity. All sets s_i^k are bounded. All sets b_i^k are bounded because $b_i^k = s_i^k$. The sets c_i^k are bounded because $c_i^k = \emptyset$. The factorization of the Boltzmann distribution is polynomially bounded for the JRP as a whole.

5 Experimental Results

We perform experiments on the single-product lotsizing problem and the dynamic joint replenishment problem. The analysis of the previous chapters indicates that both problems are decomposable. State-of-the-art EDA solve decomposable problems in polynomial time. We conduct an experimental scalability analysis to assess the degree of the polynom and how the constant factor depends on the problem instance.

The Hierarchical Bayesian Optimization Algorithm (hBOA) [34, 32] is used in all experiments because it is a state-of-the-art EDA. It has been used for solving complicated problems from computer science and physics previously, see [33]. An introduction is available in [32]. hBOA performs selection by restricted tournament replacement (RTR). In RTR, a specified percentage of the population is iteratively replaced by the best individual out of a randomly chosen subset of individuals of size w (called the window size). We chose to replace 50% of the population. After some initial testing, the window size for RTR was set to 4, independent of the size of the problem l.

5.1 Single-Product Case

We implemented fitness function (11) for hBOA. Solutions were encoded on the binary string as explained in Sect. 4.1. In every generation of the algorithm, the feasibility of the population was maintained by setting the first bit of a solution to 1.

Scalability theory for hBOA assumes BBs of constant and bounded size. In order to assess whether the scalability results for the lotsizing problem are in accordance with scalability theory, we propose two test problems with constant demand and *constant* BB-size in Table 1. In the optimal solution of the first test problem, BBs are of size 2. In the optimal solution of the second test problem, BBs are of size 6. Instead of using 2 and 6, another pair of different sizes could also have been chosen.
We propose a second test problem with seasonal demand in Table 2 where BBs of the optimal solution have *varying* sizes from 2 to 6.

For all problems, we varied the problem size T from 6 to 60 in steps of size 6, resulting in 10 problem sizes per problem. For each problem size, the optimal costs was computed with the solver XPRESS MP as listed in the tables. Then, we derived the minimal population size that hBOA requires to solve the problem to optimality using a bisection method. An instance was assumed to be solved to optimality, if in at least 27 out of 30 independent consecutive runs of hBOA,

Table 1. Test problems with constant BB-sizes for the single-product lotsizing problem

BB-size	Optimal costs	c	h	$z_t\ \forall t$
2	$\frac{T}{6} \cdot 5400$	15000	8	100
6	$\frac{T}{6} \cdot 27000$	1500	8	100

Table 2. Test problem with varying BB-sizes for the single-product lotsizing problem

BB-size	Optimal costs	c	h	z_t
$2-6$	$T=6:3680$	1000	2	$z=$ (100,140,180,140,120,110,
	$T=12:6100$			80,50,30,50,80,90,
	$T=18:9600$			100,140,180,140,120,110,
	$T=24:12080$			80,50,30,50,80,90,
	$T=30:15680$			100,140,180,140,120,110,
	$T=36:18160$			80,50,30,50,80,90,
	$T=42:21680$			100,140,180,140,120,110,
	$T=48:24160$			80,50,30,50,80,90,
	$T=54:27760$			100,140,180,140,120,110,
	$T=60:30240$			80,50,30,50,80,90)

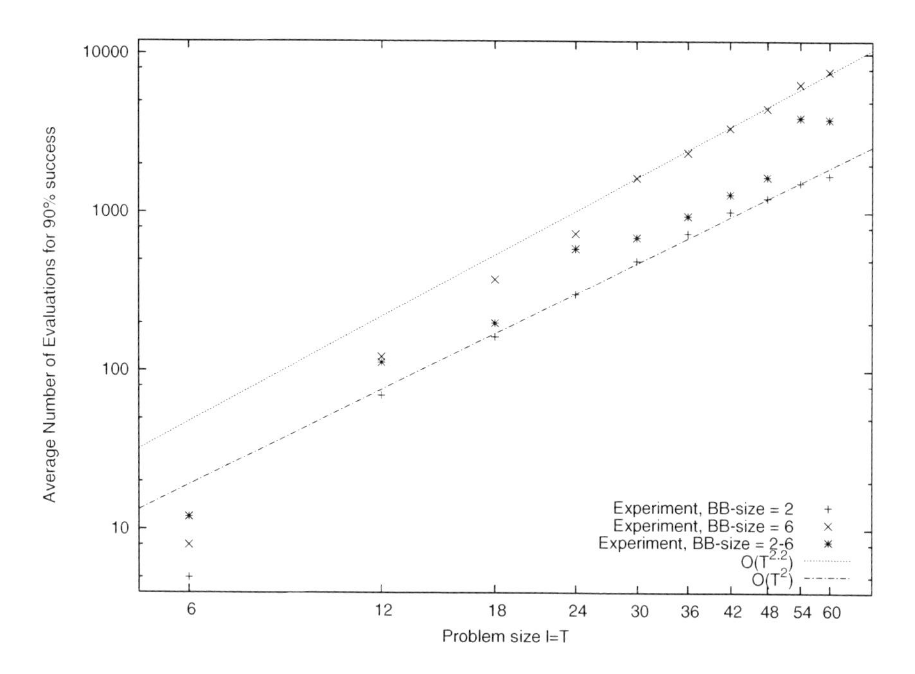

Fig. 4. Scalability results for the single-product lotsizing problem

the entire population converged towards the optimal solution. For the minimally required population size, the number of fitness evaluations was averaged over the number of successful runs.

Figure 4 illustrates how the average number of fitness evaluations depends on the problem size l. Additionally, the average number of evaluations has been approximated by a function of the form $O(l^r)$, where r was set such that experimental results were fitted accurately, emphasizing more stable experimental results for larger problems. Smaller instances were neglected due to high volatility of the results. Straight lines in the plot indicate polynomial scalability.

As can be seen in Fig. 4, the average number of fitness evaluations grows with a low-order polynomial depending on the problem size $l = T$. For a BB-size of 2, r has been set to $r = 2.0$, for a BB-size of 6, r has been set to $r = 2.2$. hBOA scales within these bounds, if the BB-size is in between $2 \leq$ BB-size ≤ 6.

Scalability theory for hBOA on non-hierarchical problems predicts that the number of fitness evaluations needed to reliably solve decomposable problems of bounded order grows with a low-order polynomial depending on the problem size with respect to the problem size l [32]. Our estimate of r for a BB-size of 6 lies slightly above quadratic scalability. Still, hBOA succeeds in solving single-product lotsizing problems whitin low-order polynomial time.

The size of the constant factor of the polynomial that approximates the experimental results grows with the size of the BBs. This means that hBOA needs more time to solve instances where large batches are ordered, compared to instances where smaller batches are ordered. If batches are not sized identically, the constant factor grows with the size of the largest batch ordered in the optimal solution. Note, that the size of the batches is unknown in advance because they depend on the parameter settings. We do not perform a detailed study of problem difficulty.

5.2 Dynamic Joint Replenishment Problem

We implemented fitness function (12) in hBOA. Solutions were represented as described in Sect. 4.2. Feasibility of all solutions was maintained as follows in each generation. We assume positive demand. All bits $T*k \; \forall \; k = 0, 1, 2, \ldots, K-1$ were set to 1, if they were 0. Additionally, whenever $\sum_{k=1}^{K} x_{(k*T)+j} = 0$ for any $j = 0, 1, 2, \ldots, T-1$, then x_j was set to 0. This means that if none of the real product is ordered at time-point j, the artificial product is not ordered as well.

We conduct scalability analysis for $K = 2$ and $K = 6$ products. For each case, we propose in Table 3 two test problems with constant demand and constant BB-size, resulting in 4 problems in total. For both $K = 2$ and $K = 6$, a problem with BB-size of 2 and a problem with BB-size of 6 is designed. Just like in the single-product case, we expect that a problem instance with seasonal demand and varying BB-size between 2 and 6 would scale up inside the bounds of these problems.

Table 3. Test problems with constant BB-sizes for the JRP

	BB-size	Optimal costs	$z_t^k \; \forall \; t, k$	$h^k \; \forall \; k$	c^o	$c^k \; \forall \; k$
2	2	$\frac{l}{18} \cdot 11400$	100	8	2000	100
Products	6	$\frac{l}{18} \cdot 3260$	10	5	1560	100
6	2	$\frac{l}{42} \cdot 2100$	10	5	100	50
Products	6	$\frac{l}{42} \cdot 9100$	10	5	1600	500

By varying the number of time-points T between 6 and 60 with a step size of 6, we obtained 10 problem instances for the 2 products case. The scalability analysis for two products spans problem sizes from 18 to 180 bits. For the 6 products case, T was varied from 6 to 30 in steps of 6, yielding 5 problem instances. This was necessary due to limited computational resources available. The scalability analysis for $K = 6$ products spans problem sizes from 42 to 210 bits.

For each instance, we obtained the optimal costs using the solver XPRESS MP as listed in Table 3. We derived the minimal population size that hBOA required to solve the problem to optimality using a bisection method. A problem instance was solved to optimality, if in at least 27 out of 30 consecutive and independent runs of hBOA, the optimal solution was found. For the minimal required population size, the average number of fitness evaluations was averaged over all successful runs.

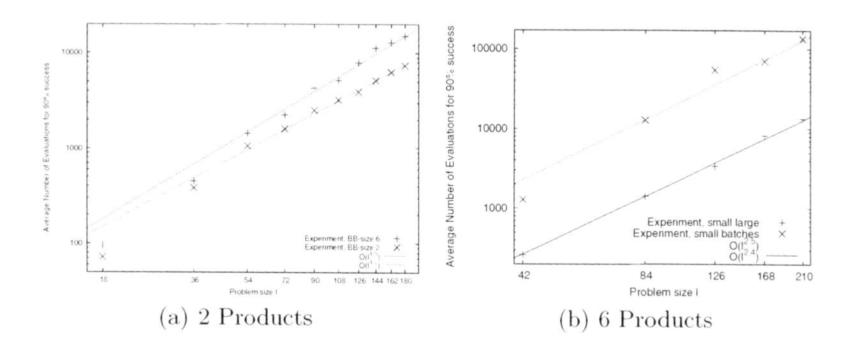

(a) 2 Products (b) 6 Products

Fig. 5. Scalability results for hBOA on the joint replenishment problem

Figure 5 illustrates how the average number of fitness evaluations depends on the size of the problem for $K = 2$ and $K = 6$ products. In addition, the average number of evaluations has been approximated by a function of the form $O(l^r)$, where r was set such that experimental results were fitted accurately, again emphasizing larger problem sizes. Both plots have a log-log scale. Straight lines in the plots indicate polynomial scalability.

For the case that the number of products $K = 2$, r has been set to $r = 1.9$ (BB-size of 6) and $r = 1.7$ (BB-size of 2). For the 6 products case, $r = 2.5$ (BB-size of 6) and $r = 2.4$ (BB-size of 2). Thus, hBOA succeeds in solving decomposable instances of the JRP in low-order polynomial time. For the two-products case, the constant factor grows with the size of the BBs. Like in the single-product case, more time is needed to solve instances of the JRP where larger batches are ordered optimally, compared to instances of the JRP where smaller batches are ordered.

The scalability results from the 6-products case illustrated in Fig. 5(b) yield counterintuitive results. hBOA scales polynomially for small and large BB-sizes, but obviously the problem set with a BB-size of 2 is harder to solve and requires more fitness evaluations than the corresponding instance with a BB-size of 6. This result can be explained with the population sizing model from [14] which states that the required population size scales inversely proportional to the signal-to-noise ratio defined as σ_{bb}/d. Noise σ_{bb} denotes the standard deviation of fitness values from all solutions that include the best BB and indicates the amount of fitness variability in the instance. The signal d denotes the difference between the mean fitness $\bar{f}_1$ of solutions that contain the best BB and the mean fitness $\bar{f}_2$ of solutions that contain the second-best BB. The signal-to-noise ratio for the JRP instance with 6 products and 6 periods ($l = 42$) is exemplarily presented in Table 4. Note that, for a BB-size of 2, the average costs of solutions that contain the second-best BB is lower than that of solutions that contain the best BB. This deception is not present in the case of BB-size 6, rendering the instance with smaller BB-size harder to solve than the instance with larger BBs.

Table 4. Signal-to-noise ratio for JRP with six products

	$\bar{f}_1$	$\bar{f}_2$	d	σ_{bb}	Signal-to-noise ratio
BB-size 6	20107.81	20257.80	-150.01	1717.19	-0.09
BB-size 2	2864.06	2823.44	40.62	309.92	0.13

6 Conclusions and Outlook

Decomposability of fitness functions is well-understood and frequently assumed in theoretical GA and EDA literature. State-of-the-art EDAs reliably solve decomposable problems in a low-order polynomial number of fitness evaluations depending on the problem size. This success makes it a tempting idea to apply EDA to real-world problems. It is essential to bridge the gap between theoretical work that focuses on solving decomposable problems and applied work that focuses on solving problems of practical interest. However, the complexity of the real world makes a direct adaption of theoretical concepts a stiff task and it is rarely known which problems are decomposable.

In this chapter, we demonstrated that decomposability is of practical relevance and valid for certain problems in inventory management. The decomposability of single-product lotsizing and the dynamic joint replenishment problem was analyzed. The results indicated that these lotsizing problems are indeed decomposable into subproblems of bounded complexity. We conducted a scalability analysis that showed that a state-of-the-art EDA can reliably solve the problems in a low-order polynomial number of fitness evaluations depending on the problem size.

The results are promising and reveal the potential of EDA applications in inventory management. However the complexity of real-world lotsizing problems is higher due to capacity constraints and bill of material structures. This provides interesting areas for future research.

References

1. Aksoy, Y., Erenguc, S.S.: Multi-item inventory models with coordinated replenishments. International Journal of Operations Management 8(1), 63–73 (1988)
2. Arkin, E., Joneja, D., Roundy, R.: Computational complexity of uncapacitated multi-echelon production planning problems. Operations Research Letters 8, 61–66 (1989)
3. Boctor, F.F., Laporte, G., Renaud, J.: Models and algorithms for the dynamic joint replenishment problem. International Journal of Production Research 42(13), 2667–2678 (2004)
4. Bosman, P.A.N., Thierens, D.: Learning probabilistic models for enhanced evolutionary computation. In: Jin, Y. (ed.) Knowledge Incorporation in Evolutionary Computation, pp. 147–176. Springer, Berlin (2004)
5. de Bodt, M., Gelders, L., van Wassenhove, L.: Lot sizing under dynamic demand conditions: A review. Engineering Costs and Production Economics 8, 165–187 (1984)

6. Deb, K., Goldberg, D.E.: Analysing deception in trap functions. In: Whitley, L.D. (ed.) Foundations of Genetic Algorithms 2, pp. 93–108. Morgan Kaufmann, San Mateo (1993)
7. Erenguc, S.S.: Multiproduct dynamic lot-sizing model with coordinated replenishments. Naval Research Logistics 35, 1–22 (1988)
8. Federgruen, A., Tzur, M.: A simple forward algorithm to solve general dynamic lot sizing models with n periods in O(n log n) or O(n) time. Management Science 37, 909–925 (1991)
9. Goldberg, D.E.: Genetic Algorithms in Search, Optimization, and Machine Learning. Addison-Wesley, Reading (1989)
10. Goldberg, D.E.: *The Design of Innovation: Lessons from and for Competent Genetic Algorithms*. Genetic Algorithms and Evolutionary Computation, vol. 7. Springer, Berlin (2002)
11. Goldberg, D.E., Deb, K., Kargupta, H., Harik, G.: Rapid, accurate optimization of difficult problems using fast messy genetic algorithms. In: Forrest, S. (ed.) Proceedings of the Fifth International Conference on Genetic Algorithms, pp. 56–64. Morgan Kaufmann, San Francisco (1993)
12. Goldberg, D.E., Korb, B., Deb, K.: Messy genetic algorithms: Motivation, analysis and first results. Complex Systems 10(5), 385–408 (1989)
13. Harik, G.: Linkage learning via probabilistic modeling in the ECGA. Technical Report 99010, IlliGAL, University of Illinois, Urbana, Illinois (1999)
14. Harik, G., Cantú-Paz, E., Goldberg, D.E., Miller, B.L.: The gambler's ruin problem, genetic algorithms, and the sizing of populations. Evolutionary Computation 7(3), 231–253 (1999)
15. Harik, G., Goldberg, D.E.: Learning linkage. In: Belew, R.K., Vose, M.D. (eds.) Foundations of Genetic Algorithms 4, pp. 247–262. Morgan Kaufmann, San Francisco (1997)
16. Holland, J.H.: Adaptation in Natural and Artificial Systems. University of Michigan Press, Ann Arbor, Michigan (1975)
17. Joneja, D.: The joint replenishment problem: New heuristics and worst case performance bounds. Operations Research 38, 711–723 (1990)
18. Kallel, L., Naudts, B., Reeves, C.R.: Properties of fitness functions and search landscapes. In: Naudts, B., Kallel, L., Rogers, A. (eds.) Theoretical Aspects of Evolutionary Computing, pp. 175–206. Springer, Berlin (2001)
19. Kao, E.P.C.: A multi-product dynamic lot-size model with individual and joint setup costs. Operations Research 27, 279–289 (1979)
20. Kargupta, H.: SEARCH, Polynomial Complexity, and the Fast Messy Genetic Algorithm. PhD thesis, University of Illinois, Urbana, Illinois (1995)
21. Larrañaga, P., Lozano, J.A. (eds.): *Estimation of Distribution Algorithms: A New Tool for Evolutionary Computation*. Genetic Algorithms and Evolutionary Computation, vol. 2. Kluwer Academic Publishers, Norwell (2001)
22. Lozano, J.A., Larrañaga, P., Inza, I., Bengoetxea, E. (eds.): *Towards a New Evolutionary Computation*. Studies in Fuzziness and Soft Computing, vol. 192. Springer, Berlin (2006)
23. Mahnig, T.: Populationsbasierte Optimierung durch das Lernen von Interaktionen mit Bayes'schen Netzen. PhD thesis, University of Bonn, Sankt Augustin, GMD Research Series No. 3/2001
24. Mandl, F.: Statistical Physics, 2nd edn. The Manchester Physics Series. Wiley, Chichester (1988)
25. Mühlenbein, H., Höns, R.: The estimation of distributions and the mimimum relative entropy principle. Evolutionary Computation 13(1), 1–27 (2005)

26. Mühlenbein, H., Mahnig, T.: FDA – A scalable evolutionary algorithm for the optimization of additively decomposed functions. Evolutionary Computation 7(4), 353–376 (1999)
27. Mühlenbein, H., Mahnig, T.: Evolutionary algorithms: From recombination to search distributions. In: Kallel, L., Naudts, B., Rogers, A. (eds.) Theoretical Aspects of Evolutionary Computing, pp. 137–176. Springer, Berlin (2000)
28. Mühlenbein, H., Mahnig, T., Rodriguez, A.O.: Schemata, distributions and graphical models in evolutionary optimization. Journal of Heuristics 5, 215–247 (1999)
29. Mühlenbein, H., Paaß, G.: From recombination of genes to the estimation of distributions: I. Binary parameters. In: Ebeling, W., Rechenberg, I., Voigt, H.-M., Schwefel, H.-P. (eds.) PPSN 1996. LNCS, vol. 1141, pp. 178–187. Springer, Berlin (1996)
30. Narayanan, A., Robinson, E.P.: More on models and formulations for the dynamic joint replenishment problem. International Journal of Production Research 44(2), 297–383 (2006)
31. Ochoa, A., Soto, M.R.: Linking entropy to estimation of distribution algorithms. In: Lozano, J.A., Larrañaga, P., Inza, I., Bengoetxea, E. (eds.) Towards a New Evolutionary Computation: Advances on Estimation of Distribution Algorithms. Studies in Fuzziness and Soft Computing, vol. 192, pp. 1–38. Springer, Berlin (2006)
32. Pelikan, M.: Bayesian optimization algorithm: From Single Level to Hierarchy. PhD thesis, University of Illinois at Urbana-Champaign, Department of Computer Science, Urbana, Illinois (2002)
33. Pelikan, M., Goldberg, D.E.: Hierarchical BOA solves ising spin glasses and MAXSAT. In: Cantú-Paz, E., Foster, J.A., Deb, K., Davis, D., Roy, R., O'Reilly, U.-M., Beyer, H.-G., Standish, R., Kendall, G., Wilson, S., Harman, M., Wegener, J., Dasgupta, D., Potter, M.A., Schultz, A.C., Dowsland, K., Jonoska, N., Miller, J. (eds.) GECCO 2003. LNCS, vol. 2724, pp. 1271–1282. Springer, Berlin (2003)
34. Pelikan, M., Goldberg, D.E., Cantú-Paz, E.: BOA: The Bayesian optimization algorithm. In: Banzhaf, W., Daida, J., Eiben, A.E., Garzon, M.H., Honavar, V., Jakiela, M., Smith, R.E. (eds.) Proceedings of the GECCO 1999 Genetic and Evolutionary Computation Conference, pp. 525–532. Morgan Kaufmann, San Francisco (1999)
35. Pelikan, M., Goldberg, D.E., Cantú-Paz, E.: Linkage problem, distribution estimation, and Bayesian networks. Evolutionary Computation 8(3), 311–341 (2000)
36. Pelikan, M., Goldberg, D.E., Lobo, F.: A survey of optimization by building and using probabilistic models. Computational Optimization and Applications 21(1), 5–20 (2002)
37. Pelikan, M., Sastry, K., Butz, M.V., Goldberg, D.E.: Hierarchical BOA on random decomposable problems. In: Keijzer, M., Cattolico, M., Arnold, D., Babovic, V., Blum, C., Bosman, P., Butz, M.V., Coello Coello, C., Dasgupta, D., Ficici, S.G., Foster, J., Hernandez-Aguirre, A., Hornby, G., Lipson, H., McMinn, P., Moore, J., Raidl, G., Rothlauf, F., Ryan, C., Thierens, D. (eds.) Proceedings of the GECCO 2006 Genetic and Evolutionary Computation Conference, pp. 431–432. ACM Press, New York (2006)
38. Robinson, E.P., Gao, L.L.: A dual ascent procedure for multiproduct dynamic demand coordinated replenishment with backlogging. Management Science 42, 1556–1564 (1996)
39. Silver, E.A.: Coordinated replenishments of items under time varying demand: Dynamic programming formulation. Naval Research Logistics Quarterly 26, 141–151 (1979)
40. Thierens, D.: Scalability problems of simple genetic algorithms. Evolutionary Computation 7(4), 331–352 (1999)

41. Thierens, D., Goldberg, D.E.: Mixing in genetic algorithms. In: Forrest, S. (ed.) Proceedings of the Fifth Conference on Genetic Algorithms, pp. 38–45. Morgan Kaufmann, San Mateo (1993)
42. Veinott, A.F.: Minumum concave cost solutions of Leontief substitution models of multi-facility inventory systems. Operations Research 17, 262–291 (1969)
43. Wagelmans, A., van Hoesel, S.: Economic lot sizing: An O(n log n) algorithm that runs in linear time in the Wagner-Whitin case. Operations Research 40(S1), 145–156 (1992)
44. Wagner, H.M., Whitin, T.M.: Dynamic version of the economic lot size model. Management Science 5, 89–96 (1958)
45. Watson, R.A., Hornby, G.S., Pollack, J.B.: Modeling building-block interdependency. In: Eiben, A.E., Bäck, T., Schoenauer, M., Schwefel, H.-P. (eds.) PPSN 1998. LNCS, vol. 1498, pp. 97–106. Springer, Berlin (1998)
46. Wemmerlöv, U.: A comparison of discrete single stage lot-sizing heuristics with special emphasis on rules based on the marginal cost principle. Engineering Costs and Production Economics 7, 45–53 (1982)

Hybrid Genetic Algorithms for the Lot Production and Delivery Scheduling Problem in a Two-Echelon Supply Chain

S. Ali Torabi[1], Masoud Jenabi[2], and S. Afshin Mansouri[3]

[1] Department of Industrial Engineering, Faculty of Engineering, University of Tehran, Tehran, Iran
satorabi@ut.ac.ir

[2] Department of Industrial Engineering, Amirkabir University of Technology, Tehran, Iran
m.jenabi@aut.ac.ir

[3] Brunel Business School, Brunel University, Uxbridge, Middlesex UB8 3PH, United Kingdom
Afshin.Mansouri@brunel.ac.uk

Summary. This chapter addresses integrated production and delivery scheduling of several items in a two-echelon supply chain. A single supplier produces the items on a flexible flow line (FFL) under a cyclic policy and delivers them directly to an assembly facility over a finite planning horizon. A new mixed zero-one nonlinear programming model is developed, based on the basic period (BP) policy to minimize average setup, inventory-holding and delivery costs per unit time where stock-out is prohibited. This problem has not yet been addressed in literature. It is computationally complex and has not been solved optimally especially in real-sized problems. Two efficient hybrid genetic algorithms (HGA) are proposed using the power-of-two (PT-HGA) and non-power-of-two (NPT-HGA) policies. The solution's quality of the proposed algorithms is evaluated and compared with the common cycle approach in a number of randomly generated problem instances. Numerical experiments demonstrate the merit of the NPT-HGA and indicate that it constitutes a very promising solution method for the problem.

Keywords: Flexible flow lines, Lot production and delivery scheduling, Basic period approach, Hybrid genetic algorithm.

1 Introduction

Production management involves a set of activities which are triggered by customer orders and/or forecasted demands to determine an appropriate production schedule in a production facility (i.e. indicating the starting and finishing times of required manufacturing operations). One of the key issues in production planning and scheduling is the lot-sizing problem. This problem arises where significant setup times/costs are to be incurred when switching from one product

A. Fink and F. Rothlauf (Eds.): Advances in Computational Intelligence, SCI 144, pp. 253–275, 2008.
springerlink.com

to another in the production sequence. Therefore, the economies of scale requires a lot production strategy (i.e. producing the products in a lot-by-lot manner).

The economic lot scheduling problem (ELSP) deals with lot sizing and scheduling of several items with static demands over an infinite planning horizon at a single capacitated facility, where only one item can be produced at a time [3].

Nowadays, there is a firm tendency to develop integrated models in research community for simultaneously cost-effective planning of different activities in supply chains. Among them, integrated production and delivery planning between adjacent supply parties is of particular interest, which can reduce the total logistics-related costs considerably. The economic lot and delivery-scheduling problem (ELDSP) is a variant of the ELSP where a supplier produces several items for an assembly facility and delivers them in a static condition [8]. The objective is to minimize the average transportation, setup and inventory holding costs per unit time across the supply chain without backlogging.

Most of the contributions reported in the literature dealing with static demand have focused on three cyclic schedules (i.e. a schedule that is repeated periodically). These include the common (or rotation) cycle approach, the basic period approach and the time varying lot sizes approach [22].

The common cycle approach restricts the products' cycle times to an equal length. The main advantage is that a feasible schedule (if any) can be found; because it is always possible to find a cycle time in which one lot of each product can be produced which satisfies the corresponding demands during the cycle time.

The basic period approach allows different cycle times for different products, but restricts each product's cycle time to an integer multiple of a time period called basic period. In the course of this approach, even finding a feasible schedule becomes a complex problem, given the number of production runs of each product per global cycle (i.e. the least common multiple of individual cycle times). This is the main disadvantage of the basic period approach, but in general, it provides better solutions compared to the common cycle approach.

The time varying lot sizes approach allows multiple runs for each product at each cycle with different lot sizes over a cyclic schedule, and guarantees finding a feasible schedule if one exists. Modeling a problem with this approach is more complicated than the other approaches, but usually gives better solutions.

The ELSP was first introduced by Eilon [3] who formulated it under the common cycle policy. Bomberger [1] considered different cycles for products, where each cycle time must be an integer multiple of a basic period (BP) which is long enough to meet the demand of all items. The production frequency of each product is then determined as a multiple of the selected BP. Elmaghraby [4] provided a review of the various contributions to ELSP and an improvement upon the BP approach which is called the extended basic period (EBP) method. In the course of this policy, the items are allowed to be loaded on two BPs simultaneously and at the same time relax the requirement that the basic period should be large enough to accommodate such simultaneous loading. Furthermore, several authors have extended the ELSP to multi-stage production systems under common cycle or basic period production policies (for example, see [5], [6], [16–20], and [22, 23]).

Hahm and Yano [8] introduced the economic lot and delivery scheduling problem to the single-product case. They also extended the problem to the multiple-product case under the common cycle and nested schedule cases respectively [9, 10]. Khouja [15] studied the ELDSP for a supplier that uses a volume flexible production system. Jensen and Khouja [13] studied the ELDSP under a common cycle approach considering some simplified assumptions. Based on these assumptions, they interpreted the sequencing phase as an optimization problem to minimize the weighted flow time. Moreover, they assumed that there was sufficient time within the common cycle time (T) to setup and produce the required products. Under these conditions, they developed an optimal polynomial time algorithm for the problem. Torabi et al. [23] considered the ELDSP in flexible flow lines (typical flow lines with at least one machine at each stage) under the common cycle approach over a finite planning horizon. They developed an effective HGA to obtain near optimal solutions.

The complexity of the ELSP and ELDSP even under the simplifying assumptions (like the single-stage production facility and zero setup cost) have been shown to be NP-hard [12]. The difficulty arises from the fact that two or more items compete for the available facility. As such, several heuristic/meta-heuristic approaches have been proposed in literature. Among them, the evolutionary-based approaches presented by Khouja et al. [14], and Yao and Huang [25] under basic period policy are two cases dealing with single-stage production systems.

It is noteworthy that most of the previous researches in this field assume the planning horizon to be infinite. However, the planning horizon is rarely infinite in real-world situations. In practice, the length of the finite horizon is usually determined by a mid-range contract between the supplier and the assembler. At the end of such a planning horizon, either the scheduling problem will cease to exist, or a new cyclic schedule needs to be derived for the next horizon based on a new contract. Moreover, it is common to consider a finite planning horizon where no significant fluctuations exist in the demand rate. Hence it could be argued that the assumption of infinite planning horizon considerably restricts the applicability of the proposed contributions. To cope with this shortcoming, some research papers have considered the finite horizon case (e.g., [16, 23]).

To the best of our knowledge, no research has been reported on the ELDSP in flexible flow lines under basic period approach over a finite planning horizon. Flexible flow lines can be found in a vast number of industries including the automotive, chemical, cosmetics, electronics, food, packaging, paper, pharmaceutical, printing, textile, wood-processing and semiconductor industries [21]. They are particularly common in the process industry. In this study, we address the finite horizon ELDSP in flexible flow lines (FH-ELDSP-FFL) under a basic period approach. A new mixed zero-one nonlinear program has been developed to determine optimal solutions concerning (i) an assignment of products in basic periods, (ii) an assignment of products to machines in the stages with multiple parallel machines, (iii) a sequence of products on each machine, (iv) lot sizes and (v) an optimal production schedule for each production run.

To solve this problem, it is assumed that the cycle time of each product i, denoted by T_i, is an integer multiple k_i of a basic period F; i.e., $T_i = k_i F$ for all i. In addition, the basic period F needs to be determined such that the planning horizon PH be an integer multiple of a global cycle HF. In other words, $PH = rHF$ where r is an integer and H denotes the least common multiple (LCM) of the k_i's. Two hybrid genetic algorithms are suggested based on two variants of BP approach: the power of two policy (i.e. restricting the time multiples to powers of 2) and non-power of two $(1, 2, 3, \ldots)$ time multiples.

The rest of this chapter is organized as follows. The problem formulation is presented in Sect. 2. Section 3 explains the proposed HGAs. Section 4 developes an efficient procedure for determining upper bounds on product's cycle time coefficients. An efficient feasibility test for capacity checking along with an iterative repair procedure for infeasible solutions is proposed in Sect. 5. Computational experiments are provided in Sect. 6. Finally, Sect. 7 is devoted to concluding remarks.

2 Problem Formulation

The following assumptions are considered for the problem formulation:

- Parallel machines at each stage are identical.
- Machines are continuously available and each machine can only process one product at a time.
- At the stages with parallel machines, each product is processed entirely on one machine.
- Setup times/costs in the supplier's production system are sequence independent.
- The production sequence at each basic period for each machine at each stage is unique and determined by solution method.
- The supplier incurs linear inventory holding costs on semi-finished products.
- Both the supplier and the assembler incur linear holding costs on end products.
- Preemption/lot-splitting is not allowed.

Moreover, the following notations are used for the problem formulation:

Parameters

n number of products
m number of work centers (stages)
m_j number of parallel machines at stage j

$M_{k'j}$ k'-th machine at stage j
d_i demand rate for product i
p_{ij} production rate of product i at stage j
s_{ij} setup time of product i at stage j
sc_{ij} setup cost of product i at stage j
h_{ij} inventory holding cost per unit of product i per unit time between stages j and $j+1$
h_i inventory holding cost per unit of final product i per unit time
A transportation cost per delivery
PH planning horizon
M a large real number

Decision variables

σ_k production sequence vector for basic period k (which contains a sub-set of products to be produced by the supplier)
$\sigma_{kk'j}$ production sequence vector for machine k' at stage j related to the basic period k (which contains a sub-set of products to be produced by the supplier)
r number of production cycles over the finite planning horizon
$n_{kk'j}$ number of products assigned to machine $M_{k'j}$ in basic period k
F basic period
b_{ij} starting time for processing product i at stage j
k_i cycle time coefficient of product i

$$x_{ilk'kj} = \begin{cases} 1 & \text{if product } i \text{ at basic period } k \text{ is assigned to the } l\text{-th position of} \\ & \text{machine } M_{k'j} \\ 0 & \text{otherwise} \end{cases}$$

It should be noted that the global cycle time is equal to the the least common multiple of the k_i variables, i.e., $H = \text{LCM}(k_1, k_2, \ldots, k_n)$. Moreover, the production cycle time for product i (T_i), the production lot size of product i (Q_i) and the processing time for a lot of product i at stage j (pt_{ij}) can be computed as follows:

$$T_i = k_i F \tag{1}$$

$$Q_i = d_i T_i \tag{2}$$

$$pt_{ij} = Q_i / p_{ij} = d_i k_i F / p_{ij} \tag{3}$$

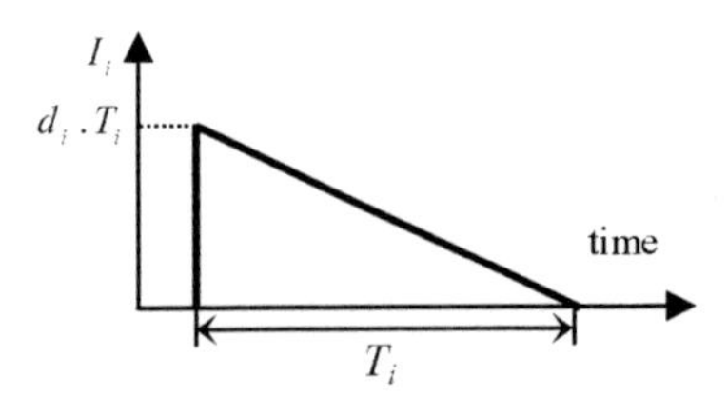

Fig. 1. Inventory level of final product i at the assembler in one cycle

Moreover, at stages with only one machine, the value of m_j and index k' are equal to one. Since the monetary value of the items increases as they are moving through production stages, the h_{ij} values are non-decreasing; i.e., $h_{i,j-1} \leq h_{ij}$.

The objective function of the problem (P) includes three elements: inventory holding costs, setup costs, and transportation costs. The setup and transportation cost per unit time could be easily computed as follows:

$$C = \sum_{i=1}^{n} \sum_{j=1}^{m} \frac{sc_{ij}}{k_i F} + \frac{A}{F} \tag{4}$$

The inventory holding cost is incurred at both the supplier and the assembler. Figure 1 shows the inventory curve of final product i in one cycle at the assembly facility.

It could be derived from the figure that the average inventory holding cost of products per unit time at the assembly facility is equal to: $F/2 \sum_{i=1}^{n} h_i k_i d_i$.

Two types of inventory (i.e., work-in-process (WIP)) and finished product inventories are considered for the supplier. Figures 2 and 3 show the amount of WIP inventory of product i between two successive stages $j-1$ and j, and the inventory level of final product i, respectively.

Based on Fig. 2, the average WIP inventory of product i between two successive stages $j-1$ and j per unit time is calculated as follows:

$$\begin{aligned} I_{i,j-1} &= \frac{1}{T_i} \left\{ \frac{d_i T_i}{2} \frac{d_i T_i}{p_{i,j-1}} + d_i T_i \left(b_{ij} - b_{i,j-1} - \frac{d_i T_i}{p_{i,j-1}} \right) + \frac{d_i T_i}{2} \frac{d_i T_i}{p_{ij}} \right\} \\ &= d_i \left(b_{ij} + \frac{d_i k_i F}{2 p_{ij}} - b_{i,j-1} - \frac{d_i k_i F}{2 p_{i,j-1}} \right) \end{aligned} \tag{5}$$

Therefore, the total WIP inventory holding cost for all products per unit time at the supplier amounts to:

$$TC_{WIP} = \sum_{i=1}^{n} \sum_{j=2}^{m} h_{i,j-1} d_i \left\{ b_{ij} + \frac{d_i k_i F}{2 p_{ij}} - b_{i,j-1} - \frac{d_i k_i F}{2 p_{i,j-1}} \right\} \tag{6}$$

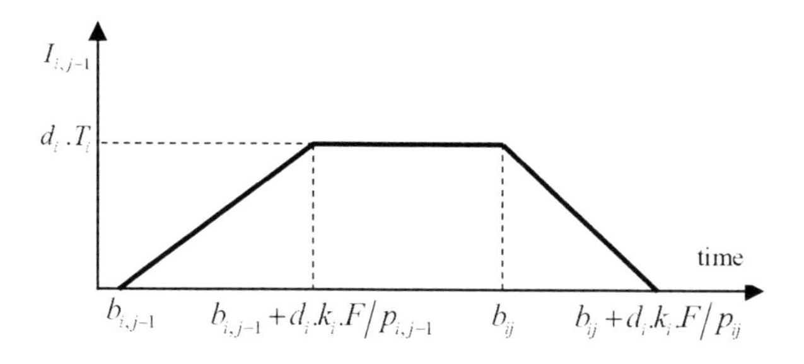

Fig. 2. WIP between stages $j-1$ and j at the supplier

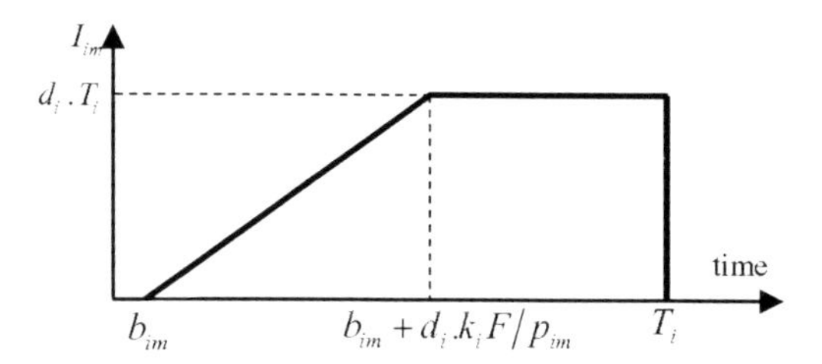

Fig. 3. Final product inventory at the supplier

Moreover, it could be derived from Fig. 3 that the average inventory of final product i per unit time is:

$$\begin{aligned} I_{im} &= \frac{1}{T}\left\{\frac{d_iT_i}{2}\frac{d_iT_i}{p_{im}} + d_iT_i\left(T_i - b_{im} - \frac{d_iTi}{p_{im}}\right)\right\} \\ &= d_i\left(1 - \frac{d_i}{2p_{im}}\right)k_iF - d_ib_{im} \end{aligned} \tag{7}$$

Thus, the total inventory holding cost for all final products per unit time amounts to:

$$TC_{FI} = \sum_{i=1}^{n} h_id_i\left(1 - \frac{d_i}{2p_{im}}\right)k_iF - \sum_{i=1}^{n} h_id_ib_{im} \tag{8}$$

Finally, the total cost per unit time (i.e., objective function of problem P) can be computed as follows:

$$\begin{aligned} TC = \frac{A}{F} &+ \sum_{i=1}^{n}\sum_{j=1}^{m}\frac{sc_{ij}}{k_iF} + \sum_{i=1}^{n} F\left[\frac{h_id_ik_i}{2}\left(3 - \frac{d_i}{p_{im}}\right)\right. \\ &\left. + \frac{d_i^2}{2}\sum_{j=2}^{m} h_{i,j-1}\left(\frac{1}{p_{ij}} - \frac{1}{p_{i,j-1}}\right)\right] \\ &+ \sum_{i=1}^{n}\sum_{j=2}^{m} h_{i,j-1}d_i(b_{ij} - b_{i,j-1}) - \sum_{i=1}^{n} h_id_ib_{im} \end{aligned} \tag{9}$$

Considering this objective function as well as logical relationships between the variables of problem P, the following mixed zero-one nonlinear model is developed to obtain an optimal solution of the problem.

Problem P:

$$\begin{aligned}\text{Min } Z = & \frac{A}{F} + \sum_{i=1}^{n}\sum_{j=1}^{m}\frac{sc_{ij}}{k_i F} \\ & + \sum_{i=1}^{n} F\left[h_i \frac{d_i k_i}{2}\left(3 - \frac{d_i}{p_{im}}\right) + \frac{k_i d_i^2}{2}\sum_{j=2}^{m} h_{i,j-1}\left(\frac{1}{p_{ij}} - \frac{1}{p_{i,j-1}}\right)\right] \\ & + \sum_{i=1}^{n}\sum_{j=2}^{m} h_{i,j-1} d_i (b_{ij} - b_{i,j-1}) - \sum_{i=1}^{n} h_i d_i b_{im}\end{aligned} \tag{10}$$

subject to:

$$b_{i,j-1} + \frac{d_i k_i F}{p_{i,j-1}} \leq b_{ij}; i = 1, \ldots, n; j = 2, \ldots, m \tag{11}$$

$$\begin{gathered} b_{i,j} + \frac{d_i k_i F}{p_{i,j}} + s_{ij} - b_{uj} \leq M(2 - x_{ilk'kj} - x_{u(l+1)k'kj}); \\ i = 1, \ldots, n; u \neq i; j = 1, \ldots, m; k' = 1, \ldots, m_j; \\ l < n; k = 1, \ldots, H; H = \text{LCM}(k_1, \ldots, k_n) \end{gathered} \tag{12}$$

$$\begin{gathered} \sum_{i=1}^{n} x_{ilk'kj} \leq 1; \\ j = 1, \ldots, m; k' = 1, \ldots, m_j; l = 1, \ldots, n; k = 1, \ldots, H; H = \text{LCM}(k_1, \ldots, k_n) \end{gathered} \tag{13}$$

$$\begin{gathered} \sum_{i=1}^{n} x_{i(l+1)k'kj} \leq \sum_{u=1}^{n} x_{ulk'kj}; \\ i = 1, \ldots, n; j = 1, \ldots, m; k = 1, \ldots, m_j; \\ l < n; k = 1, \ldots, H; H = \text{LCM}(k_1, \ldots, k_n) \end{gathered} \tag{14}$$

$$\sum_{k'=1}^{m_j}\sum_{l=1}^{n}\sum_{k=1}^{k_i} x_{ilk'kj} = 1; i = 1, \ldots, n; j = 1, \ldots, m \tag{15}$$

$$\sum_{k'=1}^{m_j}\sum_{l=1}^{n} x_{ilk'(t+bk_i)j} = \sum_{k'=1}^{m_j}\sum_{l=1}^{n} x_{ilk'(t+(b+1)k_i)j}; \tag{16}$$

$$i = 1,\ldots,n; j = 1,\ldots,m; t = 1,\ldots,k_i; b = 0,\ldots,\frac{H}{k_i} - 2$$

$$\sum_{k'=1}^{m_j}\sum_{l=1}^{n} x_{ilk'kj} = \sum_{k'=1}^{m_j}\sum_{l=1}^{n} x_{ilk'k,j+1}; \tag{17}$$

$$i = 1,\ldots,m; j = 1,\ldots,m; j < m; k = 1,\ldots,H$$

$$b_{ij} \geq s_{ij} - M\left(1 - \sum_{k'=1}^{m_j} x_{i1k'kj}\right); \tag{18}$$

$$j = 1,\ldots,m; i = 1,\ldots,n; k = 1,\ldots,H; H = \mathrm{LCM}(k_1,\ldots,k_n)$$

$$b_{im} + \frac{d_i k_i F}{p_{im}} \leq F; i = 1,\ldots,n \tag{19}$$

$$HFr = PH; H = \mathrm{LCM}(k_1,\ldots,k_n) \tag{20}$$

$$r \geq 1, \text{and integer} \tag{21}$$

$$F \geq 0; b_{ij} \geq 0 \forall i,j; x_{ilk'kj} = \{0,1\}; \forall i,l,k',j,k. \tag{22}$$

Constraints (11) state that no product can be processed on a given stage before its completion of its preceding stage. Constraints (12) guarantee that no product can be processed before the completion of its predecessor product in the production sequence ($\sigma_{kk'j}$). Constraints (13) enforce that at most one product is being processed at each position of each machine. Constraints (14) state that one product can be assigned at a position of machine $M_{k'j}$ only when another product has previously been assigned at the preceding position of the same machine. Constraints (15) ensure the assignment of product i to one of the first k_i basic periods and imply that at each stage, each assigned product has a unique position in the sequence of one machine. Constraints (16) determine the assignment of products in appropriate basic periods during the H basic periods. Constraints (17) denote that if product i has been assigned to the basic period k at stage j, it must be assigned to this basic period at all stages. Constraints (18) show that if product i is the first product in the sequence vector of one machine at stage j, its processing cannot be started before the corresponding set up operation. Constraints (19) ensure that the resulting schedule is cyclic so that the process completion time for each product at the final stage is less than or equal to a basic cycle time F. Constraint (20) implies that the planning horizon

PH is an integer multiplier of HF, where $H = \text{LCM}(k_1, \ldots, k_n)$, and F is the basic period length. Constraints (21) indicate that r is an integer number greater than or equal to one. Finally, Constraints (22) preserve the non-negativity of the variables.

In order to solve this model, the k_i values need to be determined at first. Based on these values and through solving problem P, the corresponding optimal basic period, assignments, sequence vectors and the production and delivery schedule of products might be obtained.

3 Proposed Hybrid Genetic Algorithms

During the last three decades, there has been a growing interest in solving the complex combinatorial problems using genetic algorithms (GA). Introduced by Holland [11], GA works based on the mechanism of natural selection and genetics. It starts with an initial set of solutions, called population. Each solution in the population is called a chromosome (or individual), which represents a point in the search space. The chromosomes are evolved through successive iterations, called generations, by genetic operators (selection, crossover and mutation). In a GA, a fitness value is assigned to each individual according to a problem-specific objective function. Generation by generation, the new individuals, called offspring, are created and compete with chromosomes in their parents in the current population to form a new population. To improve solution quality and to escape from converging to local optima, various strategies of hybridization have been suggested [2], [23]. In designing a hybrid genetic algorithm (HGA), the neighborhood search (NS) heuristic usually acts as a local improver into a basic GA loop.

In the proposed HGAs in this chapter, each solution is characterized by a set of k_i multipliers and the value of basic period F. One of the challenges besides the cost minimization is to generate feasible schedules. For this, a capacity feasibility test has been proposed which identifies and converts the infeasible solutions to feasible schedules. This issue will be discussed in Sect. 5.

3.1 Chromosome Representation

The chromosome structure for the proposed HGAs is binary strings representing the k_i values. Each k_i multiplier is represented by a particular part of a chromosome. For instance, the first u_1 bits are used to encode k_1 and the segment of chromosome from the $(u_1 + 1)$-th bit to the $(u_1 + u_2)$-th bit represents k_2 and so on. In order to represent all possible values of k_i for each item i, an upper bound (see Sect. 4) need to be set for the value of k_i. In the power of two policy where $k_i = 2^{v_i}$, the upper bound is to be set on the value v_i. A mapping scheme is also required to decode segments of binary strings into their corresponding integer (i.e., k_i values). The following equations are used to map a binary string consisting of u_i bits to their associated k_i for the power of two and non-power of two cases, respectively:

$$\langle b_{u_i} b_{u_i-1} \dots b_1 \rangle_2 = \left(\sum_{j=1}^{u_i} b_j 2^{j-1} \right)_{10} = (v_i)_{10} \Rightarrow k_i = 2^{v_i} \tag{23}$$

$$\langle b_{u_i} b_{u_i-1} \dots b_1 \rangle_2 = \left(\sum_{j=1}^{u_i} b_j 2^{j-1} \right)_{10} = (k_i)_{10} \tag{24}$$

3.2 Determining the σ_k Vectors

Finding feasible solutions as for the assignment and sequence of the products in different basic periods (i.e., σ_k vectors), is not easy. This section derives simple necessary and sufficient conditions in order to have a non-empty set of feasible solutions.

Given a vector of multipliers $k_i; i = 1, \dots, n$, the procedure starts with creating a vector say V' by sorting the products in ascending order of k_i. Ties are broken by sorting the products having the same multiplier k_i in a descending order of ρ_i where:

$$\rho_i = \sum_{j=1}^{m} \frac{k_i d_i}{p_{ij}}, i = 1, \dots, n \tag{25}$$

Each product i in the vector V' is assigned to the basic period t within the first k_i periods of the global cycle H which minimizes:

$$\max_{k=t,t+m_i,\dots} \left\{ \max_{j=1,\dots,m} \left(\sum_{u \in \sigma_k} \frac{k_u d_u}{p_{uj}} + \frac{k_i d_i}{p_{ij}} \right) \right\} \tag{26}$$

Finally, for each k, $k = 1, \dots, H$, the sequence of products within σ_k is determined so that if $i, u \in \sigma_k$ and i is before u in V', then i is also before u in σ_k.

3.3 Determining the $\sigma_{kk'j}$ Vectors

The first available machine (FAM) rule [23] has been employed to assign and sequence the products of each basic period to the machines of different stages. According to this procedure, for any given permutation vector V, the products are assigned to the machines of the first stage by using the FAM rule (if $m_1 > 1$). In the subsequent stages, the products are first sequenced in the increasing order of their completion times at the preceding stage. The products are then assigned to the machines at the current stage according to the FAM rule.

3.4 Initial Population

The initial population of the binary chromosomes is generated at random. Infeasible solutions are then identified via feasibility test and converted to feasible ones.

3.5 Evaluation Function

Each chromosome in the population represents a potential solution to the problem. The evaluation function assigns a real number to the individuals relating to their fitness. In our case, the fitness value for each individual is obtained by solving the non-linear problem $P1$ once the $\sigma_{kk'j}$ vectors are determined. Problem $P1$ is derived from problem P by substituting the variables by their values in the corresponding $\sigma_{kk'j}$ vectors. Incidentally, $\sigma_{kk'j}(i)$ denotes the i-th product in the associated sequence vector of machine $M_{k'j}$ in basic period k.

Problem $P1$:

$$\begin{aligned}\text{Min } Z = \frac{A}{F} &+ \sum_{i=1}^{n}\sum_{j=1}^{m}\frac{sc_{ij}}{k_iF} \\ &+ \sum_{i=1}^{n}\left[h_ik_i\frac{d_i}{2}\left(3-\frac{d_i}{p_{im}}\right)+\sum_{j=2}^{m}h_{i,j-1}k_i\frac{d_i^2}{2}\left(\frac{1}{p_{ij}}-\frac{1}{p_{i,j-1}}\right)\right]F \\ &+ \sum_{i=1}^{n}\sum_{j=2}^{m}h_{i,j-1}d_i(b_{ij}-b_{i,j-1})-\sum_{i=1}^{n}h_id_ib_{im}\end{aligned} \tag{27}$$

subject to:

$$b_{i,j-1}+\frac{k_iFd_i}{p_{i,j-1}} \le b_{ij}; i=1,\ldots,n; j=2,\ldots,m \tag{28}$$

$$\begin{gathered}b_{\sigma_{kk'j}(i-1),j}+\frac{k_{\sigma_{kk'j}(i-1)}Fd_{\sigma_{kk'j}(i-1)}}{p_{\sigma_{kk'j}(i-1),j}}-s_{\sigma_{kk'j}(i),j} \le b_{\sigma_{kk'j}(i),j}; \\ i=2,\ldots,n_{kk'j}; j=1,\ldots,m; k'=1,\ldots,m_j; \\ k=1,\ldots,H; H=\mathrm{LCM}(k_1,\ldots,k_n)\end{gathered} \tag{29}$$

$$\begin{gathered}b_{\sigma_{kk'j}(1),j} \ge s_{\sigma_{kk'j}(1),j}; \\ j=1,\ldots,m; k=1,\ldots,m_j; k=1,\ldots,H; H=\mathrm{LCM}(k_1,\ldots,k_n)\end{gathered} \tag{30}$$

$$b_{im}+\frac{k_iFd_i}{p_{im}} \le F; i=1,\ldots,n \tag{31}$$

$$rHF = PH; r \ge 1 \quad \text{and integer} \tag{32}$$

$$F, b_{ij} \ge 0 \quad \forall i,j. \tag{33}$$

Problem $P1$ can be solved by means of the following iterative procedure:

- Initial step: Let $r = 1$ and solve the associated linear problem. Let Z_r denote the optimal objective value.
- Iterative step: Let $r = r + 1$ and solve the corresponding linear problem. If this model has no feasible solution, stop; otherwise Let Zr+1 denote the current optimal objective value. If $Z_{r+1} < Z_r$ set $F^* = PH/rH$; otherwise the best objective value remains unchanged. Repeat the iterative step.

It is noteworthy that this procedure acts as an explicit enumeration method which ensures that the non-linear problem $P1$ could be solved to optimality.

3.6 Selection

Tournament selection is used in the proposed HGAs. It randomly chooses two chromosomes from the parent pool and selects the fitter individual with probability $1 - \varphi$ and the other one with probability φ where $0.5 < \varphi < 1$. The spouse duplication method is used to copy the selected parents in a separate set of the same size. Finally, pairs of individuals are picked randomly from the two sets for recombination through the crossover operation.

3.7 Crossover Operator

The main purpose of crossover is to exchange genetic material between randomly selected parents with the aim of producing better offspring. In this research, the two-point crossover is used wherein two cut points are selected at random among the selected chromosomes. The middle segment of the parents are exchanged to generate two descendants (see Fig. 4).

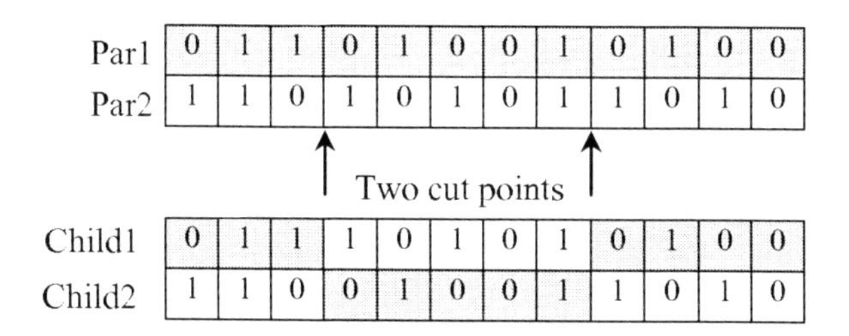

Fig. 4. Two points crossover

3.8 Mutation

Mutation introduces random variation into the population. Mutation is used to produce small perturbations on chromosomes to promote the diversity of the population. Most genetic algorithms incorporate a mutation operator mainly to

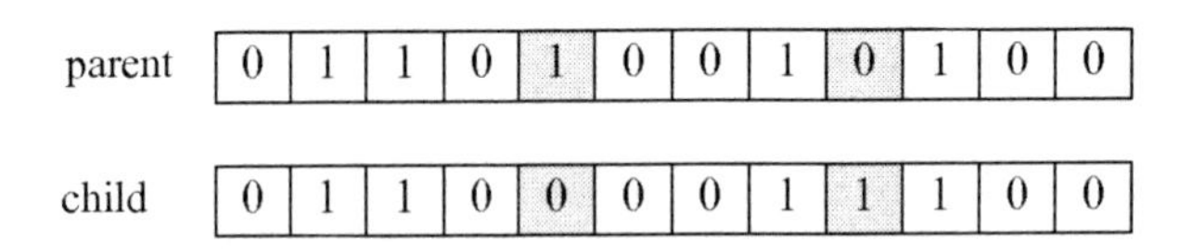

Fig. 5. Swap mutation

avoid convergence to local optima in the population and to recover lost genetic materials. A small portion of individuals (*mut_size*) is selected at random from the enlarged population (i.e. parents in the current population and offsprings generated by crossover operation) for mutation. In the proposed HGAs, swap mutation has been used as mutation operator. Fig. 5 illustrates an example of this operator.

3.9 Local Improver

A local improvement procedure is employed in the proposed HGAs based on an iterative neighborhood search (NS). The neighborhood search is applied to the offspring generated via crossover or mutatation operators. The search continues until a given chromosome is replaced by an elite (dominating) neighbor within a predetermined reasonable time. The inversion operator is used in the NS in which the genes between two randomly selected positions are inversed. Fig. 6 is an example of this operator.

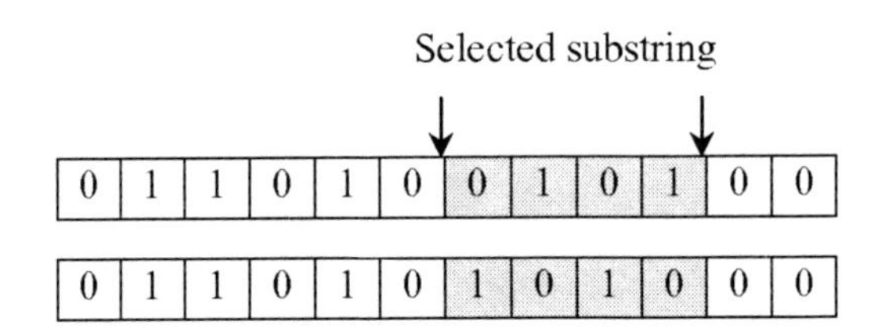

Fig. 6. Inversion operator

3.10 Population Replacement

Chromosomes for the next generation are selected from the enlarged population. Once the offspring are generated (through crossover and mutation operations) and improved by neighborhood search, the improved offspring are added to the current population. Therefore, the number of individuals in the enlarged population will be equal to: *pop_size* + *cross_size* + *mut_size*.

Individuals of the enlarged population are sorted according to their fitness. Sixty percent of the next population is filled by the top-ranked chromosomes of the enlarged population. The remaining chromosomes are selected randomly from the unselected chromosomes of the enlarged population.

3.11 Termination Criteria

The HGAs stop once a pre-determined number of generations (*max_gen*) or non-improving generations (*max_nonimprove*) has been executed.

4 Upper Bounds on k_i Values

In order to represent all possible and feasible for k_i multipliers, an upper bound is determined for each k_i. In the PT-HGA, an upper bound on k_i is derived through determining an upper bound for the objective value of each product i. The Following procedure describes how an upper bound for each $k_i(k_i^{UB})$ is computed:

Step1: For each product i, let $k_i = 1$ and calculate $\sum_j d_i/p_{ij}$. Arrange the products in ascending order of these values. Assign the products to the machines and sequence them at all stages via FAM rule. Finally, find the corresponding common cycle solution and its objective function, TC_{cc}.

Step2: Calculate the cost share of each product i using this formula:

$$TC^i_{cc} = TC_{cc} - \sum_{u=1,u\neq i}^{n} TC^u_{cc}.$$

It should be noted here that $TC^i_{BP} \leq TC^i_{cc}$ (see [24]), where TC^i_{BP} is the cost share of product i under basic period approach. The TC^i_{BP} values are determined in the next step.

Step3: Assume there is only one product (say product i) with the following objective function:

$$\begin{aligned} TC^i_{BP} = &\sum_{j=1}^{m} \frac{sc_{ij}}{k_i F} \\ &+ k_i F\left[\frac{h_i d_i}{2}\left(3 - \frac{d_i}{p_{im}}\right) + \frac{d_i^2}{2}\sum_{j=2}^{m} h_{i,j-1}\left(\frac{1}{p_{ij}} - \frac{1}{p_{i,j-1}}\right)\right] \\ &+ \sum_{j=2}^{m} h_{i,j-1} d_i (b_{ij} - b_{i,j-1}) - h_i d_i b_{im} \end{aligned} \quad (34)$$

It is obvious that in order to obtain the optimal solution, the starting times need to be determined such that $(b_{ij} - b_{i,j-1}) - b_{im}$ is minimized. Moreover, the smallest feasible value of $b_{ij} - b_{i,j-1}$ is equal to $k_i F d_i / p_{i,j1}$ and the largest feasible value of b_{im} is equal to $k_i F(1 - d_i/p_{im})$. Therefore, the best value of TC^i_{BP} is as follows:

$$TC_{BP}^{i} = \sum_{j=1}^{m} \frac{sc_{ij}}{k_i F} + k_i F \underbrace{\left[\frac{h_i d_i}{2} \left(1 + \frac{d_i}{p_{im}} \right) + \frac{d_i^2}{2} \sum_{j=2}^{m} h_{i,j-1} \left(\frac{1}{p_{ij}} + \frac{1}{p_{i,j-1}} \right) \right]}_{H_i} \tag{35}$$

Finally, for a given value of F, an upper bound on k_i, denoted by k_i^{UB} can be derived by using the following equations:

$$TC_{BP}^{i} \leq TC_{cc}^{i} \Rightarrow \sum_{j=1}^{m} \frac{sc_{ij}}{k_i F} + k_i F H_i \leq TC_{cc}^{i} \Rightarrow k_i^2 F^2 H_i - k_i F TC_{cc}^{i} + \sum_{j=1}^{m} sc_{ij} \leq 0 \tag{36}$$

$$k_i^{UB} = \frac{TC_{cc}^{i} + \sqrt{(TC_{cc}^{i})^2 - 4H_i \left(\sum_{j=1}^{m} sc_{ij} \right)}}{2H_i F_{min}}; \tag{37}$$

in nonpower of two case

$$\Rightarrow v_i^{UB} = \log_2 \left\lceil \frac{TC_{cc}^{i} + \sqrt{(TC_{cc}^{i})^2 - 4H_i \left(\sum_{j=1}^{m} sc_{ij} \right)}}{2H_i F_{min}} \right\rceil; \tag{38}$$

in power of two case

To determine the minimum value of F, F_{min}, assume that F must be large enough so that at least one product with $k_i = 1$ can be produced meanwhile. Consequently, F_{min} is obtained from the following equation:

$$\max_{i=1,\dots,n} \left\{ \sum_{j=1}^{m} s_{ij} + \sum_{j=1}^{m} \frac{d_i F}{p_{ij}} \right\} \leq F \Rightarrow F \geq \max_{i=1,\dots,n} \left\{ \frac{\sum_{j=1}^{m} s_{ij}}{1 - \sum_{j=1}^{m} d_i / p_{ij}} \right\} \tag{39}$$

5 Feasibility Test and Repair Procedure

A simple test for capacity feasibility can be carried out for a given chromosome and its related k_i, σ_k and $\sigma_{kk'j}$ vectors. To do so, the process completion times of the products for all H basic periods are first calculated. The following procedure is used for this purpose.

for $k = 1, \ldots, H$
  for each $i \in \sigma_k$
    for $j = 1, \ldots, m$
      process1$=\sum_{u \in \sigma_{kk'j}} k_u d_u / p_{uj}$, product u is before i at basic period k on machine $M_{k'j}$.
      process2$=\sum_{u \in \sigma_{kk',j-1}} k_u d_u / p_{u,j-1} + k_i d_i / p_{i,j-1}$, product u is before i at basic period k on machine $M_{k',j-1}$.
      fin$= \max\{$*process1*,*process2*$\} + k_i d_i / p_{ij}$
    end
    ft_{ik} =*fin*
  end
  $ft_k = \max_i\{ft_{ik}\}$
end

If the related completion time is greater than or equal to 1 in at least one basic period, the corresponding chromosome is infeasible. Otherwise, it is feasible. In other words, if at least one of the ft_k values, $k = 1, \ldots, H$, is greater than or equal to 1, this solution is infeasible.

To convert an infeasible solution into a feasible one, the following iterative repair procedure is proposed, based on the k_i values modifications.

Step 1: Choose the basic period with maximal value of ft_k, say basic period $k1$.

Step 2: From among the products of basic period $k1$, select the product with the largest process time $(\max_i\{\sum_{j=1,\ldots,m} k_i d_i / p_{ij}\}; i \in \sigma_{k1})$, say product i.

Step 3: If $k_i \neq 1$; set $k_i = k_i - 1$ (in the case of power of two, if $v_i \neq 0, v_i = v_i - 1$), and obtain σ_k and $\sigma_{kk'j}$ vectors for this new set of multipliers. If this solution is feasible, stop, otherwise go to *step1*. If $k_i = 1$ or $v_i = 0$; select the product with the next largest process time and go to *step3*.

It is noteworthy that all chromosomes obtained via genetic operators (crossover, mutation and local improver) are checked using the aforementioned feasibility test.

6 Computational Experiments

To verify the efficiency of the proposed algorithm in terms of the solution quality and the required computation time, a number of numerical experiments were implemented. The experiments were executed on a PC with an Intel Pentium IV 1800 MHz CPU. The HGAs were coded on MATLAB 6.5. Moreover, LINGO 6.0 optimization software was used to solve the mixed zero-one non-linear models.

6.1 Parameter Setting

The parameters of the HGAs were tuned empirically in the course of some initial experiments. These values were found to be effective in terms of solution quality and computation time: population size *pop_size*=n, maximum number of generations *max_gen*=$m \times n$, maximum number of non-improving generations *max_nonimprove*=n, crossover probability P_c=0.8, mutation probability P_m=0.2, and tournament selection parameter φ=0.7.

6.2 Data Sets

Data sets were randomly generated using these uniform distributions: $d_i \sim U(100, 1000)$, $p_{ij} \sim U(5000, 15000)$, $s_{ij} \sim U(0.01, 0.15)$, $h_{i1} \sim U(1, 10)$, $A \sim U(10000, 20000)$. In order to have non-decreasing h_{ij} values in successive stages (considering the fact that value is being added to the products as they are processed through the stages), the h_{ij} values for the stages $j \geq 2$ were generated using a recursive formula as: $h_{ij} = h_{i,j-1} + U(1, 3)$. In order to reflect the potential correlation between sc_{ij} and s_{ij} values, the corresponding sc_{ij} for each randomly generated s_{ij} was generated with this equation: $sc_{ij} = 15000 \times s_{ij} + 1000 \times U(0, 1)$. Finally, the number of parallel machines at each stage was randomly set to either 1 or 2.

6.3 Performance Evaluation

To evaluate the efficiency of the proposed solution methods, nine test problems of different size were considered. For each size, 20 problem instances were generated at random. The problem instances were subdivided into small-size problems (with 4 and 5 products with 2 and 3 stages) and medium to large-size problems (having 5 and 10 products with 2, 5 and 10 stages). For small problems, the solutions of the proposed algorithms were compared with those of LINGO. For medium and large problems, the results of the proposed algorithms were compared with lower bounds. An index $\lambda = (TC - LB)/LB$ is used for this purpose where TC is the total cost of a problem instance obtained by the proposed algorithm and LB is the associated lower bound.

In order to calculate the LB for a given problem, the following equation needs to be minimized:

$$\begin{aligned} Z = \frac{A}{F} &+ \sum_{i=1}^{n} \sum_{j=1}^{m} \frac{sc_{ij}}{k_i F} \\ &+ \sum_{i=1}^{n} \left[h_i k_i \frac{d_i}{2} \left(3 - \frac{d_i}{p_{im}} \right) + \sum_{j=2}^{m} h_{i,j-1} k_i \frac{d_i^2}{2} \left(\frac{1}{p_{ij}} - \frac{1}{p_{i,j-1}} \right) \right] F \qquad (40) \\ &+ \sum_{i=1}^{n} \sum_{j=2}^{m} h_{i,j-1} d_i (b_{ij} - b_{i,j-1}) - \sum_{i=1}^{n} h_i d_i b_{im} \end{aligned}$$

It could be concluded that if the b_{ij} values are determined so that the $(b_{ij} - b_{i,j-1})$ terms are minimized, then the above equation will be minimized. According to constraint (11), the minimum value for $(b_{ij} - b_{i,j-1})$ is $d_i k_i F / p_{i,j-1}$

and the maximum feasible value for b_{im} is $F(1-k_i d_i/p_{im})$, where F is calculated using this formula:

$$F = \sqrt{\frac{A + \sum_{j=1}^{m} sc_{ij}/k_i}{\sum_{i=1}^{n} k_i \left[\frac{h_i d_i}{2}\left(1+\frac{d_i}{p_{im}}\right) + \frac{d_i^2}{2}\sum_{j=2}^{m} h_{i,j-1}\left(\frac{1}{p_{ij}}+\frac{1}{p_{i,j-1}}\right)\right]}} \quad (41)$$

As a result, the LB can be computed as follows:

$$\begin{aligned} LB = {} & \frac{A}{F} + \sum_{i=1}^{n}\sum_{j=1}^{m}\frac{sc_{ij}}{k_i F} \\ & + \sum_{i=1}^{n} k_i F \left[h_i \frac{d_i}{2}\left(1+\frac{d_i}{p_{im}}\right) + \frac{d_i^2}{2}\sum_{j=2}^{m} h_{i,j-1}\left(\frac{1}{p_{ij}}+\frac{1}{p_{i,j-1}}\right)\right] \end{aligned} \quad (42)$$

6.4 Results

Table 1 represents the results for the small-sized problem instances. The results for the medium and large-sized problem instances are presented in Tables 2 and 3 respectively.

In Table 1, the quality of the solutions found by the proposed algorithms is compared with the solutions found using a LINGO 6.0 optimization software. Columns 2 and 3 show the number of times out of 20 instances where NPTHGA and PTHGA respectively have found a better solution compared to LINGO. The figures indicate that the proposed algorithms outperform LINGO in most cases. The main reason for this could be attributed to the nonlinearity of the mathematical model which enforces LINGO to stop searching once a local optimum has been found instead of the global optimum. The average superiority of the proposed algorithms over LINGO in terms of cost reduction or distance to the LINGO's final results are represented in the columns 4 and 5. As the averages indicate, the solutions' quality found by the NPTHGA and PTHGA was 6.3 and 4.3 percent better than the solutions of LINGO, respectively. Moreover, the figures in columns 6–8 concerning the execution times show that both proposed algorithms were considerably faster than LINGO. Furthermore, they indicate that PTHGA requires less CPU time compared to NPTHGA in finding the final solutions. However, considering the better performance of the NPTHGA over PTHGA in terms of quality, this extra execution time might be justified.

It should be noted that for solving the small-sized problem instances by LINGO, this parameter was set empirically, since it was not possible to specify the smallest possible upper bound for big M in the mathematical model. According to our experiments, $M = 1000$ would yield good results. For the values very larger than 1000, LINGO reports the problem as being infeasible.

For the medium and large size problem instances, performance ratio λ have been calculated and used as a measure to compare the proposed algorithms. In Table 2 the performances of the proposed algorithms are compared with each

Table 1. Results for small size test problems

(1)	(2)	(3)	(4)	(5)	(6)	(7)	(8)
4x2	18	17	4.80	3.44	2736.98	35.64	34.43
4x3	17	16	6.08	5.05	5684.45	75.24	53.47
5x2	17	16	6.82	5.35	5770.77	81.11	54.69
5x3	19	18	11.57	9.07	9737.29	170.96	133.65
Average			6.3	4.3			

Column headings:

(1): problem size (n×m)

(2): number of times that the NPTHGA's solution was better than the LINGO's solution

(3): number of times that the PTHGA's solution was better than the LINGO's solution

(4): the average superiority of NPTHGA compared to LINGO (%) in terms of cost reduction

(5): the average superiority of PTHGA compared to LINGO (%) in terms of cost reduction

(6): average CPU time for LINGO (s)

(7): average CPU time for NPTHGA (s)

(8): average CPU time for PTHGA (s)

Table 2. Results for medium and large size test problems (comparisons with LB)

Problem size (n×m)	Avg. performance ratio of NPTHGA (%)	Avg. performance ratio of PTHGA (%)	Avg. CPU time of NPTHGA (s)	Avg. CPU time of PTHGA (s)	Cost comparison of PTHGA vs. NPTHGA (%)
5×5	8.01	10.26	270.45	149.34	28.08
5×10	18	18.86	825.59	787.66	4.77
10×2	6.04	7.84	988.86	636.36	29.8
10×5	15.29	16.4	1869.1	1825.9	7.25
10×10	23.06	23.63	2234.52	2126.58	2.47

Table 3. Results for medium and large size test problems (comparisons with common cycle approach)

problem size (n×m)	Avg. cost reduction in NPTHGA's solution compared to common cycle approach	Avg. cost reduction in PTHGA's solution compared to common cycle approach	Cost comparison of the PTHGA vs. NPTHGA
5×5	7.59	5.43	39.77
5×10	4.63	4.58	10.92
10×2	6.92	5.13	34.89
10×5	5.27	5.16	2.13
10×10	3.84	3.72	3.22

other. The results indicate that the solution quality of the NPTHGA averages 18.19% better than the solution quality of the PTHGA.

Table 3 reports the average improvement in total cost by using the proposed algorithms compared to the common cycle approach (i.e. the case where $k_i = 1$

is assumed for all products). These results reveal that the average reduction in the total cost of solutions found by PTHGA and NPTHGA are 4.8% and 5.65% respectively, when compared with the common cycle approach. These results show the advantage of using a basic period policy instead of a common cycle in the problem addressed in this chapter.

7 Concluding Remarks

In this chapter, the basic period approach has been used to solve the economic lot and delivery-scheduling problem in a two-echelon supply chain, in which a supplier produces several items in a flexible flow line system and delivers them to an assembler on a cyclic basis over a finite planning horizon. A new mixed zero-one nonlinear model has been developed to solve the problem to optimality. Due to the computational complexity, providing an optimal solution is impossible especially for medium and large size problem instances. For this, two hybrid GAs (called NPTHGA and PTHGA) have been developed based on different variants of basic period approach.

In the numerical experiments, the quality of the proposed hybrid GAs was compared with LINGO on small size problems. For medium and large-size problems, the hybrid GAs were compared with reference to a lower bound. Computational results indicate that using a basic period policy instead of a common cycle approach yields better solutions. Furthermore, the NPTHGA outperforms the PTHGA with respect to the solution quality, but the PTHGA outperforms the NPTHGA with respect to the computation time.

Acknowledgement

This study was supported by the University of Tehran under the research grant No. 8109920/1/01. Also, Afshin Mansouri was supported in part by EPSRC under grant EP/D050863/1 as well. The authors are grateful for this financial supports.

References

1. Bomberger, E.E.: A dynamic programming approach to the lot size scheduling problem. Management Science 12, 778–784 (1966)
2. Cheng, R., Gen, M.: Parallel machine scheduling problems using memetic algorithms. Computers and Industrial Engineering 33, 761–764 (1997)
3. Eilon, S.: Scheduling for batch production. Institute of Production Engineering Journal 36, 549–579 (1957)
4. Elmaghraby, S.E.: The economic lot scheduling problem: review and extensions. Management Science 24, 587–598 (1978)
5. El-najdawi, M., Kleindorfer, P.R.: Common cycle lot size scheduling for multi-product, multi-stage production. Management Science 39, 872–885 (1993)

6. El-najdawi, M.: Multi cyclic flow shop scheduling: An application for multi-product, multi-stage production processes. International Journal of Production Research 39, 81–98 (1997)
7. Fatemi Ghomi, S.M.T., Torabi, S.A.: Extension of common cycle lot size scheduling for multi-product, multi-stage arborscent flow-shop environment. Iranian Journal of Science and Technology, Transaction B 26, 55–68 (2002)
8. Hahm, J., Yano, C.A.: The economic lot and delivery-scheduling problem: The single item case. International Journal of Production Economics 28, 235–252 (1992)
9. Hahm, J., Yano, C.A.: The economic lot and delivery-scheduling problem: The common cycle case. IIE Transactions 27, 113–125 (1995)
10. Hahm, J., Yano, C.A.: The economic lot and delivery scheduling problem: Models for nested schedules. IIE Transactions 27, 126–139 (1995)
11. Holland, J.H.: Adaptation in Natural and Artificial Systems, 2nd edn. University of Michigan / MIT Press (1992)
12. Hsu, W.L.: On the general feasibility test of scheduling lot sizes for several products on one machine. Management Science 29, 93–105 (1983)
13. Jensen, M.T., Khouja, M.: An optimal polynomial time algorithm for the common cycle economic lot and delivery scheduling problem. European Journal of Operational Research 156, 305–311 (2004)
14. Khouja, M., Michalewicz, Z., Wilmot, M.: The use of genetic algorithm to solve the economic lot size scheduling problem. European Journal of Operational Research 110, 509–524 (1998)
15. Khouja, M.: The economic lot and delivery-scheduling problem: Common cycle, rework, and variable production rate. IIE Transactions 32, 715–725 (2000)
16. Ouenniche, J., Boctor, F.F.: Sequencing, lot sizing and scheduling of several components in job shops: The common cycle approach. International Journal of Production Research 36, 1125–1140 (1998)
17. Ouenniche, J., Boctor, F.F.: The multi-product, economic lot-sizing problem in flow shops: the powers-of-two heuristic. Computers and Operations Research 28, 1165–1182 (2001)
18. Ouenniche, J., Boctor, F.F.: The two-group heuristic to solve the multi-product, economic lot-sizing and scheduling problem in flow shops. European Journal of Operational Research 129, 539–554 (2001)
19. Ouenniche, J., Boctor, F.F.: The G-group heuristic to solve the multi-product, sequencing, lot-sizing and scheduling problem in flow shops. International Journal of Production Research 39, 89–98 (2001)
20. Ouenniche, J., Bertrand, J.W.M.: The finite horizon economic lot sizing problem in job shops: The multiple cycle approach. International Journal of Production Economics 74, 49–61 (2001)
21. Quadt, D., Kuhn, H.: Conceptual framework for lot-sizing and scheduling of flexible flow lines. International Journal of Production Research 43, 2291–2308 (2005)
22. Torabi, S.A., Karimi, B., Fatemi Ghomi, S.M.T.: The common cycle economic lot scheduling in flexible job shops: The finite horizon case. International Journal of Production Economics 97, 52–65 (2005)
23. Torabi, S.A., Fatemi Ghomi, S.M.T., Karimi, B.: A hybrid genetic algorithm for the finite horizon economic lot and delivery scheduling in supply chains. European Journal of Operational Research 173, 173–189 (2006)

24. Yao, M.J., Elmaghraby, S.E.: On the economic lot scheduling problem under power-of-two policy. Computers and Mathematics with Applications 41, 1379–1393 (2001)
25. Yao, M.J., Huang, J.X.: Solving the economic lot scheduling problem with deteriorating items using genetic algorithms. Journal of Food Engineering 70, 309–322 (2005)

Author Index